The Concise Guide to
American Aircraft
of World War II

David Mondey

**CHARTWELL
BOOKS, INC.**

First published in Great Britain by
Hamlyn Publishing Group Limited in 1982
an imprint of Reed Consumer Books Limited
Michelin House, 81 Fulham Road, London SW3 6RB
and Auckland, Melbourne, Singapore and Toronto

This edition published in 1994 by
Chartwell Books, Inc.
A division of Book Sales, Inc.
Raritan Center
114 Northfield Avenue
Edison, NJ 08818

A CIP catalogue record for this book is available
from the British Library

ISBN 0 78580 147 2

Produced by Mandarin Offset
Printed and bound in China

CONTENTS

Aeronca L-3 Grasshopper

Aeronca L-3B Grasshopper of a US Army liaison/observation unit.

History and Notes

The US Army Air Corps had been slow to appreciate the value of light aircraft for employment in an observation/liaison role, but information received from Europe in late 1940, where World War II was already more than a year old, highlighted their usefulness. Consequently, in 1941 the US Army began its own evaluation of this category of aircraft, obtaining four commercial lightplanes from each of three established manufacturers, namely Aeronca, Piper and Taylorcraft. For full field evaluation larger numbers of these aircraft were ordered shortly afterwards, to be deployed in the US Army's annual manoeuvres which were to be held later in the year. It took very little time for the service to appreciate that these lightweight aeroplanes had a great deal to offer, both for rapid communications and in support of armed forces in the field.

The name Aeronca Aircraft Corporation had been adopted in 1941 by the company established in late 1928 as the Aeronautical Corporation of America. One of its most successful products was the Model 65 high-wing monoplane, developed to meet commercial requirements for a reliable dual-control tandem two-seat trainer. The four of these aircraft supplied initially to the USAAC became designated YO-58, and these were followed by 50 O-58, 20 O-58A and 335 O-58B aircraft, serving with the USAAF (established on 20 June 1941). In the following year the O (Observation) designation was changed to L (Liaison), and the O-58, O-58A and O-58B designations became respectively L-3, L-3A and L-3B. An additional 540 aircraft were delivered as L-3Bs and 490 L-3Cs were manufactured before production ended in 1944. The designations L-3D/-3E/-3F/-3G/-3H/-3J were applied to civil Model 65s with varying powerplant installations, which were pressed into military service when the United States became involved in World War II.

Most L-3s were generally similar, with small changes in equipment representing the variation from one to another. All shared the welded steel-tube fuselage/tail unit with fabric covering, and wings with spruce spars, light alloy ribs and metal frame ailerons, all fabric-covered. Landing gear was of the non-retractable

Aeronca L-3 Grasshopper

tailwheel type, with the main units divided and incorporating oleo-spring shock-absorbers in the side vees.

With the requirement for a trainer suitable for glider pilots, Aeronca developed an unpowered version of the Model 65. This retained the wings, tail unit and aft fuselage of the L-3, but introduced a new front fuselage providing a third seat forward for an instructor, the original tandem seats being used by two pupils: all three occupants had similar flying controls and instruments. A total of 250 of these training gliders was supplied to the USAAF under the designation TG-5, and three supplied to the US Navy for evaluation were identified as LNR. Production of Aeronca liaison aircraft continued after the war, with planes supplied to the USAF under the designation L-16.

Specification

Type: two-seat light liaison and observation monoplane
Powerplant: one 65-hp (48-kW) Continental O-170 flat-four piston engine
Performance: maximum speed 87 mph (140 km/h); cruising speed 46 mph (74 km/h); service ceiling 10,000 ft (3 050 m); range 200 miles (322 km)
Weights: empty 835 lb (379 kg); maximum take-off 1,300 lb (590 kg)
Dimensions: span 35 ft 0 in (10.67 m); length 21 ft 0 in (6.40 m); height 7 ft 8 in (2.34 m); wing area 158 sq ft (14.68 m²)
Armament: none
Operators: USAAF, USN for evaluation

Allied LRA-1

History and Notes

During the interwar years many military services carried out some evaluation of gliders with a view to employing them in a training or cargo-carrying role. In Germany, where gliders had been developed extensively following the limitations imposed by the Versailles Treaty, they had not only been used for an initial stage of pilot training, but had been adopted for the rapid deployment of airborne troops, their equipment and supplies. The early success of gliders in such a role, during the preliminary gambits of World War II, gave new impetus to such evaluations being carried out in the USA.

The US Navy had shown intermittent interest in the possible use of gliders from as long ago as 1920, but German westward attacks against the European Allies, at the beginning of May 1940, added a new sense of urgency. Amongst the US Navy's projects was a flying boat transport glider with very clean lines, of which two prototypes were built by the Allied Aviation Corporation. This company had been established in January 1941 to fabricate moulded plywood structures for use by aircraft manufacturers, and these two all-wooden prototypes (11647-11648) represented the company's first attempt to build a complete aircraft.

The design was unique, with a large two-step flying boat hull to accommodate a crew of two and 10 troopers. Jettisonable wheeled landing gear was provided to offer amphibious capability. The tail unit and flying control surfaces were entirely conventional; the unusual feature was the low-set cantilever monoplane wing, the wing roots and inner wing section resting on the water to provide stability when afloat. In addition, provision had been made for the installation of a small engine to enable the craft to manoeuvre on the water. Four additional prototypes had been ordered from Bristol Aeronautical Corporation, only two of which were built, and 100 production examples ordered from each manufacturer were cancelled in 1943 when the US Navy decided that its requirements could be satisfied more effectively by powered aircraft.

The two prototypes built by Allied Aviation were designated XLRA-1 in service use, the X signifying experimental, L glider, R cargo, and A the manufacturer.

Specification

Type: flying boat transport glider
Dimensions: span 72 ft 0 in (21.95 m); length 40 ft 0 in (12.19 m); height 12 ft 3 in (3.73 m)
Operator: USN, for evaluation only

A fascinating design exercise, the Allied XLRA-1 was designed to provide the US Navy with the means to assault-land troops.

Beech Model 17

History and Notes

Known widely as the Staggerwing, the Beech Model 17 was first flown in 1934 and was the first design to appear from the factory Walter Beech opened after he left the Travel Air organisation. Of light alloy construction with fabric covering, the Beech 17, with its comfortable enclosed four/five-seat cabin, quickly became established in the United States civil market and 424 had been built by the time the US government took the nation into World War II in 1941.

The initial production variant was the B17L with a 225-hp (168-kW) Jacobs L-4 radial engine, followed by the B17B with the more powerful 285-hp (213-kW) Jacobs L-5, while the B17R and B17E, which were introduced in 1935, were both powered by Wright engines, the former by the 420-hp (313-kW) R-975 and the latter by the 285-hp (213-kW) R-760. Later developments included the D17S with a 450-hp (336-kW) Pratt & Whitney R-985 Wasp Junior and the F17D, powered by a 330-hp (246-kW) Jacobs L-6.

Comfort and performance were also among the requirements for the transportation of senior US military officers, and the Beech 17 was evaluated in this role, initially by the US Navy with one C17R powered by a (336-kW) Wright R-975 and given the military designation JB-1. Ten examples of the D17S were acquired in 1939, with the designation GB-1, later supplemented by a small number of impressed civil aircraft and well over 300 GB-2s, many of which were among 105 Beech 17 Travellers supplied to Britain under Lend-Lease and used principally by the Royal Navy.

The US Army Air Corps purchased three D17S aircraft in 1939, designating them YC-43. A production order for 27 UC-43s was placed in 1942; these were powered by the R-985-AN-1 Wasp Junior and incorporated some interior and equipment changes which resulted in an increase in gross weight to 4,700 lb (2 132 kg). Two subsequent orders for 75 and 105 UC-43s brought the total USAAF procurement to 207 aircraft. As in the case of the US Navy, at the outbreak of hostilities many civil Beech 17s were impressed and given designations in the range UC-43A to -43H, -43J and -43K. The former civil model numbers were D17R, D17S, F17D, E17B, C17R, D17A, C17B, B17R, and D17W respectively.

Production for the civil market recommenced in 1945 and approximately 90 C17S models had been built by 1948 when the line closed. This version was powered by the 450-hp (336-kW) Pratt & Whitney R-985.

Specification

Type: four/five-seat liaison and communications aircraft

Powerplant (UC-43): one 450-hp (336-kW) Pratt & Whitney R-985-AN-1 Wasp Junior radial piston engine

Performance: maximum speed 198 mph (319 km/h) at 5,000 ft (1 525 m); cruising speed 170 mph (274 km/h) at 5,000 ft (1 525 m); service ceiling 20,000 ft (6 095 m); range 500 miles (805 km)

Weights: empty 3,085 (1 399 kg); maximum take-off 4,700 lb (2 123 kg)

Dimensions: span 32 ft 0 in (9.75 m); length 26 ft 2 in (7.98 m); height 10 ft 3 in (3.12 m); wing area 296 sq ft (27.50 m²)

Armament: none

Operators: Brazilian Air Force, Chinese Air Force, RN, USAAC/USAAF, USN

A boost to Beech Model 17 production came with the US entry into the war, when large numbers of UC-43 Traveller communications aircraft were ordered.

Beech Model 18

History and Notes

The first civil Beech Model 18 flew on 15 January 1937, and initial deliveries soon began, though only in relatively modest numbers. The key to the type's eventual success (over 9,000 built in 32 years of production) was the Model 18D of 1939, which introduced 330-hp (246-kW) Jacobs L-6 engines, which provided improved performance without an increase in fuel consumption.

The type commended itself to the US forces as a utility transport, and the first US Army Air Corps order, placed in 1940, was for 11 aircraft to be designated C-45 in their role as staff transports, as which they were generally similar to the civil Model B18S with the R-985 radial engine. Subsequent procurement covered 20 C-45A utility transports, and 223 C-45B transports with minor interior and equipment changes. Some of these aircraft were supplied to the UK under Lend-Lease, being designated Expediter I in RAF service. The USAAF designations C-45C, C-45D and C-45E were applied to two impressed Model B18S aircraft, two AT-7s completed as transports and six similarly completed AT-7Bs. The major, and final, production variant for the USAAF was the C-45F: this had a slightly lengthened nose, and production totalled 1,137. Lend-Lease C-45Fs served with the RAF and FAA as Expediter IIs, and with the RCAF as Expediter IIIs. All the foregoing C-45 designations became UC-45 designations in 1943.

In 1941 the Beech AT-7 Navigator derivative was introduced for navigator training. This had positions for three trainees and a dorsal astrodome. Some 577 were built, followed by six AT-7As with floats and a ventral fin. Nine winterised AT-7Bs were followed by 549 of the AT-7C version with different engines. Another 1941 arrival was the AT-11 Kansan bombing/gunnery trainer. This was fitted with round rather than square windows, had a small bomb bay and nose bombardier's position, and was armed with two 0.3-in (7.62-mm) guns, one in the nose and the other in a dorsal turret. Production totalled 1,582, including 36 completed as AT-11A navigation trainers.

Last of the USAAF's wartime variants was the photo-reconnaissance F-2: the first 14 were impressed Model B18S aircraft with cabin-mounted mapping cameras; then came 13 F-2As converted from C-45As with four cameras; and finally 42 F-2Bs converted from UC-45Fs with extra camera ports in the fuselage sides. In June 1948 the survivors became RC-45As, and at the same time the AT series dropped the A prefix, and CQ-3 drone-director conversions from UC-45Fs became CD-45Fs.

US Navy and US Marine Corps procurement of the Model 18 also totalled over 1,500 examples. The initial JRB-1 was equivalent to the F-2, the JRB-2 was a transport, and the JRB-3 and -4 were similar to the C-45B and UC-45F. The SNB-1, -2 and -3 were retrospective designations to cover aircraft similar to the AT-11, -7 and -7C respectively. The three last variants were the SNB-2H air ambulance, the SNB-2P photographic aircraft and the SNB-3Q was an electronic countermeasures trainer.

During 1951-2 all surviving USAF UC-45, AT-7 and AT-11 aircraft were remanufactured to zero-time condition as C-45Gs with the R-985-AN-3 engine and an autopilot, or as C-45Hs with the R-985-AN-14B engine and no autopilot. US Navy SNB-2s, -2Cs and -2Ps were remade as SNB-5 and -5P. In 1962 the remaining SNB aircraft became TC-45J training and RC-45J photographic aircraft.

Specification

Type (AT-11): navigation and bombing trainer
Powerplant: two 450-hp (336-kW) Pratt & Whitney R-985-AN-1 radial piston engines
Performance: maximum speed 215 mph (346 km/h); service ceiling 20,000 ft (6 095 m); range 850 miles (1 368 km)
Weights: empty 6,175 lb (2 801 kg); maximum take-off 8,727 lb (3 959 kg)
Dimensions: span 47 ft 8 in (14.53 m); length 34 ft 2 in (10.41 m); height 9 ft 8 in (2.95 m); wing area 349.0 sq ft (32.42 m²)
Armament: two 0.3-in (7.62-mm) machine-guns plus up to 10 100-lb (45-kg) bombs

The Beech Model 18 found extensive employment with the US forces during and after World War II. The model illustrated is a C-45G, the designation given to all C-45Fs, RC-45As, T-7s and T-11s remanufactured after the war to an improved standard.

Bell P-39 Airacobra

History and Notes

Conventional in its external appearance, the Bell P-39 was unique among US Army fighter aircraft of World War II in its powerplant installation, and was also the US Army's first single-seat fighter to be provided with tricycle type landing gear. These latter features were imposed by the desire to mount heavy armament in the nose, which reflects the general interest of all nations during the middle and late 1930s in developing fighter aircraft with good forward firepower.

In early 1935, executives of the Bell Aircraft Corporation had been present at a demonstration of the American Armament Corporation's T9 37-mm cannon. Impressed by what they had seen, they instigated the design of a fighter aircraft which would include a T9 cannon firing through the propeller hub, as well as two 0.50-in (12.7-mm) machine-guns mounted in the fuselage nose and synchronised to fire between the rotating propeller blades. The decision to locate the cannon to fire through the propeller hub meant that the engine had to be mounted within the fuselage, directly above the rear half of the low-set monoplane wing, with the propeller driven by an extension shaft which passed beneath the cockpit floor. In turn, this engine position, virtually over the aircraft's centre of gravity, highlighted the desirability of introducing a tricycle type landing gear, an installation presenting few problems as there was adequate room in the fuselage nose to accommodate the nosewheel unit.

The concept was sufficiently attractive to the US Army Air Corps to win an order for a single XP-39 prototype on 7 October 1937, and this flew for the first time on 6 April 1938. Twelve months later, following extensive evaluation by the US Army, 12 YP-39s were ordered for a wider service test, plus a single YP-39A without a turbocharger for the Allison V-1710 engine. Following service evaluation of the XP-39, the National Advisory Committee for Aeronautics (predecessor of the National Aeronautics and Space Administration) carried out a study of this prototype, recommending changes which included the provision of fairing doors for the mainwheel units; a lower-profile cockpit canopy; resiting of the engine air intake and coolant radiators; and deletion of the turbocharger. The original prototype, modified to this configuration under the designation XP-39B, was test flown and demonstrated improved performance. As a result, a decision was made to delete the turbocharger from all future aircraft: the 13 preproduction prototypes were completed to XP-39B standard, provided with two additional 0.30-in (7.62-mm) machine-guns in the fuselage nose, and began service trials under the designation YP-39.

With the initial designation P-45, the new fighter was ordered into production on 10 August 1939, the first contract being for 80 aircraft; the designation reverted to P-39 before the first of these was delivered. The first 20 of these, completed to XP-39B standard, were designated P-39C, but the remaining 60 each received two more 0.30-in (7.62-mm) machine-guns (all four then being mounted in the wings), introduced self-

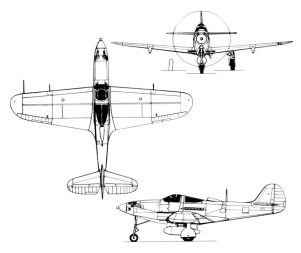

Bell P-39Q Airacobra

sealing fuel tanks, and had provision for a 500-lb (227-kg) bomb or a 75-US gallon (284-litre) fuel drop tank to be carried beneath the fuselage: these changes brought the designation P-39D.

The first large order, for 369 P-39Ds, was placed in September 1940, and initial deliveries of these began about seven months later, the first export Airacobras ordered by a British Purchasing Commission beginning to come off the production line at about the same time. British orders totalled 675 aircraft similar to the P-39D, but with the 37-mm cannon replaced by one of 20-mm, and the six 0.30-in (7.62-mm) machine-guns replaced by an equal number of 0.303-in (7.7-mm) calibre.

Airacobras began to reach the UK in July 1941, and in September of that year No.601 Squadron exchanged its Hawker Hurricanes for these new aircraft. Immediately they were introduced into service, the full implication of the decision to delete the turbocharger was appreciated for the first time, the aircraft having an inadequate rate of climb and high-altitude performance unacceptable for deployment in the European theatre. Only about 80 of the total order entered service with the RAF, equipping only No. 601 Squadron, which exchanged them for Supermarine Spitfires in March 1942. Of the balance, more than 250 were supplied to the Russian air force under a British aid scheme; about 200 in Britain were transferred to the USAAF in Britain in late 1942, and about 200 were repossessed by the US Army Air Force in America after the USA entered World War II in December 1941. These ex-British Airacobras were designated P-400s in USAAF service.

Constructed in large numbers, a total of 9,558 Airacobras being built before production ended, there were no major design changes in the several variants which followed. The P-39F, of which 229 were built, succeeded the P-39D into production. Generally similar, it differed by having an Aeroproducts hydraulically

Bell P-39 Airacobra

Bell Airacobra Mk I of No. 601 Squadron, RAF, in October 1941.

Bell P-400 Airacobra of the 67th Fighter Squadron, 35th Fighter Group, USAAF, based in New Caledonia during 1942.

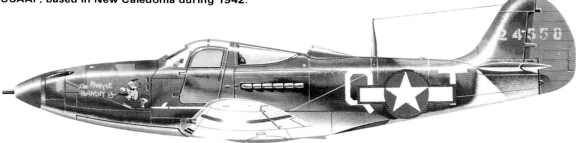

Bell P-39L Airacobra of the 93rd Fighter Squadron, 81st Fighter Group, USAAF, based in Tunisia during 1943.

Bell P-39N Airacobra of the 4° Stormo, Italian Co-Belligerent Air Force, based in Yugoslavia during 1944.

Bell P-39Q Airacobra of Major Aleksandr Pokryshin, Red Air Force, on the southern part of the Eastern Front during 1943-4.

Bell P-39 Airacobra

operated constant-speed propeller instead of the Curtiss type of the earlier models. The P-39J, of which 25 were built, had a different version of the Allison V-1710 engine, while the P-39K (210 built) and P-39L (250 built) differed in detail equipment and by installation of the more powerful V-1710-63 engine; the former had an Aeroproducts, the latter a Curtiss propeller. The P-39M (240 built) had the lower-power V-1710-83 engine and an increased-diameter propeller. The final production versions, the P-39N and P-39Q, were built in large numbers for supply to the Russian air force under Lend-Lease. To improve performance the P-39N carried less fuel and armour, and the P-39Q could be identified easily by two underwing fairings, each housing a 0.50-in (12.7-mm) machine-gun in place of the four wing-mounted 0.30-in (7.62-mm) guns.

Total P-39 production was 9,558, of which 4,773 (primarily P-39D/-39N/-39Q) were supplied to Russia. Variants included three XP-39E experimental aircraft with laminar-flow wings, and small numbers of TP-39F and RP-39Q two-seat trainers. Seven P-39s were supplied to the US Navy for use as target drones under the designation F2L. The US Navy had shown interest in the type for use as a ship-based fighter, and a single XFL-1 Airabonita was produced with tailwheel type landing gear, strengthened fuselage, and an arrester hook. First flown on 13 May 1940, it failed to enter production after carrier trials had proved unsatisfactory.

Although deletion of the turbocharger had limited the Airacobra's potential as a fighter aircraft, it was used most successfully in North Africa in late 1942 in the ground-attack role, and was deployed widely in the Pacific theatre by the USAAF. Until 1944, when more potent fighters began to enter service, the P-39 together with the Curtiss P-40 represented the main first-line equipment of the USAAF's fighter squadrons. A small number of P-39s were used by the Portuguese air force, acquired after these aircraft had force-landed in Portugal, about 150 were supplied to Free French forces in the latter stages of the war, and a similar number to the Italian co-belligerent air force.

Specification

Type: single-seat monoplane fighter/fighter-bomber
Powerplant (P-39M): one 1,200-hp (895-kW) Allison V-1710-83 inline piston engine
Performance: maximum speed 386 mph (621 km/h) at 9,500 ft (2 895 m); cruising speed 200 mph (322 km/h); service ceiling 36,000 ft (10 970 m); range 650 miles (1 046 km)
Weights: empty 5,610 lb (2 545 kg); maximum take-off 8,400 lb (3 810 kg)
Dimensions: span 34 ft 0 in (10.36 m); length 30 ft 2 in (9.19 m); height 11 ft 10 in (3.61 m); wing area 213 sq ft (19.79 m²)
Armament: one 37-mm T9 cannon, two 0.50-in (12.7-mm) machine-guns and four 0.30-in (7.62-mm) machine-guns, plus provision for one 500-lb (227-kg) bomb
Operators: FFAF, ICoAF, (Portuguese Air Force), RAF, Soviet Air Force, USAAF, USN

Designed as an interceptor, the Bell P-39 Airacobra's forte proved to lie at the opposite end of the altitude spectrum as a ground-attack machine.

Bell P-59 Airacomet

History and Notes

Details of development of the turbojet engine in the UK, stemming from the work of Frank Whittle (later Sir Frank), was transferred as routine to the United States as part of an agreed interchange of technological progress, intended to speed the end of World War II. In the USA the General Electric Company, which had wide experience of the design, development and construction of industrial turbines, dating back to before the start of the 20th century, was chosen initially to proceed with the development of national aircraft gas turbines based on the Whittle engine. Because of the Bell Aircraft Corporation's geographical location, in relation to the General Electric plant, this company was chosen to design and build a fighter aircraft to be powered by the first American-built gas turbine.

Realising that early engines would develop only limited thrust, Bell elected for a twin-engine installation, with one engine carried on each side of the fuselage, beneath the wings. The configuration selected was that of a mid-wing monoplane, and wide-span main landing gear units were mounted under the wings, well outboard of the engines and retracting inward into the wing, the nosewheel unit retracting aft into the

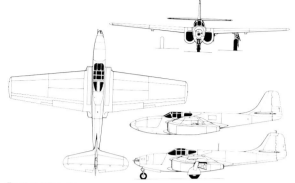

Bell P-59B Airacomet (upper side view: XP-59A)

fuselage nose. In other respects the design was conventional, with care taken to ensure a fairly high tailplane position so that it would be clear of the efflux from the turbojets.

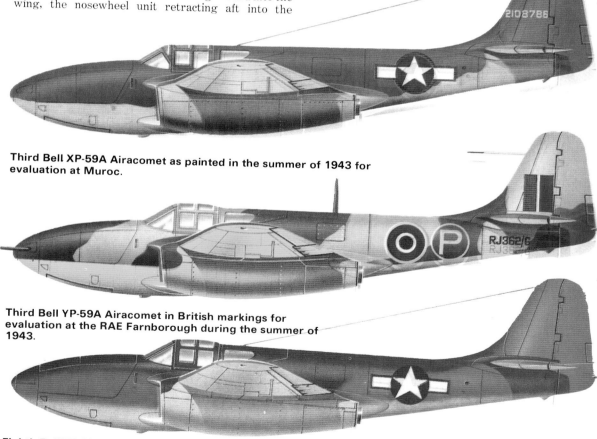

Third Bell XP-59A Airacomet as painted in the summer of 1943 for evaluation at Muroc.

Third Bell YP-59A Airacomet in British markings for evaluation at the RAE Farnborough during the summer of 1943.

Eighth Bell YP-59A Airacomet as painted for evaluation by the US Navy in late 1943 at Naval Air Station Patuxent River.

Bell P-59 Airacomet

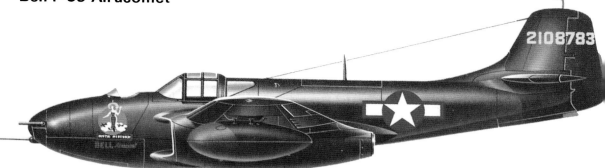

Thirteenth Bell YP-59A Airacomet, named Mystic Mistress, with open second cockpit for the use of a drone operator in early 1945.

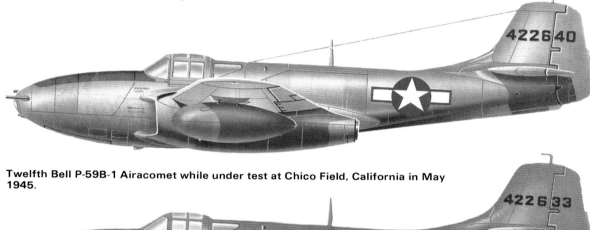

Twelfth Bell P-59B-1 Airacomet while under test at Chico Field, California in May 1945.

Fifteenth Bell P-59B-1 Airacomet, named *Reluctant Robot,* flown as a pilotless drone in 1944-5.

The first XP-59A, powered by two 1,250-lb (567-kg) thrust General Electric Type I-A turbojets, was flown for the first time from Muroc Dry Lake on 1 October 1942. Two more XP-59As were built, followed by a batch of 13 YP-59As for test and evaluation. The majority of these latter aircraft, delivered during 1944, were each powered by two 1,650-lb (748-kg) thrust General Electric I-16 (later J31) turbojets. The 20 P-59A and 30 P-59B Airacomets which followed had J31-GE-3 and J31-GE-5 engines respectively, the P-59Bs also having increased fuel capacity.

Flown for test and evaluation by the USAAF's 412th Fighter Group, a specially formed trials unit, the P-59 was found to have inadequate performance and proved an indifferent gun platform. As a result no further examples were built.

Specification
Type: single-seat jet fighter
Powerplant (P-59B): two 2,000-lb (907-kg) thrust General Electric J31-GE-5 turbojets
Performance: maximum speed 409 mph (658 km/h) at 35,000 ft (10 670 m); cruising speed 375 mph (604 km/h); service ceiling 46,200 ft (14 080 m); range 400 miles (644 km)
Weights: empty 8,165 lb (3 704 kg); maximum take-off 13,700 lb (6 214 kg)
Dimensions: span 45 ft 6 in (13.87 m); length 38 ft 1½ in (11.62 m); height 12 ft 0 in (3.66 m); wing area 385.8 sq ft (35.84 m²)
Armament: one 37-mm M4 cannon and three 0.50-in (12.7-mm) machine-guns mounted in the nose
Operator: USAF

Bell P-63 Kingcobra

History and Notes

At a fairly early stage in the development of the Bell P-39 Airacobra, work had been carried out to enhance the performance of this aircraft by the introduction of aerodynamic improvements. Three experimental aircraft were built, each utilising the basic fuselage of the P-39D, to which was added a new laminar-flow wing with square wingtips and a revised tail unit. In fact, each of the three XP-39Es, as these aircraft were designated, had a different tail unit. It was planned originally to power the prototypes with the Continental Aviation and Engineering Corporation's I-1430 12-cylinder inverted-vee piston engine, which had demonstrated a power output in excess of 2,000 hp (1 491 kW). However, Allison V-1710 engines of little more than half of that horse power were installed, presumably because of unreliability of the Continental engine. Testing of the XP-39Es began in February 1942 and, proving satisfactory, the type was ordered into production under the designation P-76. Some 4,000 aircraft were to be built at Bell's Marietta, Ohio, facility but were cancelled only three months later.

It was decided, instead, to build a larger and more powerful version for utilisation in a close-support fighter/fighter-bomber role, and the research and design development which had been carried out for the XP-39E was used in finalising the design of what was to become known as the P-63 Kingcobra. In its layout this latter aircraft was generally similar to the P-39, but apart from being larger and with the V-1710 engine

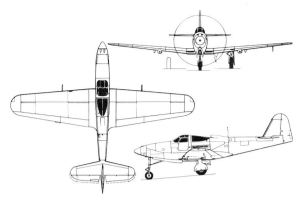

Bell P-63A Kingcobra (dashed lines: P-63A-6)

more powerful than those installed in all but the P-39K and P-39L production aircraft, efforts had to be made to render this new development more suitable for the close-support role regarded as its primary mission.

Two prototypes were ordered by the US Army Air Corps in June 1941 under the designation XP-63, and these made their first flights on 7 December 1942 and 5 February 1943, both powered by the 1,325-hp (988-kW) Allison V-1710-47 engine.

The Bell P-63 Kingcobra was based on the P-39D, and the initial P-63A-1 had an armament of one 37-mm nose cannon and two 0.50-in (12.7-mm) wing guns.

Bell P-63 Kingcobra

Both aircraft were lost in an early stage of their test programme, resulting in the construction of a third prototype, the XP-63A, first flown on 26 April 1943 and powered by a V-1710-93 engine with a war emergency rating of 1,500 hp (1 119 kW). It was planned subsequently to flight-test this prototype with a Packard-Merlin V-1650-5 engine installed, under the designation XP-63B, but this did not happen.

The performance of the XP-63A was found to be satisfactory, and the type was ordered into production in September 1942. Initial deliveries of P-63As began in October 1943, and by the time production ended in 1945 more than 3,300 Kingcobras had been built in several versions. By far the majority, something in excess of 2,400, were supplied to the USSR under Lend-Lease, and about 300 went to the Free French Armée de l'Air. Very few of the total production of P-63 close-support fighters/fighter-bombers were delivered to the USAAF, and so far as is known no Kingcobras were used operationally by that service.

Equipment of production batches varied considerably, resulting in many sub-types. The first production P-63A-1s had V-1710-93 engines, a nose-mounted 37-mm M4 cannon and two 0.50-in (12.7-mm) machine-guns in underwing fairings; other sub-types had two additional 0.50-in (12.7-mm) guns mounted in the fuselage nose. P-63A-1s and -5s could accommodate a 75-US gallon (284-litre) or 175-US gallon (662-litre) drop tank, or a 522-lb (237-kg) bomb beneath the wing centre-section; P-63A-6s had underwing racks for two similar bombs or additional fuel; and P-63A-10s could mount three air-to-ground rockets beneath each wing. The weight of defensive armour, intended primarily to give protection from ground weapons, increased progressively from 87.7 lb (39.8 kg) on the P-63A-1 to 236.3 lb (107.2 kg) on the P-63A-10.

The P-63A was succeeded on the production line by the P-63C with the V-1710-117 engine, this offering with water injection an emergency war rating of 1,800 hp (1 342 kW). A distinctive identification feature of the P-63C was provided by the introduction of a small ventral fin. Other variants included a single P-63D with V-1710-109 engine, a bubble canopy, and increased wing span; 13 P-63Es, all that had been produced of 2,930 on order when contracts were cancelled at the war's end, and which were generally similar to the P-63D except for a reversion to the standard cockpit canopy; and two P-63Fs, a version of the P-63E with a V-1710-135 engine and modified tail surfaces.

One other unusual version of the Kingcobra was built extensively (in excess of 300) for use by the USAAF in a training programme involving the use of live ammunition. Developed from the P-63A, all armour and armament was removed, and the external surface of the wings, fuselage and tail unit were protected externally by the addition of a duralumin alloy skin

Compared with the P-63A-1, the Bell P-63A-9 Kingcobra carried the same gun armament (with increased 37-mm ammunition stowage) but had provision for a pair of 522-lb (237-kg) bombs under the outer wing panels.

Bell P-63 Kingcobra

weighing some 1,500 lb (680 kg). Other protection included the installation of bulletproof glass in windscreen and cockpit side and upper windows, the provision of a steel grille over the engine air intake and steel guards for the exhaust stacks, and the use of a propeller with thick-walled hollow blades. All of these precautions were to make it possible for the aircraft to be flown as a target that could withstand, without significant damage, the impact of frangible bullets. When a hit was made by an attacking aircraft a red light blinked to confirm the accuracy of the weapon being fired against it.

The first five of these target aircraft were designated RP-63A-11; the 95 RP-63A-12s which followed had increased fuel tankage; the next production version, with the V-1710-117 engine, became designated RP-63C (200 built); and the final version was the RP-63G (32 built), these having the V-1710-135 engine. Although never flown as pilotless drone aircraft, the designations of these three versions were changed

Although widely produced for the USAAF, the Bell P-63 Kingcobra was most extensively deployed in Russia, which received more than 2,400.

subsequently to QF-63A, QF-63C, and QF-63G respectively.

Specification
Type: single-seat close-support fighter/fighter-bomber and target aircraft
Powerplant (P-63A): one 1,325-hp (988-kW) Allison V-1710-93 inline piston engine
Performance: maximum speed 410 mph (660 km/h) at 25,000 ft (7 620 m); cruising speed 378 mph (608 km/h); service ceiling 43,000 ft (13 110 m); range with maximum weapon load and internal fuel 450 miles (724 km); ferry range with maximum internal and external fuel 2,200 miles (3 541 km)
Weights: empty 6,375 lb (2 892 kg); maximum take-off 10,500 lb (4 763 kg)
Dimensions: span 38 ft 4 in (11.68 m); length 32 ft 8 in (9.96 m); height 12 ft 7 in (3.84 m); wing area 248 sq ft (23.04 m²)
Armament: one 37-mm M4 cannon, two wing-mounted and two nose-mounted 0.50-in (12.7-mm) machine-guns, plus up to three 522-lb (237-kg) bombs
Operators: FFAF, Soviet Air Force, USAAF

Boeing B-17 Flying Fortress

Boeing B-17D of the 14th Squadron, 19th Bombardment Group, based at Clark Field, Luzon (Philippine Islands) in December 1941.

History and Notes

Although it is sometimes introduced as the most famous of all US World War II aircraft, there are many who will argue that Boeing's B-17 Flying Fortress ranks equally with several other superb machines which became available to the US Army at just the right moment. The North American P-51 Mustang has its ardent advocates for pride of place in the USAAF's wartime armoury, but it was a child of war, conceived to live, fight and endure in the battle-torn skies of Europe. The origin of the Fortress was very different, its gestation long and troubled.

In the first few years after World War I the US Army Air Corps' Brigadier General William ('Billy') Mitchell began his campaign in favour of strategic bombing, demonstrating (perhaps inconclusively) the ascendancy of bomber over battleship in July 1921 and September 1923 by the destruction of captured or obsolete warships anchored at sea. His burning belief in air power led to a bitter campaign, against the US Navy initially, but later involving also the US Army. In the last month of 1925 'Billy' Mitchell was court-martialled and suspended from the service: he resigned very soon after this verdict, so that he could continue his campaign for the creation of the air force which he believed was needed by the USA. World War II was to prove him right in his ideas for in 1946, 10 years after his death, he was elevated to the rank of the nation's heroes by the posthumous award of the Congressional Medal of Honor.

Although Mitchell had been discredited in 1925, there were many of his former colleagues who were less outspoken but nevertheless believed in the concept of air power. With Mitchell no longer there to provide support and encouragement, the efforts of this small steering nucleus were necessarily slow. More far sighted, in some ways, were the nation's aircraft manufacturers. Boeing, for example, began work in 1930 on its Models 214 and 215, twin-engined developments of its revolutionary Model 200 Monomail civil airliner. Built as a private venture these were ordered in small numbers as Y1B-9 and YB-9, but the first significant order for monoplane bombers went to the Glenn L. Martin Company for 48 twin-engined B-10 bombers.

Deliveries of production B-10s began in June 1934, and in a changed climate of opinion the US Army had issued a month earlier its specification for an even

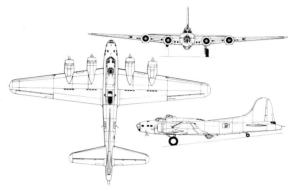

Boeing B-17E Flying Fortress

more advanced multi-engined bomber, able to haul a bomb load of 2,000 lb (907 kg) over a range of between 1,020 miles (1 640 km) and, optimistically, 2,200 miles (3 540 km), at speeds of between 200 and 250 mph (322 and 402 km/h). So far as the US Army was concerned, 'multi' meant more than one engine but Boeing, invited to submit its proposal for this requirement, elected to use four engines to power its Model 299, on which design work was initiated in mid-June 1934.

For Boeing the Model 299, built as a private venture, was a make or break gamble. Hitherto the company had built aircraft in only 'penny packet' numbers: the failure of the B-9 to win a worthwhile order had forced economies of near desperation upon Boeing, with its work force split in half and working two weeks on and two weeks off. Unless the Model 299 entered production in significant numbers the company faced, at the least, a very bleak prospect. Not surprisingly, every effort was devoted to the success of the project; every employee knew that he or she had an important contribution to make if the company was to survive.

The US Army specification had stipulated that the prototype should be available for test in August 1935, and however impossible this target had seemed in mid-1934, it became reality on 16 July 1935 when the Model 299 was rolled out of its hangar at Boeing Field, Seattle, for its first introduction to the press. Headlines on the following day announced the new '15-ton Flying Fortress', and seizing upon the name the company had it registered as the official name of its Model 299. Contrary to popular belief, this was not because of its

Boeing B-17 Flying Fortress

defensive armament, but because it was procured as an aircraft which would be operated as a mobile flying fortress to protect America's coastline, a concept which needs some explanation.

USAAC protagonists of air power were still compelled to step warily, despite procurement of the B-10 bomber, for the US Navy had the most prestigious support in the corridors of power and was determined to keep the upstart US Army in its place. Even if strategic bombers were required, efforts must be made to prevent the US Army acquiring such machines. The USAAC was, however, quite astute when needs be and so, with tongue in cheek, succeeded in procuring 13 YB-17s, the original service designation of the Fortress, for coastal defence. However, this explanation anticipates the story.

On 28 July 1935 the Model 299 flew for the first time: just over three weeks later it was flown non-stop to Wright Field, Ohio, to be handed over for official test and evaluation. The 2,100-mile (3 380-km) flight had been made at an average speed of 252 mph (406 km/h), a most impressive performance which augured well for the future. The elation of the Boeing company was understandable, especially with confirmation that initial trials were progressing well. On 30 October 1935 hopes were dashed with the news that the prototype had crashed on take-off. Subsequent investigation was to prove that the attempt to take-off had been made with the controls locked, and in view of the satisfactory testing prior to this accident, the USAAC decided on the procurement of 13 YB-17s (later Y1B-17s), plus one example for static testing.

The prototype (X13372) which had crashed at Wright Field was powered by four 750-hp (559-kW) Pratt & Whitney R-1690-E Hornet radial engines. The cantilever monoplane wings were in a low-wing configuration, the wing section at the root so thick that it was equal to half the diameter of the circular-section fuselage; and wide-span trailing-edge flaps were provided to help reduce take-off and landing speeds. Landing gear was of the electrically retractable tailwheel type. Armament comprised five machine-guns, and a maximum bomb load of 4,800 lb (2 177 kg) could be carried in the fuselage bomb bay.

The initial Y1B-17 (36-149) flew for the first time on 2 December 1936, and differed from the prototype by having 930-hp (694-kW) Wright GR-1820-39 Cyclone

Two Boeing B-17Fs of the 322nd Bombardment Squadron from the 8th Air Force's 91st Bombardment Group which claimed the 8th Air Force's highest number of 'kills' (420 aircraft).

Boeing B-17 Flying Fortress

radials, accommodation for a crew of nine, and minor changes in detail. Twelve were delivered between January and August 1937, equipping the USAAC's 2nd Bombardment Group at Langley Field, Virginia. The thirteenth aircraft went to Wright Field for further tests and after one of the Y1B-17s survived without damage the turbulence of a violent storm, it was decided that the static test example would, instead, be completed as an operational aircraft. Designated Y1B-17A, this aircraft (37-369) was provided with 1,000-hp (746-kW) GR-1820-51 engines each fitted with a Moss/General Electric turbocharger (supercharger powered by a turbine driven by exhaust gases). It flew for the first time on 29 April 1938, and subsequent testing by the USAAC gave convincing proof of the superiority of the turbocharged engine over those which were normally aspirated, and such engines were to become standard on all future versions of the Fortress.

The utilisation of the Y1B-17s, designated B-17 in service with the 2nd Bombardment Group, did little to improve relations between the US Army and US Navy. When three of the force were used to stage an 'interception' of the Italian liner *Rex* some 750 miles (1 207 km) out in the Atlantic, to demonstrate that the USAAC was more than capable of defending the nation's coastline, it sparked a row which dispersed the air power disciples from General Headquarters Air Force (GHQAF) to other commands, where they were remote from each other and potential influential supporters. Orders for additional B-17s had to be reduced after it had been underlined by Major General Stanley D. Embrick that ' . . . the military superiority of a B-17 over the two or three smaller aircraft which could be procured with the same funds has yet to be established.' This helps explain why, despite the growing war clouds in Europe, the USAAC had less than 30 B-17s when Hitler's forces invaded Poland on 1 September 1939.

The order for Y1B-17s was followed by a contract for 39 B-17Bs, more or less identical to the Y1B-17A prototype with turbocharged engines. The first of these flew on 27 June 1939, and all had been delivered by March 1940. In 1939 the B-17C was ordered, the first of the 38 on contract making its first flight on 21 July 1940. They differed by having 1,200-hp (895-kW) R-1820-65 engines, and by an increase from five to seven machine-guns.

A Douglas-built Boeing B-17G unloads over its target. Easily seen are the type's four main defensive turrets (chin, dorsal, ventral and tail) each with a pair of 0.50-in (12.7-mm) guns.

Boeing B-17 Flying Fortress

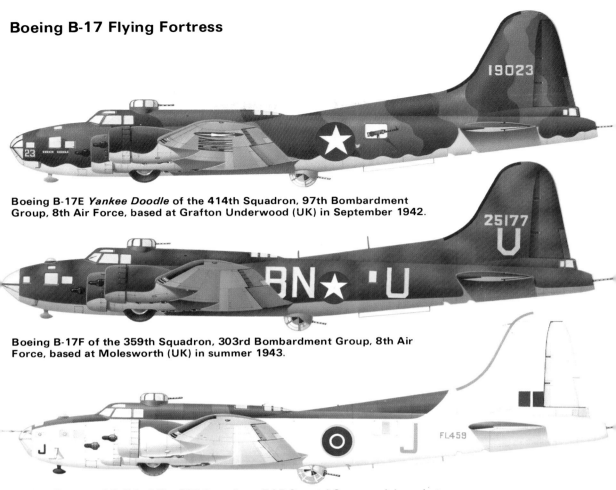

Boeing B-17E *Yankee Doodle* of the 414th Squadron, 97th Bombardment Group, 8th Air Force, based at Grafton Underwood (UK) in September 1942.

Boeing B-17F of the 359th Squadron, 303rd Bombardment Group, 8th Air Force, based at Molesworth (UK) in summer 1943.

Boeing Fortress Mk IIA of No. 220 Squadron, RAF Coastal Command, based at Ballykenny (Northern Ireland) in late 1942.

The B-17C was the first version of this bomber to be supplied to the RAF in Great Britain, which designated the 20 examples received in early 1941 as Fortress Is. Equipping No. 90 Squadron, they were used operationally for the first time on 8 July 1941 when aircraft launched a high-altitude (30,000 ft / 9 145 m) attack on Wilhelmshaven. In the 26 attacks made on German targets during the next two months the Fortress Is proved unsatisfactory, although there was American criticism of the way in which they had been deployed. Nonetheless, their use in daylight over German territory had proved that their operating altitude was an inadequate defence in itself, and so they needed more formidable defensive armament, for Messerschmitt Bf 109E and 109F fighters had little difficulty in intercepting them at heights of up to 32,000 ft (9 750 m). Until improvements in the Fortress were made, or means found of deploying them more effectively, they were withdrawn from operations over Europe.

With the end of 1941 drawing near, the USA was soon to become involved in World War II, initially in the Pacific theatre, but following the containment of the initial explosion of Japanese expansion it was

decided that the Allies would first concentrate their efforts on bringing about a speedy conclusion of the war in Europe. Thus, large numbers of B-17s which otherwise would have found employment in the Far East were instead to equip the USAAF's 8th Air Force in Britain. Those allocated to serve with the Anglo-American Northwest African Air Forces were later to become part of the US 15th Air Force.

In 1940 Boeing received an order for 42 B-17Ds. These differed little from the B-17C, but as a result of early reports of combat conditions in Europe were provided with self-sealing tanks and additional armour for protection of the crew, and these were delivered during 1941. The B-17E which followed was the first version to benefit from the RAF's operational experience with its Fortress Is. A major redesign provided a much larger tail unit to improve stability at high altitude, and to overcome the criticism of inadequate defence 13 machine-guns were mounted in one manual and two power-operated turrets, radio compartment, waist stations and in the nose. Of the 512 of this version built under two contracts, the first flew on 5 September 1941. B-17Es were the first to serve with the 8th Air

Boeing B-17 Flying Fortress

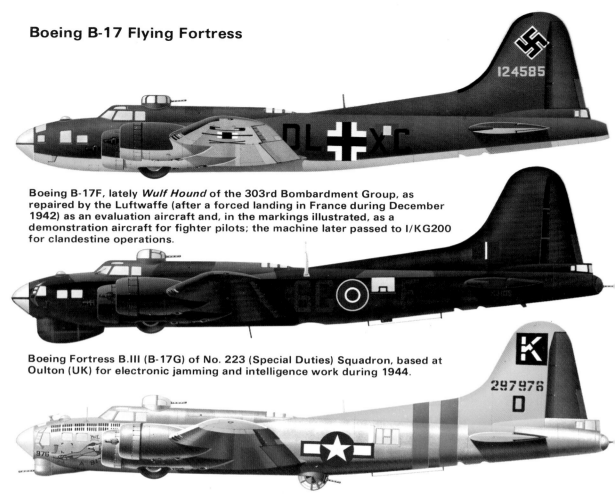

Boeing B-17F, lately *Wulf Hound* of the 303rd Bombardment Group, as repaired by the Luftwaffe (after a forced landing in France during December 1942) as an evaluation aircraft and, in the markings illustrated, as a demonstration aircraft for fighter pilots; the machine later passed to I/KG200 for clandestine operations.

Boeing Fortress B.III (B-17G) of No. 223 (Special Duties) Squadron, based at Oulton (UK) for electronic jamming and intelligence work during 1944.

Boeing B-17G *A Bit O'Lace* of the 447th Bombardment Group, 8th Air Force, based at Rattlesden (UK) in early 1945.

Force in Europe, with deliveries beginning in July 1942. They were used operationally for the first time by the 97th Bombardment Group, 12 aircraft being detailed for a daylight attack on Rouen on 17 August, with fighter escort provided by RAF Supermarine Spitfires.

The B-17F, of which the first flew on 30 May 1942, was the first version to be built in large numbers: Boeing produced 2,300 at Seattle, and further construction of 1,105 came from Douglas (605) and Lockheed Vega (500). Major changes included a redesigned nose, and strengthened landing gear to cater for a higher gross weight. Other changes included increased fuel capacity, the introduction of additional armour, provision of external bomb racks beneath the inner wings and, on late production aircraft, the introduction of R-1820-97 engines.

The B-17Es and B-17Fs became used extensively by the 8th Air Force in Europe, but in two major operations against German strategic targets, on 17 August and 14 October 1943, a total of 120 aircraft were lost. Clearly the Fortresses could not mount an adequate defence, no matter how cleverly devised was the box formation in which they flew. The hard truth was that without adequate long-range fighter escort they were very vulnerable to attack during mass daylight operations. Many of the losses were attributed to head-on attack, and the final major production version was planned to offset this shortcoming.

Thus the B-17Gs had a 'chin' turret housing two 0.50-in (12.7-mm) machine-guns mounted beneath the fuselage nose, which meant that this version carried a total of 13 0.50-in (12.7-mm) guns. To increase the aircraft's operational ceiling, later production examples had an improved turbocharger for their R-1820-97 engines. B-17G production totalled 8,680, built by Boeing (4,035), Douglas (2,395), and Lockheed Vega (2,250).

Although used most extensively in Europe and the Middle East, B-17s were operational in every area where US forces were fighting. In the Pacific theatre they offered invaluable service for maritime patrol, reconnaissance, and conventional and close-support bombing. A number of variants were also produced or converted for special purposes and operations, and details of these follow. Although almost 13,000 B-17s

Boeing B-17 Flying Fortress

were built, only a few hundred B-17Gs were retained in USAAF service after the end of the war, and these were soon made redundant.

B-17H: small number of air-sea rescue aircraft with search radar and droppable lifeboat; later designated SB-17G

B-40: bomber escort, originating from XB-40 converted from a B-17F; four TB-40 trainers were built, and some YB-40s carried up to 30 guns; not operationally successful

BQ-7: pilotless flying-bomb, radio controlled, from which crew of two parachuted after setting it on course; inaccurate and little used

CB-17G and VB-17G: B-17Gs equipped as staff transports

DB-17P: drone directors

F-9: photo-reconnaissance versions with different camera installations producing F-9A, -9B and -9C variants; other designations for PR aircraft included FB-17 and RB-17G

QB-17L and QB-17N: target drones

TB-17G: special duty trainers

XB-38: one aircraft equipped experimentally with Allison V-1710-89 engines

XC-108: transport conversion to accommodate 38 passengers

XC-108A: cargo transport with freight door on port side

XC-108B: experimental fuel tanker

YC-108: VIP transport

Specification

Type: 10-seat long-range medium bomber/reconnaissance aircraft

Powerplant (B-17G): four 1,200-hp (895-kW) turbocharged Wright R-1820-97 Cyclone radial piston engines

Performance: maximum speed 287 mph (462 km/h) at 25,000 ft (7 620 m); cruising speed 182 mph (293 km/h); service ceiling 35,800 ft (10 850 m); range with 6,000-lb (2 722-kg) bomb load 2,000 miles (3 219 km)

Weights: empty 36,135 lb (16 391 kg); maximum take-off 65,500 lb (29 710 kg)

Dimensions: span 103 ft 9 in (31.62 m); length 74 ft 4 in (22.66 m); height 19 ft 1 in (5.82 m); wing area 1,420 sq ft (131.92 m²)

Armament: 13 0.50-in (12.7-mm) machine-guns, plus up to 17,600 lb (7 983 kg) of bombs

Operators: RAF, USAAC/USAAF

5 Grand was the 5,000th Boeing-built B-17 (a B-17G in this instance), and its metal surface was signed by thousands of Boeing employees. The hatch in the rear fuselage side (and one on the other side) mounted a single machine-gun.

Fast Woman was a B-17F-40-BO, built in Seattle by
Boeing and attached to the 359th Bomb Squadron of
the 303rd Bomb Group, 8th Air Force. The aeroplane is shown as it appeared when the group began
operations in Europe from its base at Molesworth in
Huntingdonshire in January 1943.

Boeing B-29 Superfortress

History and Notes

That there were senior officers of the US Army Air Corps who were well aware of the need to procure long-range strategic bombers had been made clear in the entry dealing with the Boeing B-17. In addition to the long and drawn out process of getting B-17s into squadron service, the USAAC had also initiated procurement of more potent aircraft, ordering a prototype XBLR-1 (Experimental Bomber Long Range-1) from Boeing, which was built and flown as the XB-15. A competitive XBLR-2 (later XB-19) was ordered from Douglas Aircraft Company, and after it and the XB-15 had been evaluated, both were put into 'cold storage' until more powerful engines became available.

The outbreak of war in Europe in 1939 made it essential that USAAC planners should at least talk about long-range bomber projects, and the initial identification of such was VHB (very heavy bomber). When it seemed likely that such an aircraft might have to be deployed over the vast reaches of the Pacific Ocean the identification VLR (very long-range) seemed more apt, and it was the VLR project which General Henry H. ('Hap') Arnold, head of the USAAC, got under way at the beginning of 1940.

Requests for Proposals were sent to five US aircraft manufacturers on 29 January 1940: in due course design studies were submitted by Boeing, Consolidated, Douglas and Lockheed, these being allocated the respective designations XB-29, XB-32, XB-31 and XB-30. Douglas and Lockheed subsequently withdrew from the competition, and on 6 September 1940 contracts were awarded to Boeing and Consolidated (Convair) for the construction and development of two (later three) prototypes of their respective designs.

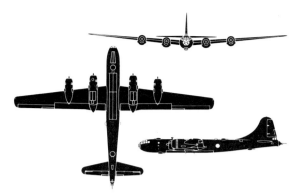

Boeing B-29 Superfortress

Convair's XB-32 Dominator was the first to fly, on 7 September 1942, but extensive development delayed its entry into service.

Boeing, because of the company's foresight, was much further along the design road in 1940, and being able to convince the USAAC that they would have production aircraft available within two or three years, had received orders for more than 1,500 before a prototype was flown. The reason for the advanced design state of Boeing's proposal was due to the fact that as early as 1938 the company had offered to the USAAC its ideas for an improved B-17, with a pressurised cabin to make high-altitude operations less demanding

The 'wing' under the fuselage of this Bell-built Boeing B-29 Superfortress indicates that the aircraft is fitted with APQ-7 bombing radar for adverse-weather operations.

Boeing B-29 Superfortress

on the crew. While there was then no requirement for such an aircraft, the US Army encouraged Boeing to keep the design updated to meet the changing conditions of war. This was reflected by the designs identified as Models 316, 322, 333, 334 and 341. The design for the XB-29 was a development of the Model 341, designated Model 345, and the first of the prototypes made its maiden flight on 21 September 1942.

The USAAC's specification had called for a speed of 400 mph (644 km/h), so the XB-29 had a high aspect ratio cantilever monoplane wing mid-set on the circular-section fuselage. Because such a wing would entail a high landing speed, the wide-span trailing-edge flaps were of the Fowler type which effectively increased wing area by almost 20 per cent, thus allowing a landing to be made at lower speed. Electrically-retractable tricycle landing gear was provided and, as originally proposed by Boeing, pressurised accommodation was included for the flight crew. In addition, a second pressurised compartment just aft of the wing gave accommodation to crew members who, in the third XB-29 and production aircraft, sighted defensive gun turrets from adjacent blister windows. The crew and aft compartments were connected by a crawl-tunnel which passed over the fore and aft bomb bays. The tail gunner was accommodated in a pressurised compartment, but this was isolated from the other crew positions. The powerplant consisted of four Wright R-3350 Cyclone twin-row radial engines, each with two General Electric turbochargers mounted one in each side of the engine nacelle. The 16 ft 7 in (5.05 m) diameter four-blade metal propellers were of the constant-speed and fully-feathering type.

Prototype production was followed by 14 YB-29

Compared with the B-29, the Boeing B-29A had a wing of increased span, different engines and a four-gun front dorsal turret. The model was built only at the Renton plant.

service test aircraft, the first of these flying on 26 June 1943. Deliveries of YB-29s began almost immediately to the 58th Very Heavy Bombardment Wing (VHBW), a unit which had been established on 1 June in advance of the first flight. B-29 production was the largest aircraft manufacturing project undertaken in the USA during World War II, with literally thousands of sub-contractors supplying components or assemblies to the four main production plants: Boeing at Renton and Wichita; Bell at Marietta, Georgia; and Martin at Omaha, Nebraska.

Deliveries of production B-29s started in the autumn of 1943, and these began to equip the 58th VHBW so that it could proceed with training and get groups ready for operational service. One of the tricky questions was where to send the units initially, for the Allied/US agreement to end the war in Europe first would suggest their deployment against Germany and German-occupied territories. However, as 1943 was nearing its end, the situation in the Far East suggested that they could be used more effectively in that area, and the decision was made to send them to operate in the theatre for which they had been designed.

On 4 April 1944, the 20th Air Force was established to operate the B-29s, but as at that time no island bases were available from which the B-29s could strike at the Japanese home islands, preparations had already been made for them to operate initially from bases in China. Something like half a million Chinese farmers and

29

Boeing B-29 Superfortress

The Boeing B-29 Superfortress combined in one package all the new tools of the strategic bombing trade: a high-altitude, stable platform; advanced optical and radar bombsights; heavy, automated and centrally-controlled defensive firepower; very long range; and in relation to the range, a heavy bombload. Designed to perform precision bombing from high altitudes, the B-29 was successful only after it had been employed in a way totally different from that conceived by its proponents. The Superfortress fleets, operating by night and at low altitudes, dumped tons of incendiaries on the highly combustible Japanese cities, so turning them to ash. *The Big Stick* was one of the B-29s operating from the island bases in the Marianas group. It was allocated to the 500th Bomb Group of the 73rd Bomb Group (Very Heavy), 20th Air Force.

Boeing B-29 Superfortress

peasants laboured with simple hand tools to create four airfields for the B-29s in the Chengtu area of Szechwan province, and the first aircraft landed at Kwanghan air base on 24 April 1944. By 10 May all four bases were operational and the first attack against a Japanese home island target, the Imperial Iron and Steel Works at Yawata, Kyushu, was made by 77 aircraft of XX Bomber Command on 15 June 1944. There were many problems to these operations from the Chinese bases, not least of which was logistics: about 150 B-29s were used continually to haul essential fuel and supplies to Kunming, over the Himalayan 'hump' from India, thus making it possible for 100 B-29s to remain operational. But it was not until the establishment of bases on Saipan, Guam and Tinian in the Marianas that the major B-29 offensive could be launched against Japan.

The first of XXI Bomber Command's Superfortresses landed on Saipan's Isley Field on 12 October 1944; Tinian's first airstrip was operational in late December; and that on Guam on 2 February 1945. But the answer to the question of how to employ the B-29s most effectively was not resolved until the night of 9/10 March 1945, when 334 aircraft flying from Guam, Saipan and Tinian set out to attack Tokyo, some 1,600 miles (2 575 km) distant. When they returned they had recorded the most devastating air attack ever made, with 83,793 people dead, 40,918 injured and 1,008,005 rendered homeless.

This was to be the continuing pattern for XXI Bomber Command, while XX Bomber Command reduced Formosa's towns and docks to little more than rubble. It remained only for the B-29s *Enola Gay* and *Bock's Car*, of the 393rd Bombardment Squadron, to drop the world's only operational atomic bombs over Hiroshima and Nagasaki on 6 and 9 August 1945 respectively to bring World War II to a close. In these closing stages

XXI Bomber Command B-29s had dropped some 160,000 tons (162 560 tonnes) of bombs on Japanese targets, averaging 1,193 tons (1 212 tonnes) per day during the last three months: the USAAC's VLR project was justified.

B-29 production totalled 1,644 from Boeing's Wichita plant, with 668 built by Bell and 536 by Martin. The Renton plant produced only B-29As, with slightly increased span and changes in fuel capacity and armament: production continued until May 1946 and totalled 1,122 aircraft.

Specification

Type: 10-seat long-range strategic bomber/reconnaissance aircraft
Powerplant (B-29): four 2,200-hp (1 641-kW) Wright R-3350-23-23A/-41 Cyclone 18 turbocharged radial piston engines
Performance: maximum speed 358 mph (576 km/h) at 25,000 ft (7 620 m); cruising speed 230 mph (370 km/h); service ceiling 31,850 ft (9 710 m); range 3,250 miles (5 230 km)
Weights: empty 70,140 lb (31 815 kg); maximum take-off 124,000 lb (56 245 kg)
Dimensions: span 141 ft 3 in (43.05 m); length 99 ft 0 in (30.18 m); height 29 ft 7 in (9.02 m); wing area 1,736 sq ft (161.27 m²)
Armament: two 0.50-in (12.7-mm) machine-guns in each of four remotely-controlled power-operated turrets, and three 0.50-in (12.7-mm) guns or two 0.50-in (12.7-mm) guns and one 20-mm cannon in the tail turret, plus a bomb load of up to 20,000 lb (9 072 kg)
Operator: USAAF

Maintenance of Boeing B-29s in the primitive airfield conditions of China proved a constant problem, and was most marked with complex items such as the R-3350 engines.

Boeing B-314

History and Notes

As early as January 1935, Pan American Airways had signified to the US Bureau of Air Commerce its wish to establish a transatlantic service and, despite its ownership of the large Martin M-130 and Sikorsky S-42 long-range four-engined flying boats, the airline wanted a new aircraft for the route.

Boeing submitted a successful tender to the Pan American specification and a contract for six Boeing 314s was signed on 21 July 1936. The manufacturer used features of the earlier XB-15 heavy bomber, adapting the wing and horizontal tail surfaces for its 82,500-lb (37 421-kg) gross weight flying boat, which could accommodate up to 74 passengers in four separate cabins. The engines were not the 1,000-hp (746 kW) Pratt & Whitney R-1830 Twin Wasps of the XB-15, but 1,500-hp (1 119-kW) Wright GR-2600 Double Cyclones which gave the machine a maximum speed of 193 mph (311 km/h). The fuel capacity of 4,200 US gallons (15 898 litres) conferred a maximum range of 3,500 miles (5 633 km); some of the fuel was stored in the stabilising sponsons which also served as loading platforms.

The first Boeing 314 took off on its maiden flight on 7 June 1939, this original version having a single fin and rudder, later replaced by twin tail surfaces to improve directional stability. These proved to be inadequate, and the original centreline fin was restored, without a movable rudder. The aircraft was awarded Approved Type Certificate 704 and entered transatlantic airmail service on 20 May 1939, passenger service commencing on 28 June. At that time the 314 was the largest production airliner in regular passenger service.

Pan American ordered another six aircraft which were designated Model 314A, improved by the installation of 1,600-hp (1 193-kW) Double Cyclones with larger-diameter propellers, an additional 1,200 US gallons (4 542 litres) of fuel capacity, and a revised interior. The first 314A flew on 20 March 1941 and delivery was complete by 20 January 1942. Five of the original order were retrospectively converted to 314A standard in 1942. Three of the repeat order were sold, before delivery, to BOAC for transatlantic service and operation on the Foynes-Lagos sector of the wartime 'Horseshoe Route'.

Of Pan American's nine 314/314As, four were requisitioned by Army Transport Command and given the military designation C-98. They were little used, however, and in November 1942 one was returned to the airline. The other three were transferred to the US Navy to join two acquired direct from Pan American; the airline provided crews for the US Navy's B-314 operations and the aircraft were partially camouflaged but operated with civil registrations.

BOAC and Pan American terminated Boeing 314 services in 1946 and the surviving aircraft were sold to American charter airlines.

Specification

Type: long-range flying boat transport
Powerplant (B-314A): four 1,600-hp (1 193-kW) Wright R-2600 Cyclone 14 radial piston engines
Performance: maximum speed 193 mph (311 km/h) at 10,000 ft (3 050 m); cruising speed 183 mph (295 km/h); service ceiling 13,400 ft (4 085 m); range 3,500 miles (5 633 km)
Weights: empty 50,268 lb (22 801 kg); maximum take-off 82,500 lb (37 421 kg)
Dimensions: span 152 ft 0 in (46.33 m); length 106 ft 0 in (32.31 m); height 27 ft 7 in (8.41 m); wing area 2,867 sq ft (266.34 m²)
Armament: none
Operators: USAAF, USN

This was one of four Boeing 314As impressed by the Army Transport Command under the designation C-98. One boat was returned to Pan American in November 1942, while the remaining three (including this boat) became part of the US Navy's B-314 fleet (five boats).

Boeing C-73

History and Notes

On 8 February 1933, Boeing flew the prototype of a new civil airliner which was identified by the company as its Model 247. This had derived via the design of the single-engined civil Model 200 Monomail and the twin-engined Model 214 (US Army designation B-9) bomber, each of which had a cantilever monoplane wing.

A revolutionary aircraft, it has since become regarded as a prototype for the modern airliner, for it was a clean cantilever low-wing monoplane of all-metal construction with twin-engined powerplant, retractable landing gear, and accommodated a pilot, co-pilot, stewardess and 10 passengers. With one engine inoperative it could climb and maintain altitude with a full load, and introduced a new feature for a civil transport aircraft by being equipped with pneumatic de-icing boots on wing, tailplane and fin leading edges to prevent ice accretion from reaching a dangerous level.

Sixty examples of the Model 247 were ordered 'off the drawing board' to re-equip the Boeing Air Transport System, shortly to become a major limb of United Air Lines, and another 15 were ordered subsequently for companies or individuals. That built for Roscoe Turner and Clyde Pangborn (to compete in the England-Australia 'MacRobertson' air race of 1934) was provided with fuselage fuel tanks instead of the standard airline cabin equipment, and introduced NACA engine cowlings (to reduce drag) and controllable-pitch propellers with optimum settings for take-off and cruising performance. These improvements were incorporated retrospectively on most airline Model 247s, thus elevating them to Model 247D standard.

When the USA became involved in World War II in late 1941, these Model 247Ds remained in airline use, and 27 of them were impressed for service with the USAAF under the designation C-73. It had been anticipated that they could be used for the carriage of cargo and troops, but it was discovered that the cabin doors were too small for this purpose. Instead, they were deployed to ferry aircrew and, later in the war,

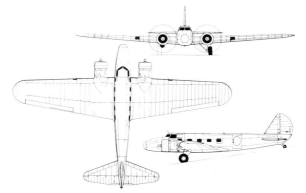

Boeing C-73 (Model 247D)

were used for training. In service they were provided with more powerful engines than the standard civil 550-hp (410-kW) Pratt & Whitney Wasp S1H-1G radials. When no longer required at the war's end they were returned to civil airline service.

Specification

Type: civil transport aircraft utilised in military role
Powerplant: two 600-hp (447-kW) Pratt & Whitney R-1340-AN-1 Wasp radial piston engines
Performance: maximum speed 200 mph (322 km/h); cruising speed 189 mph (304 km/h); service ceiling 25,400 ft (7 740 m); range 745 miles (1 199 km)
Weight: maximum take-off 13,650 lb (6 192 kg)
Dimensions: span 74 ft 0 in (22.56 m); length 51 ft 7 in (15.72 m); wing area 836 sq ft (77.66 m²)
Armament: none
Operator: USAAF

Some 27 Boeing Model 247D airliners were impressed for USAAF service, but their limited payload and small doors dictated their use for secondary tasks such as pilot ferrying before they were returned to civil use in 1945.

Boeing C-75

History and Notes

Boeing's Model 299, prototype for the military bomber aircraft which duly became the B-17 Flying Fortress, was developed in parallel with a civil version of the same aircraft which had the company designation Model 300. The basic plan was for both to have a common wing, tail unit and powerplant, but from the beginning a more spacious fuselage had been designed for the civil version. As the design progressed, however, it was decided to provide a circular-section fuselage with moderate pressurisation of 2½ lb/sq in (0.18 kg/cm²), providing a cabin altitude of 8,000 ft (2 440 m) to a height of 14,700 ft (4 480 m), so permitting the Model 307, as this final design was identified by Boeing, to operate with passengers at 20,000 ft (6 095 m), a height above much of the turbulent weather. When, in due course, the Model 307 entered airline service, this 'high-altitude' operational capability resulted in selection of the same Stratoliner.

Ten Model 307s were built, the first making its maiden flight on 31 December 1938. Unfortunately, this aircraft was lost before it could be delivered to Pan American. Of the nine which remained, three went to Pan Am, five to Transcontinental & Western Air (TWA), and one modified aircraft to Howard Hughes.

Those which had been built for TWA, as SA-307Bs, were impressed into USAAF service in 1942, receiving the designation C-75. With accommodation for 33 passengers and with a crew of five, they were operated by TWA under contract to the USAAF's Air Transport Command, as VIP transports for the highest ranking civilian and military personnel. After two and a half years service, during which these five aircraft accumulated between them approximately 3,000 transatlantic crossings, some 45,000 flight hours, and travelled about 7.5 million miles (12 million km), they were released from military service and returned to Boeing for refurbishing and conversion back to airline standards.

Specification

Type: long-range VIP transport
Powerplant: four 900-hp (671-kW) Wright GR-1820 Cyclone radial piston engines
Performance: maximum speed 246 mph (396 km/h); cruising speed 220 mph (354 km/h); service ceiling 26,200 ft (7 985 m); range 2,390 miles (3 846 km)
Weights: empty 30,000 lb (13 608 kg); maximum take-off 45,000 lb (20 412 kg)
Dimensions: span 107 ft 0 in (32.61 m); length 74 ft 4 in (22.66 m); height 20 ft 9½ in (6.34 m); wing area 1,486 sq ft (138.05 m²)
Armament: none
Operator: USAAF

The last of TWA's Boeing SA-307B Stratoliners taken over by the USAAF in 1942, this machine was later returned to civil service after refurbishment with the wings, engines and landing gear of the B-17 Flying Fortress, under the revised designation SA-307B-1.

Boeing P-26

Boeing P-26A, formerly of the USAAC's 1st Pursuit Group, seen in the markings of Escuadron de Caza, Fuerza Aerea de Guatemala, based at the Campo de la Aurora (Guatemala City) in the late 1940s.

History and Notes

Although Boeing's diminutive P-26 fighter had been retired from front-line service by the time the United States entered World War II, P-26s were among the aircraft ranged against the Japanese at Pearl Harbor, and machines of the Philippine Army Air Force's 6th Pursuit Squadron were in action as Japanese forces fought their way through the archipelago.

Work on the company-funded Boeing Model 248 began in September 1931, although the US Army Air Corps contracted to supply engines and instruments for three trials aircraft which were designated XP-636. Destined to become the first all-metal production fighter and the first monoplane to serve with the USAAC in the pursuit role, the design retained an open cockpit and, despite Boeing's experience with retractable landing gear and cantilever wings, fixed landing gear and externally-braced wings. All of these deficiencies were remedied in the Boeing 264 or YP-29, which was flown in 1934 but not put into production.

The first XP-636 was flown on 20 March 1932, and later completed an evaluation programme at Wright Field, where the second airframe had been delivered for static tests. On 25 April the third was sent to Selfridge Field, Michigan for tests with operational squadrons. Boeing subsequently received a production order for 111 P-26As, later increased to 136, which were to incorporate some improvements, including revised wing structure, the addition of flotation gear and radio; later aircraft also had higher headrests to protect the pilot in a roll-over crash. The first production P-26A made its maiden flight on 10 January 1934; the last of the 111 was delivered at the end of June 1934.

The need to reduce the landing speed of the P-26 resulted in the development of trailing-edge flaps which were fitted retrospectively to aircraft already in service, and to those still on the production line. These included the additional order for 25, completed as two P-26Bs with fuel injection-equipped Pratt & Whitney

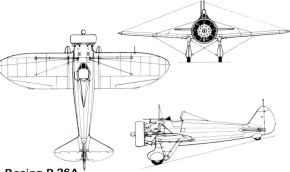

Boeing P-26A

Wasp R-1340-33 engines, and 23 P-26Cs which had minor changes to the fuel system and carburation. Many were later converted to P-26B standard.

Production was completed by 12 export Boeing Model 218s, comprising 11 from China and one for Spain; users of US surplus P-26s included Guatemala and Panama.

Specification

Type: single-seat fighter

Powerplant (P-26A/-26C): one 500-hp (373-kW) Pratt & Whitney R-1340-27 radial piston engine

Performance: maximum speed 234 mph (377 km/h) at 7,500 ft (2 285 m); cruising speed 199 mph (320 km/h); service ceiling 27,400 ft (8 350 m); range 360 miles (579 km)

Weights: empty 2,197 lb (997 kg); maximum take-off 2,955 lb (1 340 kg)

Dimensions: span 27 ft 11½ in (8.52 m); length 23 ft 7¼ in (7.19 m); height 10 ft 0½ in (3.06 m); wing area 149.5 sq ft (13.89 m²)

Armament: two fixed 0.50-in (12.7-mm) machine-guns or one 0.50-in (12.7-mm) and one 0.30-in (7.62-mm) gun.

Operators: China, Guatemala, Panama, Philippines, Spain, USAAC

Boeing XB-15/XC-105

History and Notes

However determined the majority of Americans might have been to maintain the nation's long-established policy of isolation, there were still numbers of radicals, in both the United States government and services, who realised that almost certainly the day would dawn when, for one reason or another, the USA would have to become involved in warlike activities. Given such circumstances, one of the essential weapons would be an advanced strategic bomber, and in the US Army men like Colonels Hugh Knerr and C.W. Howard were working steadily away in the 1930s to ensure, to the best of their capability, that when that moment came such a bomber would be available. Such thinking had led to the introduction into service of such bombers as the Boeing B-9, and the Martin B-10 and B-12. While it was appreciated that these did not represent the ideal, they prepared the way for the procurement of a true strategic bomber.

In 1933 came the US Army's requirement for a design study of such an aircraft: a range of 5,000 miles (8 046 km) was included in the specification to provide long-range strategic capability. Both Boeing and Martin produced design studies, but it was the former company which received the US Army's contract for construction and development of their Model 294, under the designation XB-15. When this large monoplane flew for the first time, on 15 October 1937, it was then the largest aircraft to be built in the USA.

As might be expected, it introduced a number of original features, including internal passages within the wing to permit minor engine repairs or adjustments in flight; two auxiliary power units within the fuselage to provide a 110-volt DC electrical system; sleeping bunks to allow for 'two-watch' operation; and the introduction of a flight engineer into the crew to reduce the pilot's workload. Intended to be powered by engines of around 2,000 hp (1 491 kW), which did not materialise for some years, the actual powerplant comprised four 1,000-hp (746-kW) Pratt & Whitney Twin Wasp Senior radial engines, which meant that performance was far below that estimated. Purely an experimental aircraft, it was however provided with cargo doors and flown as a cargo transport during World War II, under the designation XC-105.

Specification

Type: long-range bomber/transport
Powerplant: four 1,000-hp (746-kW) Pratt & Whitney Twin Wasp Senior radial piston engines
Performance: maximum speed 195 mph (314 km/h); service ceiling 18,900 ft (5 760 m); range 5,130 miles (8 256 km)
Weight: maximum take-off 92,000 lb (41 731 kg)
Dimensions: span 149 ft 0 in (45.42 m); length 87 ft 11 in (26.80 m); height 18 ft 0 in (5.49 m)
Armament: none as used in service as XC-105
Operator: USAAC/USAAF

A pioneering military aircraft in a number of ways, the Boeing XB-15 long-range bomber failed largely for lack of suitably powerful engines. Despite a gross weight 17,700 lb (8030 kg) greater than the later B-17F, the XB-15 had 600 hp (448 kW) less power.

Boeing XPBB-1 Sea Ranger

History and Notes

The contribution made to anti-submarine warfare (ASW) by aircraft was very limited in the early stages of World War II. As they then had none of the sophisticated avionics, equipment and weapons available to modern maritime patrol/ASW aircraft, the principal requirements were an ability to carry out extended patrols and to carry as many weapons as possible. This was to change but little until the introduction of ASV (Air to Surface Vessel) radar at a later stage of the war.

Even before the USA's entry into the war, the frequent appearance of German U-boats and surface raiders close to US coastal waters made the US Navy conscious of the fact that a long-range maritime patrol aircraft was a necessity. As a result Boeing was approached to evolve the design of a suitable aircraft, the company's resulting Model 344 proving to be the largest twin-engined flying boat to be built and flown by any of the combatant nations.

Boeing had already gained considerable experience of flying boat construction from its earlier days, culminating in the superb Model 314 'Clippers' with which Pan American Airways had inaugurated dependable services across the North Atlantic and Pacific oceans.

Boeing's design proved acceptable to the US Navy, and a contract for the construction of a prototype, under the designation XPBB-1, was awarded on 29 June 1940. This aircraft flew for the first time on 5 July 1942 and proved to be of fairly conventional design and construction. It had a wing very similar to that of its stablemate, Boeing's Model 345, the prototype of which was being built simultaneously.

Changed ideas regarding maritime patrol aircraft brought cancellation of the order for the US Navy's PBB-1 Sea Rangers, which explains why Boeing's Renton factory became dedicated to Superfortress production.

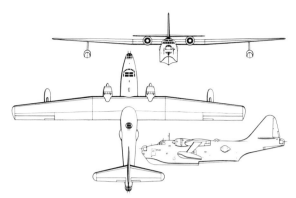

Boeing XPBB-1 Sea Ranger

Specification

Type: long-range maritime patrol/bomber flying boat
Powerplant: two 2,300-hp (1 715-kW) Wright R-3350-8 Cyclone radial piston engines
Performance: maximum speed 228 mph (367 km/h) at 14,200 ft (4 330 m); cruising speed 158 mph (254 km/h); patrol speed 127 mph (204 km/h); service ceiling 22,400 m (6 830 m); maximum range 6,300 miles (10 140 km); maximum endurance 72 hours
Weights: empty 41,531 lb (18 838 kg); maximum take-off 101,130 lb (45 872 kg)
Dimensions: span 139 ft 8½ in (42.58 m); length 94 ft 9 in (28.88 m); height 34 ft 2 in (10.41 m); wing area 1,826 sq ft (169.64 m²)
Armament: eight 0.50-in (12.7-mm) machine-guns (in bow, waist and tail positions), plus up to 20,000 lb (9 072 kg) of bombs
Operator: USN (for evaluation only)

An excellent patrol flying boat, the Boeing XPBB-1 prototype should have been followed by at least 57 production PBB-1s, but the type was made redundant by the success of landplanes.

Boeing/Stearman Model 75

History and Notes

The Stearman Aircraft Company, formed by Lloyd Stearman in 1927, became identified as the Wichita Division of the Boeing Airplane Company in 1939. As early as 1933 the company began design and construction of a new training biplane, derived from the earlier Stearman Model C; built as a private venture, this was first flown in December 1933 and, designated originally as the Stearman X70, was submitted as a contender in 1934 to meet a US Army Air Corps requirement for a new primary trainer.

The first service to show positive interest in this aircraft was the US Navy which, in early 1935, contracted for the supply of 61 Stearman Model 70s under the designation NS-1 (Trainer, Stearman, 1). These, however, received a different powerplant to that installed originally, for the US Navy had in storage a quantity of 225-hp (168-kW) Wright J-5 (R-790-8) radial engines which were specified for installation in this initial order, the company changing the model number of aircraft so equipped to Model 73. The X70 supplied for US Army evaluation was subjected to protracted testing and eventually, in early 1936, the USAAC contracted for the supply of 26 aircraft under the designation PT-13 (Primary Trainer, 13). These, powered by 215-hp (160-kW) Lycoming R-680-5 engines, were the first of the Stearman Model 75s.

This cautious approach by the US Army should not be considered as a reflection upon the capability of the new trainer. The truth of the matter was that at that period the USAAC had little money to spend on new aircraft: not only had this service to be as certain as possible that it was procuring the best available, but

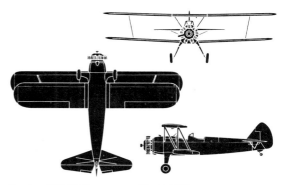

Boeing PT-13 (Stearman Model 75)

even then was only able to procure small quantities. Soon, however, the fortunes of war were to bring Boeing contracts for thousands of the Stearman-designed trainers and although, officially, the aircraft were Boeing Model 75s from 1939, they were persistantly regarded as Stearman 75s throughout the war. The name Kaydet, bestowed later by Canada, and adopted generally in reference to these aircraft, was also unofficial except in Canada.

This attractive two-seat biplane was of mixed construction, the single-bay wings being basically of wood with fabric covering, the remainder of welded steel-tube structure with mostly fabric covering.

The designation N2S-5 was given by the US Navy to the standardised Stearman Model E-75 identified by the USAAF as the PT-13D. This example has USAAF fuselage markings and US Navy tail markings.

Boeing/Stearman Model 75

Landing gear was non-retractable tailwheel type, the divided cantilever main units having cleanly faired oleo-spring shock absorbers. The powerplant was to vary considerably throughout a production run which lasted until early 1945, and during which well over 10,000 examples were built.

USAAC procurement continued with 92 PT-13As delivered from 1937, these having improved instrumentation and 220-hp (164-kW) R-680-7 engines, and by the end of 1941 the USAAF had received an additional 255 PT-13Bs with R-680-11 engines and only minor equipment changes. The designation PT-13C was allocated in 1941 to six PT-13As which were converted by the addition of equipment necessary to make them suitable for night or instrument flight. A change of powerplant, the 220-hp (164-kW) Continental R-670-5 engine installed in a PT-13A type airframe, brought the designation PT-17, and 3,519 of these were built during 1940 to meet the enormous demand for training aircraft. Eighteen PT-17s were equipped with blind-flying instrumentation under the designation PT-17A, and three with agricultural spraying equipment for pest control became PT-17Bs.

US Navy procurement during this same period included a first batch of 250 Model 75s with Continental R-670-14 engines, designated N2S-1, followed by 125 with Lycoming R-680-8 engines as N2S-2. N2S-3s, totalling 1,875, had Continental R-670-4 engines, and 99 aircraft diverted from US Army PT-17 production plus 577 similar aircraft on US Navy contracts were designated N2S-4. For the first time both the US Army and US Navy had a common model in 1942, basically the PT-13A airframe with a Lycoming R-680-17 engine, and these had the respective designations PT-13D and N2S-5. These were to be the last major production variants for the US forces, the US Army receiving 318 and the US Navy 1,450. A shortage of engines in 1940-1 had, however, produced two other designations: PT-18 and PT-18A. The first related to 150 aircraft with the PT-13A type airframe and a 225-hp (168-kW) Jacobs R-755-7 engine, and the six PT-18As were six of the PT-18s converted subsequently with blind-flying instrumentation.

The designation PT-27 applied to 300 aircraft procured by the US Army for supply under Lend-Lease to the Royal Canadian Air Force. A small number of these, and of the N2S-5s supplied to the US Navy, had cockpit canopies, cockpit heating, full blind-flying instrumentation and a hood for instrument training.

In North America the Stearman Kaydet retains an aura of nostalgia which Britons equate with such aircraft as the Avro 504 and Tutor, and de Havilland Tiger Moth, or Germans with the Bücker trainers. When declared surplus at the war's end many served with the air forces of other nations, and large numbers were converted for use as agricultural aircraft. Many remain in operation in this latter role in the 1980s, and the Kaydet is undoubtedly a collector's piece.

Specification

Type: two-seat primary trainer
Powerplant (N2S-5): one 220-hp (164-kW) Lycoming R-680-17 radial piston engine
Performance: maximum speed 124 mph (200 km/h); cruising speed 106 mph (171 km/h); service ceiling 11,200 ft (3 415 m); range 505 miles (813 km)
Weights: empty 1,936 lb (878 kg); maximum take-off 2,717 lb (1 232 kg)
Dimensions: span 32 ft 2 in (9.80 m); length 25 ft 0¼ in (7.63 m); height 9 ft 2 in (2.79 m); wing area 297 sq ft (27.59 m²)
Armament: none
Operators: RCAF, USAAC/USAAF, USMC, USN

A pair of Boeing (Stearman) N2S-2 trainers, generally similar to the US Army's PT-17s, lifts off the runway at Naval Air Station Anacostia, DC.

Brewster F2A Buffalo

Brewster F2A-2 Buffalo of VF-2 ('Flying Chiefs'), US Navy, aboard USS *Lexington* in March 1941.

History and Notes

The first monoplane fighter to equip a squadron of the US Navy, Brewster's F2A Buffalo originated from a US Navy requirement of 1936 for a new generation of carrier-based fighters. In requesting proposals from US manufacturers for such an aircraft, the US Navy indicated requirements which included monoplane configuration, wing flaps, arrester gear, retractable landing gear and an enclosed cockpit. Clearly, this specification recognised the fact that the carrier-based biplane was nearing the end of its useful life.

Proposals were received from Brewster, allocated the designation XF2A-1, Grumman (XF4F-1) and Seversky (XFN-1), but of these the only significant aircraft in the long term was the Grumman design, which was initially of biplane configuration and given serious consideration by the US Navy as an insurance policy against the possible failure of new fangled monoplanes.

A prototype of the Brewster XF2A-1 was ordered on 22 June 1936, and this flew for the first time in December 1937. While bearing a distinct family resemblance to the XSBA-1 of 1934, the new fighter

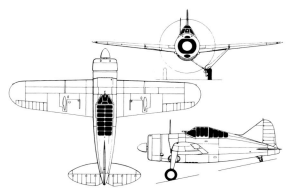

Brewster F2A-3 Buffalo

A Brewster F2A-2 Buffalo shows off its portly lines, which combined with high weight to produce only poor performance and limited manoeuvrability despite the 1,200-hp (895-kW) engine. The Buffalo's only major action in US hands was the Battle of Midway, in which the Buffalo suffered heavy losses.

Brewster B-239 flown by Sergeant H. Lampi of 2. Lentue (flight), Lentolaivue (squadron) 24, Suomen Ilmavoimat (Finnish air force), based at Tiiksjärvi in September 1942.

Brewster B-339 Buffalo Mk I of No. 453 Squadron, Royal Air Force, based at Sembawang (Singapore) in November 1941.

Brewster B-339 Buffalo Mk I of No. 21 Squadron, Royal Australian Air Force, in Dutch national markings as captured by the Japanese at Andir (Dutch East Indies) in early 1942.

appeared to be tubbier and stubbier, but a comparison of dimensions showed this to be something of an illusion. Of mid-wing monoplane configuration, it was of all-metal construction, except for fabric-covered control surfaces. Hydraulically operated split flaps were provided, and the main units of the tailwheel type landing gear retracted inward to be housed in fuselage wells. The powerplant consisted of a 950-hp (708-kW)

Wright XR-1820-22 Cyclone radial engine, driving a Hamilton three-blade metal propeller.

Service testing of the prototype began in January 1938, and on 11 June the US Navy contracted with Brewster for the supply of 54 F2A-1s. Deliveries of these started 12 months later, nine aircraft going almost immediately to equip US Navy Squadron VF-3 aboard the USS Saratoga. The available balance of 44

Brewster B-339D of 1 Afdeling (squadron), Vliegtuiggroep (group) V, ML-KNIL (Royal Netherlands Indies Army Air Corps), based at Semplak (Java) in the summer of 1941.

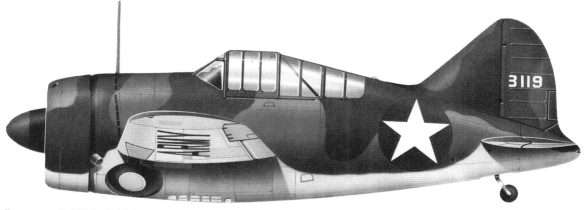

Brewster B-339D of a combined unit of the ML-KNIL (Royal Netherlands Indies Army Air Corps), based at Andir (Dutch East Indies) in March 1942.

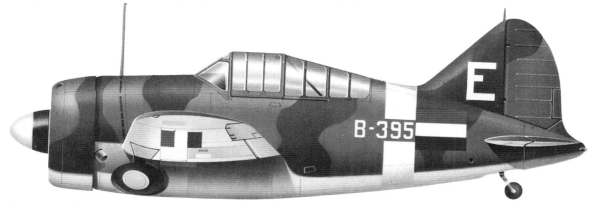

Brewster B-439 in USAAF insignia and based in Australia during mid-1942.

aircraft was, sympathetically, declared surplus to requirements and, instead, supplied to Finland which was then fighting off the might of the Soviet Union. Later equipping Nos. 24 and 26 Squadrons of the Finnish Air Regiment LeR2, they remained successfully operational until mid-1944.

An improved version was ordered by the US Navy in early 1939, this having a more powerful engine, an improved propeller and built-in flotation gear. Designated F2A-2s, these began to enter service in September 1940. They were followed by F2A-3s with more armour and a bulletproof windscreen, and these two production versions were to equip US Navy Squadrons VF-2 and VF-3, and US Marine Corps Squadron VFM-221. A number were used operationally in the Pacific but as the type was overweight, unstable and of poor

Brewster F2A Buffalo

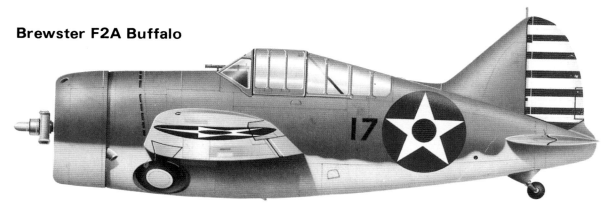

Brewster F2A-3 Buffalo of VMF-221, US Marine Corps, based at Ewa (Hawaii) in mid-1942.

manoeuvrability, it was no match for opposing Japanese fighters.

Belgian and British purchasing missions ordered 40 B-339 and 170 B-339E aircraft respectively, most of the former going to Britain after Belgium had been overrun. These orders were for land-based versions, without arrester gear and other equipment specifically for shipboard operations, but were otherwise generally similar to the F2A-3s. Of those received from the Belgian order, a small number served with Nos. 805 and 885 Squadrons of the Fleet Air Arm, the former squadron using them for support rather than combat duties during the defence of Crete.

Of those which were ordered for the RAF, which gave the type the name Buffalo, deliveries began in July 1940. No. 71 Squadron received the first of these for service trials in September, and it was realised immediately that the Buffalo's performance was totally inadequate for the type's development in the European theatre. Instead, the Buffaloes were sent to the Far East to equip the RAF's Nos. 67, 146, 243, 453 and 488 Squadrons and the Royal Australian Air Force's No. 21 Squadron to defend Singapore and the Straits Settlements. Completely unsuited to the task, the few which survived the Japanese invasion fought alongside the American Volunteer Group operating in Burma. Buffaloes with the most successful combat record were a small number of almost 100 which had been ordered

for the air arm of the Netherlands East Indies' army, which saw action in Java and Malaya. These had the Brewster model numbers B-339D and B-439. The former was similar to the B-339E, but the B-439 had a 1,200-hp (895-kW) Wright GR-1820-G205A engine.

Specification

Type: single-seat land- or ship-based fighter
Powerplant (F2A-3): one 1,200-hp (895-kW) Wright R-1820-40 Cyclone radial piston engine
Performance: maximum speed 321 mph (517 km/h) at 16,500 ft (5 030 m); cruising speed 258 mph (415 km/h); service ceiling 33,200 ft (10 120 m); range 965 miles (1 553 km)
Weights: empty 4,732 lb (2 146 kg); maximum take-off 7,159 lb (3 247 kg)
Dimensions: span 35 ft 0 in (10.67 m); length 26 ft 4 in (8.03 m); height 12 ft 1 in (3.68 m); wing area 208.9 sq ft (19.41 m²)
Armament: four 0.50-in (12.7-mm) machine-guns, plus two 100-lb (45-kg) bombs
Operators: Belgian Air Force, Finnish Air Force, Netherlands East Indies' Army, RAAF, RAF, RN, USMC, USN

Brewster F2A-2s of the US Navy peel off into a dive. Only a few US squadrons operated the type, which was soon released for export.

Brewster SBA/SBN

History and Notes

The Brewster Aeronautical Corporation was founded in the early 1930s, and for the first years of its existence concentrated on the manufacture of seaplane floats, wings and tail units under sub-contract to other manufacturers. It was not until 1934 that the company became involved in the design and prototype construction of its first aircraft, a two-seat scout-bomber required by the US Navy for service aboard the carriers USS *Enterprise* and *Yorktown* which were scheduled for launch in 1936. Designated XSBA-1, this emerged as a clean looking mid-wing monoplane of all-metal construction, except for its control surfaces which were fabric-covered. Trailing-edge flaps were provided to simplify shipboard operations. The powerplant of the prototype consisted of a single 750-hp (559-kW) Wright R-1820-4 Cyclone radial engine, and other features included hydraulically retractable tailwheel type landing gear, and internal stowage for bombs.

The XSBA-1 flew for the first time on 15 April 1936, but flight testing was to indicate that more power was necessary to provide satisfactory performance. In 1937, therefore, a 950-hp (708-kW) XR-1820-22 Cyclone engine was installed, and after tests the US Navy had no hesitation in ordering the type into production. However, at that time Brewster had inadequate production facilities, and it was decided that the 30 aircraft which the US Navy required would be built instead by the Naval Aircraft Factory (NAF), which was located at Philadelphia, Pennsylvania.

The Brewster-designed scout/bomber built by the NAF was given the designation SBN-1, and deliveries to the US Navy extended from November 1940 to March 1942. By that time more advanced designs were becoming available, and production of SBNs came to an end. These aircraft were used by US Navy Squadron VB-3 when they first entered service, and most were

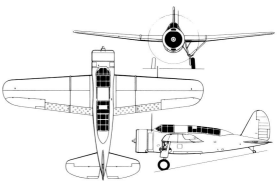

Brewster SBN-1

employed for training at a later date. Squadron VT-8 used its SBN-1s for training on board the USS *Hornet*.

Specification

Type: two-seat carrier-based scout-bomber/trainer
Powerplant (SBN-1): one 950-hp (708-kW) Wright XR-1820-22 Cyclone radial piston engine
Performance: maximum speed 254 mph (409 km/h); service ceiling 28,300 ft (8 625 m); range 1,015 miles (1 633 km)
Weight: maximum take-off 6,759 lb (3 066 kg)
Dimensions: span 39 ft 0 in (11.89 m); length 27 ft 8 in (8.43 m); height 8 ft 7 in (2.64 m); wing area 259 sq ft (24.06 m²)
Armament: one 0.30-in (7.62-mm) gun on flexible mount in rear of cockpit, plus up to 500 lb (227 kg) of bombs in internal bay
Operator: USN

The first complete aircraft of Brewster design, the XSBA-1 was relatively fast and could carry a 500-lb (227-kg) bomb, and entered production as the NAF SBN.

Brewster SB2A Buccaneer

History and Notes

With its first design, the SBA, virtually off its hands, and entering production with the Naval Aircraft Factory under the designation SBN, Brewster was able to turn to the design of an improved version. The aim was to produce a more effective scout-bomber with heavier armament, increased bomb load and higher performance; this dictated a slightly larger airframe and the installation of a more powerful engine.

A single prototype XSB2A-1 was ordered by the US Navy on 4 April 1939 and this flew for the first time on 17 June 1941. By then the company had already received several production orders, comprising 140 for the US Navy, 162 for the Netherlands, and a total of 750 for the RAF, after a British purchasing mission of 1940 was convinced of the excellence of the design. Procurement for the USAAF was also intended and the designation A-34 allocated, but the contract was cancelled before any production resulted.

The intention of providing heavier armament was realised without difficulty, the US Navy's SB2A-1 Buccaneers having eight 0.30-in (7.62-mm) machine-guns, six forward-firing and two on a flexible mounting in the aft cockpit. Unfortunately, performance was far below that anticipated and the larger, much heavier aircraft lacked manoeuvrability. Despite this, the US Navy continued to procure small numbers, acquiring 80 SB2A-2s with armament changes and 60 SB2A-3s. These latter aircraft, intended for carrier operations, featured folding wings and an arrester hook. The 162 aircraft built for the Netherlands were also taken over by the US Navy, and these were given the designation SB2A-4 and transferred to the US Marine Corps for use in a training role. They were to serve a useful purpose in establishing the US Marines' first night fighter squadron, VFM(N)-531.

Deliveries on aircraft for the RAF began in July 1942; supplied under Lend-Lease, these were identified as Bermudas in RAF service, but their performance was such that they were completely unsuitable for combat operations. As a result, the majority were converted for target towing duties, and second-line

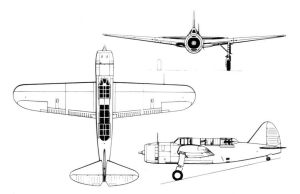

Brewster SB2A-4 Buccaneer

deployment, so far as is known, was the fate of all the 771 aircraft produced by Brewster.

Specification

Type: two-seat land- or carrier-based scout-bomber
Powerplant: one 1,700-hp (1 268-kW) Wright R-2600-8 Cyclone radial piston engine
Performance: maximum speed 274 mph (441 km/h) at 12,000 ft (3 660 m); cruising speed 161 mph (259 km/h); service ceiling 24,900 ft (7 590 m); range without bomb load 1,675 miles (2 696 km)
Weights: empty 9,924 lb (4 501 kg); maximum take-off 14,289 lb (6 481 kg)
Dimensions: span 47 ft 0 in (14.33 m); length 39 ft 2 in (11.94 m); height 15 ft 5 in (4.70 m); wing area 379 sq ft (35.21 m²)
Armament (SB2A-2): two 0.50-in (12.7-mm) fuselage-mounted machine-guns and four 0.30-in (7.62-mm) guns (two in wing and two in aft cockpit on flexible mount), plus up to 1,000 lb (454 kg) of bombs
Operators: RAF, USMC, USN

The Brewster SB2A-2 Buccaneer was similar to the SB2A-1 apart from its armament: only six guns were fitted, two of them 0.50-in (12.7-mm) weapons.

Cessna Model T-50

History and Notes

Clyde Cessna, one of America's aviation pioneers, built his first aeroplane at Enid, Oklahoma, in 1911. Fourteen years later he became a founder of the Travel Air Manufacturing Company, and in September 1927 established with Victor Roos the Cessna-Roos Aircraft Company. On 31 December 1927 this became incorporated as the Cessna Aircraft Company which, today, is one of the world's largest manufacturers of light aircraft.

The company's first twin-engined lightplane, built and flown in 1939, was a five-seat commercial transport and was typical of many very similar aircraft which became fairly common in the USA during the late 1930s. Designated T-50 by the company, it was of low-wing cantilever monoplane configuration, and was of mixed construction. Wings and tail unit were of wood, the latter with fabric covering; the fuselage, however, was a welded steel-tube structure with fabric over lightweight wooden skinning. Tailwheel type landing gear and wing trailing-edge flaps were both electrically actuated.

In 1940 the potential of this aircraft as a trainer suitable for the conversion of pilots from single-engined to twin-engined types, became apparent almost simultaneously to two North American nations. First was Canada, which required a machine of this type for the Commonwealth Joint Air Training Plan, and 550 aircraft were supplied under Lend-Lease, these being designated Crane 1A.

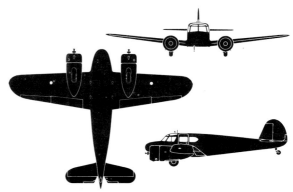

Cessna AT-17 (Model T-50)

The second requirement was for the US Army Air Corps which, in late 1940, contracted for the supply of 33 T-50s for service evaluation, allocating to them the designation AT-8. These were powered by two 295-hp (220-kW) Jacobs R-680-9 radial engines, but service trials showed that these were unnecessarily powerful for use in a two-seat trainer, and when in 1941 the first real production contracts were placed, less powerful engines by the same manufacturer were specified.

In 1942-3 the US Navy bought 67 Cessna JRC-1s for use by ferry squadrons and aircraft delivery units. The USAAF had the similar UC-78 for the same role.

Cessna Model T-50

The initial production version, designated AT-17, was equipped with Jacobs R-755-9 engines driving wooden propellers. A total of 450 was built, and these were followed into production by 223 generally similar AT-17As, which differed by having Hamilton-Standard constant-speed metal propellers. The later AT-17Bs (466) had some equipment changes, and the AT-17Cs (60) were provided with different radio for communications.

The original use of Cessna's T-50s had been in a light transport role, and in 1942 the USAAF decided that these aircraft would be valuable for liaison/communication purposes and as light personnel transports. Production of this variant totalled 1,287, the aircraft being named Bobcat and given the designation C-78, later changed to UC-78. In addition, a small number of commercial T-50s were impressed for service with the USAAF under the designation UC-78A.

The USAAF's requirement for the two-seat conversion trainers had been difficult to predict, and when it was discovered in late 1942 that procurement contracts very considerably exceeded the training requirement, Cessna was requested to fulfil the outstanding balance of AT-17Bs and AT-17Ds as UC-78B and UC-78C Bobcats respectively. Both were virtually identical, but differed from the original UC-78s by having two-blade fixed-pitch wooden propellers and some minor changes of installed equipment. Production of these two versions amounted to 1,806 UC-78Bs and 327 UC-78Cs.

In the period 1942-3, the US Navy had a requirement for a lightweight transport aircraft to carry ferry pilots between delivery points and their home bases, as well as for the movement of US Navy flight crews. The satisfactory performance of the T-50 in USAAF service induced the US Navy to procure the same type for their transport requirement, and 67 aircraft, generally similar to the UC-78, entered service under the designation JRC-1. Many examples of USAAF Bobcats remained in service for two or three years after the end of World War II.

Specification:
Type: two-seat trainer or five-seat light transport
Powerplant (UC-78): two 245-hp (183-kW) Jacobs R-755-9 radial piston engines
Performance: maximum speed 195 mph (314 km/h); cruising speed 175 mph (282 km/h); service ceiling 22,000 ft (6 705 m); range 750 miles (1 207 km)
Weights: empty 3,500 lb (1 588 kg); maximum take-off 5,700 lb (2 585 kg)
Dimensions: span 41 ft 11 in (12.78 m); length 32 ft 9 in (9.98 m); height 9 ft 11 in (3.02 m); wing area 295 sq ft (27.41 m²)
Armament: none
Operators: RCAF, USAAC/USAAF, USN

Consolidated B-24 Liberator

History and Notes
Readers with memories of World War II aircraft cannot help but recall the big, ugly, seemingly slow Liberator, characteristics which brought the nickname 'Lumbering Lib'. In the European theatre, of course, it was much overshadowed by the Boeing B-17 Flying Fortress, and to those with no detailed knowledge of military aircraft it often comes as something of a shock to learn that not only was Consolidated's Liberator built in considerably greater numbers than the B-17, but was the most extensively produced of the USA's wartime aircraft.

The Liberator's origin, like that of many US wartime aircraft, stems back to the early/mid-1930s, an era in which projects such as the Boeing XB-15 and Douglas XB-19, and development of the B-17, brought a far wider knowledge and appreciation of the 'big bomber'. The Liberator, however, represents the next generation, development of which was spurred by the tense political situation in Europe, and the growing threat of Japanese militancy as that nation overran increasingly larger areas of Manchuria. This then was the background which caused the US Army Air Corps, in January 1939, to invite Consolidated to prepare a design study for a heavy bomber with superior performance to that of the B-17: increased range, greater speed, and a higher operational ceiling were all considered to be essential.

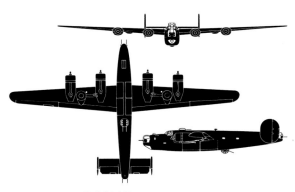

Consolidated B-24J Liberator

Consolidated wasted little time in submitting a design proposal, identifying it as their Model 32 and, as long range was paramount, it was designed round the Davis wing, first introduced on the company's Model 31 commercial flying boat design, of which a prototype was then nearing completion. Subsequently, the Model 31 was reconfigured to serve as the prototype of a military flying boat for the US Navy under the designation XP4Y-1. In reaching a decision to go ahead with prototype construction of the Model 32, the US Army almost matched the speed set by Consolidated, and seemingly they were determined to maintain this tempo for, in awarding the contract on 30 March 1939,

Consolidated B-24 Liberator

the USAAC stipulated that the construction of this prototype, designated XB-24, must be completed by the end of the year. This was achieved by Consolidated, with the first flight being made on 29 December 1939.

In terms of size the XB-24 was marginally smaller than the Fortress except in span, for the wing was just over 6 ft (1.83 m) greater in length; in terms of wing area, that of the XB-24 was approximately 26 per cent less, emphasising the high aspect ratio of the Davis wing. To ensure maximum capacity within the fuselage structure, the wing was high-mounted in shoulder-wing configuration, and to provide good low-speed handling characteristics and an acceptable landing speed, wide-span Fowler-type trailing-edge flaps were fitted. Construction of the fuselage was conventional, but deep in section to allow for installation of a bomb bay which could accommodate up to 8,000 lb (3 629 kg) of bombs stowed vertically. The bay was divided into two sections by the fuselage keel beam, this being utilised to provide a catwalk for crew transition between the fore and aft sections of the fuselage. The most unusual feature of the bomb-bay was the provision of unique 'roller shutter' doors which retracted within the fuselage when opened for attack, causing less drag than conventional bomb-bay doors that were lowered into the slipstream. The tail unit, with its easily recognisable oval-shape endplate fins and rudders, was generally similar to that which had been developed

for the Model 31 flying boat. Landing gear was of the retractable tricycle type, the free-swivelling nosewheel retracting forward into the fuselage, the main units retracting outward and upward so that the wheels were partially housed in underwing wells and faired aft by small blisters. Powerplant of the prototype comprised four wing-mounted 1,200-hp (895-kW) Pratt & Whitney R-1830-33 Twin Wasp engines.

Even before the prototype had flown, Consolidated had begun to receive orders for its new bomber. These included seven service test YB-24s and 36 B-24As for the USAAC, and 120 aircraft 'off the drawing board' for a French purchasing mission. Consolidated must have been highly relieved, therefore, when early flight tests proved successful. To meet the USAAC specification some development was necessary to achieve higher speed, but there was no doubt that the XB-24 was able to demonstrate excellent long-range capability. Furthermore, the large-volume fuselage lent itself to adaptation to fulfil other roles and, in fact, it was this versatility combined with long range which was the key to the success of the B-24.

The XB-24 was followed during 1940 by the seven

Consolidated B-24D Liberators of the 8th Air Force's 93rd Bombardment Group ('The Travelling Circus'), which made the first formation crossing of the North Atlantic in its B-24Ds during September 1942.

Consolidated B-24 Liberator

Consolidated B-24A (similar to LB-30A in RAF service) used as a USAAF transport, based at Bolling Field in October 1941.

Consolidated Liberator GR.Mk I of No. 120 Squadron, RAF Coastal Command, based at Aldergrove (Northern Ireland) in late 1942.

Consolidated Liberator GR.Mk V of No. 224 Squadron, RAF Coastal Command, in November 1942.

YB-24s for service trials, and these differed from the prototype by the provision of pneumatic de-icing boots for the leading edges of wings, tailplane and fins. But by the time that the first production aircraft began to come off the line at San Diego, France had already capitulated, and the aircraft of the French order were completed to British requirements, as specified in an order for 164 which had been placed soon after that of 120 for France; the French order was later transferred to Britain.

The RAF had allocated the name Liberator to its new bomber, this being adopted later by the USAAF, and the first of these (AM258) flew for the first time on 17 January 1941. They were, however, designated LB-30A by Consolidated, indicating Liberator to British specification, and the first six of these reached the UK during March 1941, flown directly across the North Atlantic. These initial aircraft were used as unarmed transports by BOAC, and later by RAF Ferry Command, to carry pilots and crews to fly back to Great Britain the ever increasing number of aircraft being supplied from the USA. The next batch, received in mid-1941, were to join the RAF as Liberator Is for service with Coastal Command, and these were modified in Britain to equip them with an early form of ASV (Air-to-Surface Vessel) radar, and to increase the standard armament of five 0.30-in (7.62-mm) machine-

guns to include an underfuselage gun pack, forward of the bomb bay, housing four 20-mm cannon. The Liberator I began to equip No.120 Squadron of Coastal Command in June 1941, and was the first RAF aircraft with the range and endurance to close the 'Atlantic Gap', that area of the ocean in which, until that time, sea convoys were beyond the range of air support from either North America or Great Britain.

In that same month, the USAAF began to receive its first B-24As and these, duplicating the role of the LB-30As in Britain, were allocated first to equip the Air Corps Ferrying Command, operating similar services across the North Atlantic as those of RAF Ferry Command. The first true operational bomber version, however, was the Liberator II (Consolidated LB-30), for which there was no USAAF equivalent. It differed from the Liberator I in having the fuselage nose extended 2 ft 7 in (0.79 m) by the insertion of a 'plug', to accommodate a maximum crew of 10; by the installation of Boulton Paul power-operated turrets, each housing four 0.303-in (7.7-mm) machine-guns, in mid-upper and rear fuselage positions; plus small increases in gross weight, bomb load and service ceiling. The RAF received 139 of this version, these equipping Nos. 59, 86 and 120 Squadrons of Coastal Command, and Nos. 159 and 160 Squadrons. When these two latter squadrons began operations with

Consolidated B-24 Liberator

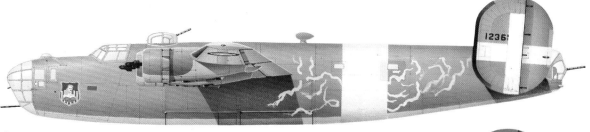

Consolidated B-24D Liberator of the 98th Bomb Group, USAAF, based at Benghazi (Libya) in February 1943 just before it landed in Sicily after a raid on Naples.

Consolidated B-24D Liberator assembly ship of the 491st Bomb Group, 8th Air Force, based at North Pickenham (UK) in autumn 1944.

Consolidated PB4Y-1 Liberator of Patrol Bomber Squadron VPB-110, US Navy, based in Devon in winter 1944.

their Liberators in the Middle East in June 1942, they were the first to deploy these aircraft in a bombing role. One aircraft of this batch (AL504) became the personal transport of Britain's prime minister, Winston Churchill, and operated under the name *Commando*.

Meanwhile, the XB-24 prototype had been modified to a new XB-24B standard, this introducing self-sealing fuel tanks and armour, but the most significant improvement was the installation of turbocharged R-1830-41 engines. This resulted in the second of the Liberator's easily identifiable features, oval-shaped nacelles, entailed by the relocation of the oil coolers in the sides of the front cowlings. With the introduction of these features, plus dorsal and tail turrets each with two 0.50-in (12.7-mm) machine-guns to supplement the original hand-held guns in beam and nose positions, nine aircraft were produced for the USAAF as B-24Cs.

They were followed by the B-24D, the first major production variant, and also the first to be employed operationally by USAAF bomber squadrons. These differed initially by the installation of R-1830-43 engines, but subsequent production batches introduced progressively changes in armament, provision of auxiliary fuel in the outer wings and bomb bay, increases in gross weight and bomb load, and in some late production examples external bomb racks below the inner wing for the carriage of two 4,000-lb (1 814-kg)

bombs. In RAF service the B-24D was designated Liberator III: Liberator IIIA identified similar aircraft supplied under Lend-Lease with US armament and equipment. Most Liberator III/IIIAs served with Coastal Command, eventually equipping 12 squadrons. A total of 122 were extensively modified in Britain, receiving ASV radar equipment including chin and retractable ventral radomes, a Leigh Light for the illumination of targets at night (especially surfaced U-boats), increased fuel capacity, but reduced armament, armour and weapon load. These were designated Liberator GR.Vs. Some were provided with small stub wings on the forward fuselage to carry eight rocket projectiles. The USAAF also operated B-24Ds in an anti-submarine role, and in 1942 the US Navy began to receive small numbers of this version as PB4Y-1s. However, at the end of August 1943 the USAAF disbanded its Anti-Submarine Command, handing over its aircraft to the US Navy in exchange for an equivalent number of aircraft of bomber configuration to be produced against outstanding US Navy orders. These ex-USAAF B-24s were also designated PB4Y-1s by the US Navy, which service was subsequently to acquire the specially-developed PB4Y-2 Privateer, which featured single tail vertical tail surfaces.

The deployment of USAAF B-24Ds in the Middle East began in June 1942, one of the first operations

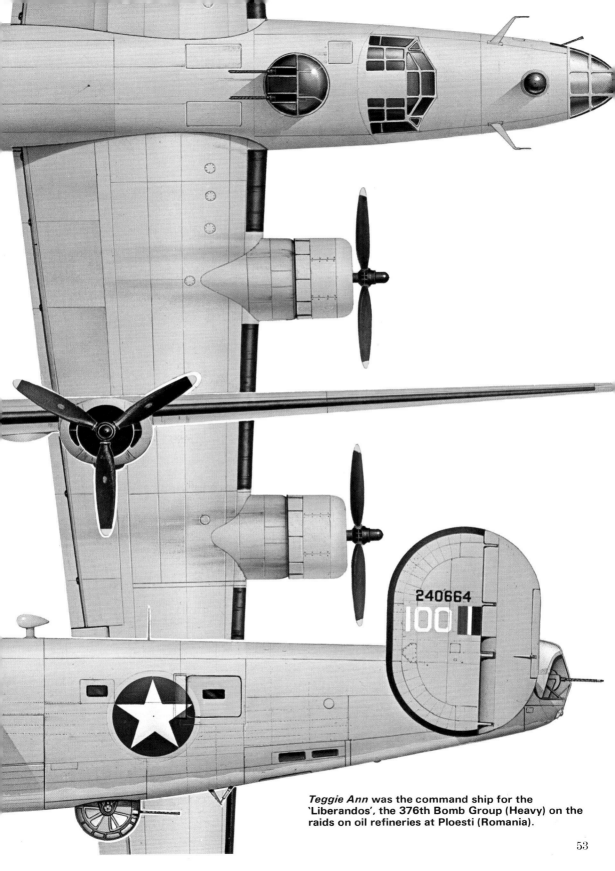

240664

100

Teggie Ann was the command ship for the
'Liberandos', the 376th Bomb Group (Heavy) on the
raids on oil refineries at Ploesti (Romania).

Consolidated B-24 Liberator

being launched by 13 aircraft of Colonel H. A. Halverson's detachment which attacked the Romanian oilfields at Ploesti on 11/12 June 1942 from the RAF base at Fayid in the Suez Canal zone. All 13 aircraft completed what the USAAF described as 'an unsuccessful attack', its only success being to alert the defences of their vulnerability. Consequently, it was a very different story on 1 August 1943, when units of the 8th and 9th Air Forces sent 177 B-24s against the same target. Although rather more successful in terms of damage caused, of the force which set out from Benghazi 55 Liberators were lost, 53 damaged, and 440 crew killed or posted missing.

By that time, of course, B-24s were being built at an enormous rate, by Consolidated at San Diego and Fort Worth, Douglas at Tulsa, and Ford with a specially built new plant at Willow Run. In mid-1942 the first transport variants began to appear, with nose and tail gun positions deleted, a large cargo door installed in the port side of the fuselage, and accommodation provided for passengers or cargo. The USAAF acquired 276 as C-87s with accommodation for a crew of five and 20 passengers; 24 similar aircraft, but provided with side windows, served with RAF Transport Command as Liberator C.VIIs; and examples flown by the US Navy were designated RY-2. Similar aircraft, but with R-1830-45 engines and equipped as VIP transports, were identified as RY-1 and C-87A by the US Navy and USAAF respectively. One special logistics version was the C-109 fuel tanker, used to ferry 2,900 US gallons (10 977 litres) of aviation fuel per load over the Himalayan 'hump', to supply Boeing B-29 Superfortresses operating from forward bases in China. An XF-7 prototype special reconnaissance version was also produced in 1943, with bomb racks removed and extra fuel tanks provided in the forward section of the bomb bay. This retained the normal defensive armament, and could also accommodate up to 11 cameras. F-7s were used extensively in the Pacific theatre, and later versions included F-7As and F-7Bs with differing camera installations.

The first production aircraft to come from the Ford plant at Willow Run were B-24Es, generally similar to the B-24D except for different propellers and minor detail changes, and this version was built also by Consolidated and Douglas, some having R-1830-65 engines. There followed the B-24G, all but the first 25 of which introduced an upper nose gun turret and had the fuselage nose lengthened by 10 in (0.25 m). These came from a new production line operated by North American Aviation at Dallas, Texas. Similar aircraft produced by Consolidated at Fort Worth, by Douglas and by Ford were designated B-24H.

The major production variant was the B-24J (6,678 built), which came from all five production lines, and which differed from the B-24H in only minor details. B-24Hs and -24Js supplied to the RAF under Lend-Lease were designated Liberator GR.VI when equipped for ASW/maritime reconnaissance by Coastal Command, or Liberator B.VI when used as a heavy bomber in the Middle East and Far East. Those used by the US Navy were identified as PB4Y-1s.

The final production versions were the B-24L, similar to the B-24D with the tail turret replaced by two manually controlled 0.50-in (12.7-mm) machine-guns, of which Consolidated San Diego built 417 and Ford 1,250; and the B-24M which differed from the B-24J in having a different tail turret. Convair built 916 of this latter version at San Diego and Ford another 1,677. Odd variants included a single B-24D provided with an experimental thermal de-icing system as the XB-24F; the XB-24K prototype of the single vertical tail version (to have been produced in large numbers as the B-24N, although only the XB-24N prototype and seven YB-24N service test aircraft were built before production ended on 31 May 1945); the single experimental XB-41 bomber escort, armed with 14 0.50-in (12.7-mm) machine-guns and converted from a B-24D; and five C-87s converted for flight engineer training under the designation AT-22 (later TB-24). Most of the USAAF's Liberators were declared surplus at the war's end, only a few remaining in service. The very last was disposed of in 1953.

From first to last, more than 18,475 Liberators had been built. In addition to those supplied to the RAF, USAAF and US Navy, others had been operated by units of the Royal Australian Air Force, Royal Canadian Air Force and South African Air Force. Nowhere had they been of greater value than in the Pacific theatre, where their long range and versatility created them 'maids of all work'. Operated extensively by the USAAF's 5th and 13th Air Forces, they fought with the US Navy and US Marines over every island step towards the Japanese home islands. In the closing stages their HE bombs or incendiaries added to the quota of destruction on Luzon, Formosa, Okinawa and, in the end, Honshu. The 'Lumbering Lib' had travelled a long and bitter route to final victory.

Specification

Type: eight/ten-seat long-range bomber/reconnaissance aircraft

Powerplant (B-24H/J): four 1,200-hp (895-kW) Pratt & Whitney R-1830-65 Twin Wasp turbocharged radial piston engines

Performance: maximum speed 290 mph (467 km/h) at 25,000 ft (7 620 m); cruising speed 215 mph (346 km/h); service ceiling 28,000 ft (8 535 m); range 2,100 miles (3 380 km)

Weights: empty 36,500 lb (16 556 kg); maximum overload take-off 71,200 lb (32 296 kg)

Dimensions: span 110 ft 0 in (33.53 m); length 67 ft 2 in (20.47 m); height 18 ft 0 in (5.49 m); wing area 1,048 sq ft (97.36 m²)

Armament: 10 0.50-in (12.7-mm) machine-guns (in nose, upper, ventral 'ball' and tail turrets, and beam positions), plus a maximum bomb load of 12,800 lb (5 806 kg) or a normal bomb load of 5,000 lb (2 268 kg)

Operators: RAAF, RAF, RCAF, USAAC/USAAF, USN

Consolidated B-24 Liberator

Consolidated B-24J Liberator of the 43rd Bomb Group, USAAF, based on Ie Shima (Okinawa) in 1945.

Consolidated Liberator GR.Mk VI of No. 547 Squadron, RAF Coastal Command, based at Leuchars (UK) in late 1944.

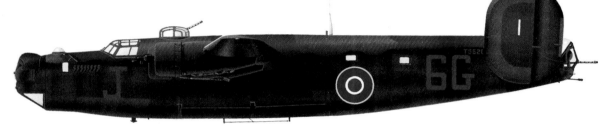

Consolidated Liberator B.Mk VI of No. 223 Squadron, No. 100 Group, RAF Bomber Command, based at Oulton (UK) in August 1944.

Consolidated Liberator B.Mk VI of No. 356 Squadron, RAF, based in the Cocos islands (Indian Ocean) during 1945.

Consolidated B-24J Liberator of the Indian air force in the late 1940s.

Consolidated (Convair) B-32 Dominator

History and Notes

For precisely the same requirement to which Boeing designed the B-29, Consolidated evolved a competing proposal, and each company was awarded a contract to build three prototypes: those ordered from Consolidated were allocated the designation XB-32.

The first prototype made its maiden flight on 7 September 1942, two weeks before the first XB-29. The second and third followed on 2 July 1943 and 9 November respectively. Like the XB-29 these featured pressurisation and remotely-controlled gun turrets, but each differed in some fairly major aspect of its configuration. The first had a rounded fuselage nose and twin fins and rudders based on those of the B-24 Liberator. The second retained this tail unit but had a modified fuselage nose with a stepped windscreen for the flight deck. The third prototype retained this fuselage design, but introduced a large single fin and rudder, and this was the basic configuration as finalised for production aircraft.

Somewhat smaller than the B-29, the B-32 was of cantilever high-wing monoplane configuration, and powered by four Wright Cyclone 18 radial engines of the same series used for the B-29. Landing gear was of the retractable tricycle type, and two cavernous bomb bays could carry 20,000 lb (9 072 kg) of bombs. Accommodation was provided for a standard crew of eight.

Consolidated were to experience extensive problems in the development of the B-32, to the extent that it was not possible to begin the delivery of production examples until November 1944, almost eight months after XX Bomber Command B-29s had been deployed on forward bases in China. Even then, production aircraft (of which 115 were built) had the intended pressurisation system and remotely-controlled gun turrets deleted.

In the final analysis, only 15 of these aircraft were to become operational before VJ-Day, these equipping the USAAF's 386th Bombardment Squadron based on Okinawa. A total of 40 TB-32s was also produced for training purposes, but with the end of the war all

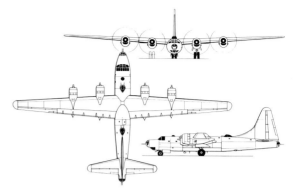

Consolidated B-32 Dominator

versions were very soon withdrawn from service.

Specification

Type: long-range strategic bomber
Powerplant: four 2,200-hp (1 641-kW) Wright R-3350-23 Cyclone radial piston engines
Performance: maximum speed 357 mph (575 km/h) at 25,000 ft (7 620 m); service ceiling 35,000 ft (10 670 m); range with maximum bomb load 800 miles (1 287 km); maximum range 3,800 miles (6 115 km)
Weights: empty 60,272 lb (27 339 kg); maximum take-off 111,500 lb (50 576 kg)
Dimensions: span 135 ft 0 in (41.15 m); length 83 ft 1 in (25.32 m); height 33 ft 0 in (10.06 m); wing area 1,422 sq ft (132.10 m²)
Armament: two 20-mm cannon (one in nose and one in tail turret) and four 0.50-in (12.7-mm) machine-guns, plus up to 20,000 lb (9 072 kg) of bombs
Operator: USAAF

Evidence of the Consolidated B-32 Dominator's failings as a strategic bomber, in comparison with the Boeing B-29, is given by the fact that of the 115 production Dominators, only 75 were bombers and the other 40 were TB-32 trainers (illustrated).

Consolidated PBY Catalina

History and Notes

With the requirement for a patrol flying boat to offer greater range and load-carrying capability than the Consolidated P2Y or Martin P3M, which were in service in the early 1930s, the US Navy contracted Consolidated and Douglas in October 1933 to build competing prototypes. Designated XP3Y-1 and XP3D-1 respectively, only a single prototype of the Douglas design was built. Consolidated's XP3Y-1, however, was to be developed to become the most extensively-built flying boat in aviation history.

Consolidated identified their design to meet the US Navy's requirement as the Model 28 and this, like the P2Y which preceded it, had a parasol-mounted wing. However, in the new design the introduction of internal bracing resulted in a wing which was virtually a cantilever, except for two small streamline struts between hull and wing centre-section on each side. Thus the Model 28 was free of the multiplicity of drag-producing struts and bracing wires which had limited the performance of earlier designs. Another innovation adding to aerodynamic efficiency was the provision of stabilising floats which, when retracted in flight, formed streamlined wingtips. The two-step hull design was very similar to that of the P2Y, but instead of strut-braced twin fins and rudders mounted high on the tailplane, the Model 28 had a clean cruciform tail unit which was a cantilever structure. Powerplant of the prototype comprised two 825-hp (615-kW) Pratt & Whitney R-1830-54 Twin Wasp engines mounted on the wing leading edges. Armament comprised four 0.30-in (7.62-mm) machine-guns and up to 2,000 lb (907 kg) of bombs.

First flown on 28 March 1935, the XP3Y-1 was soon transferred to the US Navy for service trials, which confirmed a significant improvement in performance

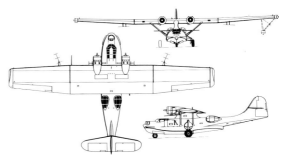

Consolidated PBY-5A Catalina (dashed lines: retractable stabilising floats)

over the patrol flying boats in service. Its extended range and improved load-carrying capability caused the US Navy to request further development to bring this new aircraft into the category of a patrol-bomber, and in October 1935 the prototype was returned to Consolidated for the necessary work to be carried out, including installation of the 900-hp (671-kW) R-1830-64 engines which had been specified for the 60 PBY-1s (a patrol-bomber designation) which had been ordered on 29 June 1935. At the same time redesigned vertical tail surfaces were introduced and the XPBY-1, as this prototype was redesignated, flew for the first time on 19 May 1936. After completing its trials, during which a record nonstop distance flight of 3,443 miles (5 541 km) was achieved, this aircraft was delivered to US Navy Squadron VP-11F during October 1936, in which month the first of the PBY-1s began to reach the

Though slow by contemporary standards, the PBY-5A more than made up for this with its reliability and range, coupled with the great tactical flexibility bestowed by the amphibian landing gear.

Consolidated PBY Catalina

squadron. Second to be equipped was Squadron VP-12, which received the first of its aircraft in early 1937.

Minor equipment changes brought the designation PBY-2 for the second production order placed on 25 July 1936, while the PBY-3s ordered on 27 November 1936 and PBY-4s on 18 December 1937 had 1,000-hp (746-kW) R-1830-66 and 1,050-hp (783-kW) R-1830-72 Twin Wasp engines respectively. All but the first examples of the PBY-4s introduced large transparent blisters over the waist gun positions, in place of sliding hatches, and these became a characteristic feature of all subsequent production aircraft.

In April 1939 the first example of the PBY-4 production aircraft was returned to the company for the installation of wheeled landing gear so that these aircraft could operate as amphibians, thus making them far more versatile. Of tricycle-type configuration, the landing gear featured a nosewheel retracting into the hull beneath the flight deck, and the main units (mounted externally on each side of the hull beneath the wing) folding to fit into small recesses. This aircraft, when completed in November 1939, emerged with the designation XPBY-5A. Testing confirmed, if confirmation was needed, the very considerable advantages of the amphibian configuration. The 33 aircraft outstanding on US Navy contracts for PBY-5s were completed as PBY-5A amphibians, and an additional 134 PBY-5As were contracted on 25 November 1940. First US Navy squadrons to be equipped with this version were VP-83 and VP-91, which received them towards the end of 1941.

Extensive service use of the PBYs had suggested that the hull would benefit from hydrodynamic improvement, and the Naval Aircraft Factory had undertaken the necessary research and development work to achieve this end, receiving an order for 156 of

A Consolidated Catalina Mk IB of No. 202 Squadron, RAF Coastal Command. This mark was essentially similar to the US Navy's PBY-5, and the antennae firmly indicate the provision of ASV radar.

these modified aircraft under the designation PBN-1 Nomad. This course was adopted in order that the design changes would not interfere with the major production coming from Consolidated. When the final production version was built by Consolidated between April 1944 and April 1945, the NAF's improvements were incorporated in a model designated PBY-6A. In addition to the hydrodynamic refinement of the hull, the NAF had also improved the lines of the wingtip floats, introduced taller vertical tail surfaces, increased fuel tankage, strengthened the wings for higher gross weight, and updated the electrical system.

From mid-1737 PBYs were introduced rapidly into service with the US Navy, and by mid-1938 14 squadrons were equipped, including five based at Pearl Harbor and three at Coco Solo; by the time that the USA became involved in World War II some 21 squadrons were equipped, 16 with PBY-5s, two with PBY-4s and three with PBY-3s.

Before this, however, interest shown by the Soviet Union resulted in an order for three aircraft and the negotiation of a licence to build the type in Russia. When these three machines were delivered they were accompanied by a team of Consolidated engineers who assisted in establishment of the Russian production facilities. Designated GST, these production aircraft were powered by Mikulin M-62 radial engines, a developed version of the M-25 licence-built Wright Cyclone which had a power rating of 900-1,000 hp (671-746 kW). The first of these GSTs began to appear in late 1939, and an unspecified number, certainly

Consolidated PBY Catalina

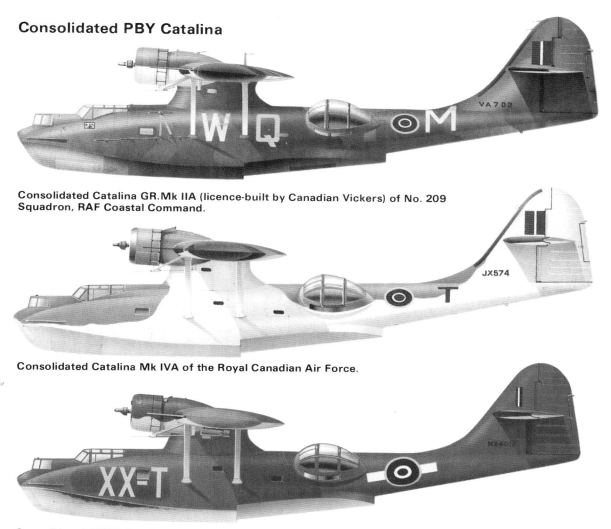

Consolidated Catalina GR.Mk IIA (licence-built by Canadian Vickers) of No. 209 Squadron, RAF Coastal Command.

Consolidated Catalina Mk IVA of the Royal Canadian Air Force.

Consolidated PBY-5 Catalina of No. 6 Squadron, Royal New Zealand Air Force, based in Fiji during 1944-5.

running into several hundreds, were built during the war for service with the Soviet navy. In addition to these home-built examples, Russia also received under Lend-Lease 137 of the PBN-1 Nomads built by the NAF and 48 of the PBY-6As.

European interest started with purchase by the British Air Ministry of a single aircraft for evaluation, this being identified by Consolidated as the Model 28-5. Flown across the Atlantic, the craft was allocated to the Marine Aircraft Experimental Establishment at Felixstowe, Suffolk in July 1939. The outbreak of war anticipated the termination of trials, but with little doubt of the excellence of the design, a first batch of 50 was ordered under the designation Catalina I, these being generally similar to the US Navy's PBY-5s except for the installation of British armament. The name Catalina had been used by Consolidated prior to receipt of the British order, and was not adopted by the US Navy until 1 October 1941.

Initial deliveries of the RAF's Catalinas began in early 1941, these entering service with Nos. 209 and 240 Squadrons of Coastal Command. They were subsequently to equip nine squadrons of Coastal Command, as well as an additional 12 squadrons serving overseas. Not surprisingly, this involved the acquisition of a considerable number of these aircraft, totalling approximately 700. The figure is only an approximate one because records vary by two or three aircraft. With the exception of 11 PBY-5As which were diverted to Britain from a US Navy order, all were non-amphibious flying boats. They comprised 100 Catalina Is equivalent to the US Navy's PBY-5, 225 Catalina IBs (PBY-5B), 36 Catalina IIAs (PBY-5), 11 Catalina IIIs (PBY-5A), 97 Catalina IVAs (PBY-5), 193 Catalina IVBs which were built by Boeing Aircraft of Canada as PB2B-1s and were generally similar to the non-amphibious PBY-5s, and 50 Catalina VIs, Boeing-built PB2B-2s which had the taller vertical surfaces first introduced on the NAF PBN-1. No Catalina Vs served with the RAF, this designation being allocated for potential supplies of NAF PBN-1s, none of which was in the event sent to Great Britain.

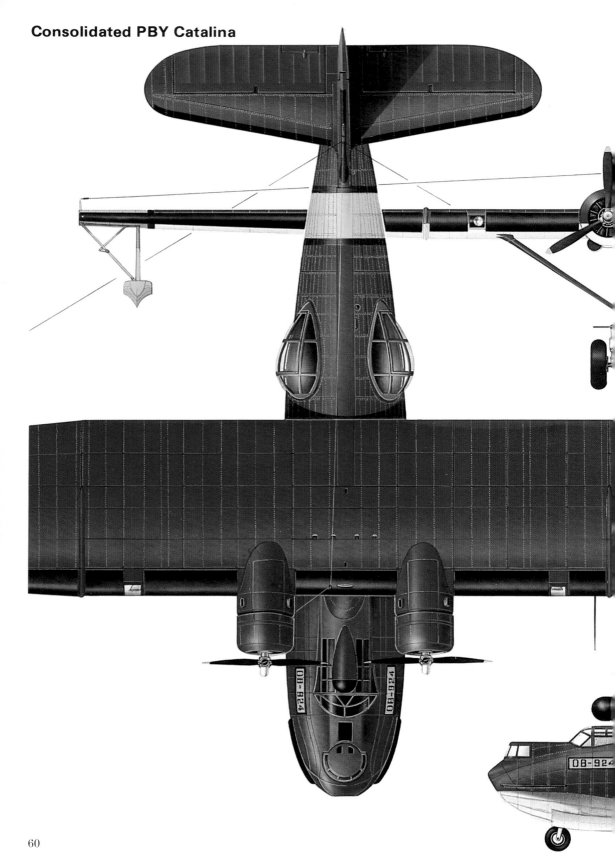

One of the more unusual aircraft handed over from the inventory of the US Army Air Force to the newly formed USAF in October 1947 was this OA-10A amphibian of the Air Rescue Service. Built during the war by Canadian Vickers, it was one of a batch of 230 which served from early 1944 on several fronts and remained in the active inventory until at least 1954. After the war they were unarmed, but were otherwise broadly similar to the PBY-5A. In the European theatre the RAF handled air/sea rescue, but from the start of 1945 the USAAF used the OA-10A in the 5th Emergency Rescue Squadron, together with the P-47 and lifeboat-dropping B-17G.

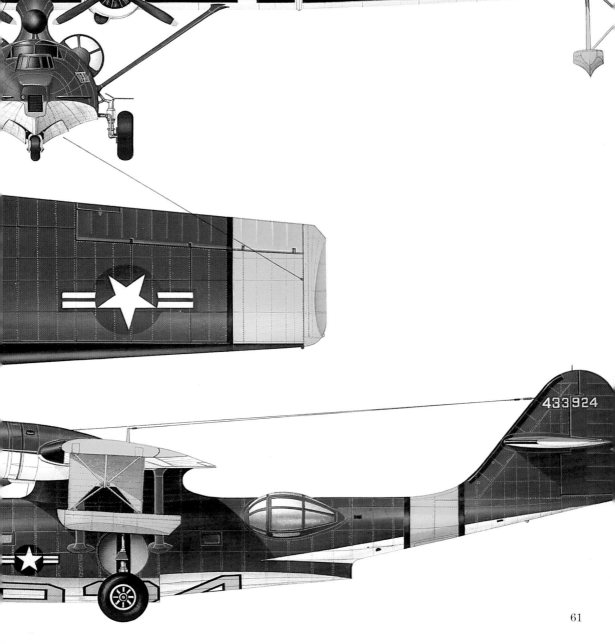

Consolidated PBY Catalina

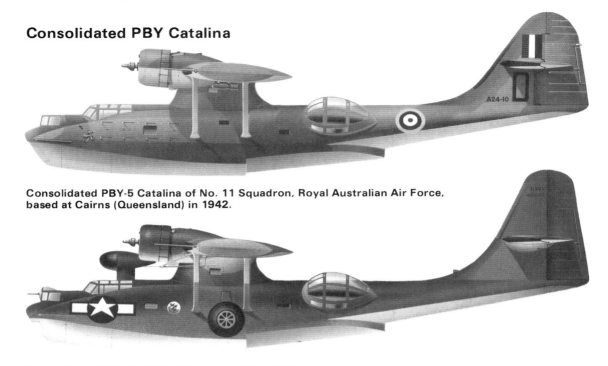

Consolidated PBY-5 Catalina of No. 11 Squadron, Royal Australian Air Force, based at Cairns (Queensland) in 1942.

Consolidated PBY-6A (PBN-1 Nomad) of the US Navy.

In RAF service the Catalina was at the centre of many notable exploits, the first coming as early as 26 May 1941 when an aircraft of No. 209 Squadron located the German battleship *Bismarck* after surface contact had been lost: it was shadowed subsequently by one of No. 240 Squadron's aircraft until surface contact was resumed. Ranging far and wide, in the Atlantic Ocean, Indian Ocean and Mediterannean Sea, they made a major contribution to the naval war. On 7 May 1945 one of No. 210 Squadron's Catalinas sank the 196th and final U-boat claimed by Coastal Command.

Soon after the receipt of Britain's first order for production aircraft, Consolidated received a French purchasing mission which, in early 1940, ordered 30 aircraft. Allocated the company's identification of Model 28-5MF, none of these was delivered before the collapse of French resistance. Other foreign orders received at about the same time covered 18 aircraft for the Royal Australian Air Force, and 48 ordered by the Dutch government for use in the Netherlands East Indies. Of these latter, 36 had been delivered before the Japanese invasion, and only nine were flown off before capture, being used as initial equipment for No. 321 Squadron in Ceylon.

Canada had its own close associations with the Catalina, both as manufacturer and customer. Under an agreement reached between the Canadian and US governments, production lines were laid down in Canada, by Boeing Aircraft of Canada at Vancouver, and by Canadian Vickers at Cartierville. Tooling for production was started in 1941: Boeing's first PB2B-1 flew on 12 May 1943, that of Canadian Vickers more than a month earlier, on 3 April 1943.

Boeing production totalled 362 aircraft, these comprising 240 PB2B-1s supplied to Australia, Britain and New Zealand; 50 PB2B-2s for Britain; 17 non-amphibious Catalinas for the RCAF, and 55 amphibians which, in Royal Canadian Air Force service, were designated Canso. Aircraft produced by Canadian Vickers totalled 379 equivalent to the PBY-5A, of which 149 were supplied to the RCAF. From the balance of 230 the US Navy planned to acquire 183 under the designation PBV-1A; in fact the US Navy received none of these, all of them being supplied to the USAAF which had previously acquired 56 PBY-5As as a direct transfer from the US Navy and which it designated OA-10. These were used throughout World War II for search and rescue missions, some carrying air-dropped lifeboats beneath each wing. The 230 aircraft built by Canadian Vickers were designated OA-10A in USAAF service, and the final production aircraft to be received were 75 PBY-6As built by Consolidated, and which were designated OA-10B.

Because there is no record of the number of Catalinas built in Russia as GSTs, it is impossible to quote any figure for total production. The best one can do is to estimate the final construction figure as one around 4,000. What is certain is the excellence of this most attractive aircraft, which was deployed in practically all of the operational theatres of World War II. It served with distinction and, in the war fought against the Japanese where the battle area was largely sea and islands, played a prominent and invaluable role. This was especially true in the first year, when Catalinas and Boeing B-17 Fortresses were the only two aircraft with the essential attribute of long range. As a result they were used in almost every possible military role until a new generation of aircraft became available in

Consolidated PBY Catalina

the Pacific theatre: such a capability meant that the Catalina rapidly gained a reputation for ability and reliability which is remembered to this day.

With the end of the war, flying boat versions were quickly retired from the US Navy, but amphibious versions remained in service for a few years: they subsequently equipped the world's smaller armed services, in fairly substantial numbers, into the late 1960s.

Specification

Type: seven/nine-seat long-range maritime patrol-bomber amphibian/flying boat
Powerplant (PBY-5A): two 1,200-hp (895-kW) Pratt & Whitney R-1830-92 Twin Wasp radial piston engines
Performance: maximum speed 179 mph (288 km/h)

at 7,000 ft (2 135 m); long-range cruising speed 117 mph (188 km/h); service ceiling 14,700 ft (4 480 m); maximum range 2,545 miles (4 096 km)
Weights: empty 20,910 lb (9 485 kg); maximum take-off 35,420 lb (16 066 kg)
Dimensions: span 104 ft 0 in (31.70 m); length 63 ft 10½ in (19.47 m); height 20 ft 2 in (6.5 m); wing area 1,400 sq ft (130.06 m²)
Armament: two 0.30-in (7.62-mm) machine-guns in bow, one 0.30-in (7.62-mm) machine-gun firing aft through a tunnel aft of the hull step and two 0.50-in (12.7-mm) machine-guns (one in each beam position), plus up to 4,000 lb (1 814 kg) of bombs or depth charges
Operators: Brazil, Netherlands East Indies, RAAF, RAF, RCAF, RNZAF, SAAF, Soviet Union, USAAF, USN

Consolidated PB2Y Coronado

History and Notes

Plans for the development of a maritime patrol bomber larger than the PBY Catalina were drawn up by the US Navy very soon after the first flight of the Catalina's XP3Y-1 prototype. The aim was to procure a patrol flying boat with increased performance and good weapon load capability, Consolidated and Sikorsky each receiving a contract for the construction of a prototype for evaluation. Sikorsky's XPBS-1, ordered on 29 June 1935, flew for the first time on 13 August 1937. Despite a number of new features (it was, for example, the first US military aircraft with both nose and tail turrets), it was the Consolidated XPB2Y-1 which, when evaluated following a first flight on 17 December 1937, was regarded as the more suitable for production. As at that time the US Navy had no funds for immediate procurement of any of these aircraft, Consolidated were to have almost 15 months in which to rectify the shortcomings revealed by initial flight tests.

Most serious of the problems was lateral instability, which the company attempted to rectify by the addition of two oval-shaped fins, mounted one each side of the tailplane. This was a move in the right direction, but stability was still far from satisfactory and was finally resolved by the design of a new tail unit, first with circular endplate fins and rudders and, finally, by endplate fins and rudders similar to those of the B-24 Liberator.

The other problem concerned the hydrodynamic performance of the flying boat's hull; fortunately, the procurement time-scale allowed Consolidated to re-design the hull, this being deeper than that of the prototype, with a much-changed nose profile.

Eventually, on 31 March 1939, the US Navy were able to order six of these aircraft, given the designation PB2Y-2 and the name Coronado, and delivery of these to US Navy Squadron VP-13 began on 31 December 1940. They were impressive aircraft, powered by four

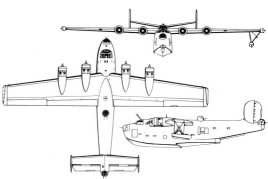

Consolidated PB2Y-3 Coronado (dashed lines: retractable stabilising floats)

1,200-hp (895-kW) Pratt & Whitney R-1830-78 Twin Wasp radial engines mounted on the high-set cantilever wing. Construction was all-metal, and interesting features included stabilising floats which retracted to form wingtips in flight, and bomb bays formed in the deep-section wing. Accommodation was provided for a crew of nine.

These PB2Y-2s were used for service trials, leading to the procurement of the PB2Y-3 Coronado, following the conversion of one of the PB2Y-2s as a prototype XPB2Y-3. They differed by having R-1830-88 engines, increased armament, and the provision of self-sealing tanks and armour. A total of 210 of this version was built, late production aircraft being equipped with ASV (Air-to-Service Vessel) radar.

Ten of the aircraft, designated PB2Y-3B, were supplied to the RAF and based initially at Beaumaris, Anglesey, intended for service with Coastal Command. Their stay there was only brief, for they were transferred to No. 231 Squadron of Transport Command, and used from June 1944 to operate freight services.

Variants in US service, converted from PB2Y-3s,

Consolidated PB2Y Coronado

included 31 PB2Y-3R transports, fitted with single-stage supercharged R-1830-92 engines; one XPB2Y-4 converted by the experimental installation of Wright R-2600 Cyclone engines; PB2Y-5s with increased fuel capacity and R-1830-92 engines; and a number of PB2Y-5H casualty-evacuation aircraft which saw service in the Pacific theatre, with military equipment removed to provide accommodation for 25 stretchers.

Specification

Type: long-range flying boat bomber
Powerplant (PB2Y-3): four 1,200-hp (895-kW) Pratt & Whitney R-1830-88 Twin Wasp radial piston engines
Performance: maximum speed 223 mph (359 km/h) at 20,000 ft (6 095 m); cruising speed 141 mph (227 km/h) at 1,500 ft (460 m); service ceiling 20,500 ft (6 250 m); range with 8,000-lb (3 629-kg)

A Consolidated PB2Y-3 Coronado patrol flying boat of the US Navy. Each of these boats cost three times as much as a PBY, which the PB2Y was designed to replace, but only limited operational service followed.

bomb load 1,370 miles (2 205 km); maximum range 2,370 miles (3 814 km)
Weights: empty 40,935 lb (18 568 kg); maximum take-off 68,000 lb (30 844 kg)
Dimensions: span 115 ft 0 in (35.05 m); length 79 ft 3 in (24.16 m); height 27 ft 6 in (8.38 m); wing area 1,780 sq ft (165.36 m²)
Armament: two 0.50-in (12.7-mm) machine-guns in each of bow, dorsal and tail turrets, and one 0.50-in (12.7-mm) gun in each of two beam positions, plus up to 12,000 lb (5 443 kg) of weapons including bombs, depth bombs and torpedoes in bomb bays
Operators: RAF, USN

Consolidated PB4Y Liberator and Privateer

History and Notes

Urgently in need of land-based bombers which could be deployed on long-range anti-shipping and anti-submarine patrols, the US Navy began in early 1942 to campaign for a supply of Consolidated B-24 Liberator bombers which could be utilised successfully in this role. There was no doubt of their suitability for such operations, for RAF Coastal Command had been using Liberator Is (basically the same as the B-24A) from the summer of 1941. These had been supplied initially to the RAF's No. 120 Squadron, and were used to help close the 'Atlantic Gap', that area of the North Atlantic which, before the Liberator became available, was the most hazardous for Allied convoys. For the first time No.120 Squadron was able to provide air support in an area that previously had been out of reach of both US

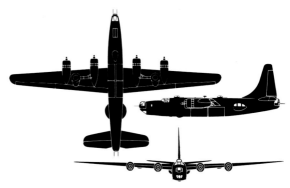

Consolidated PB4Y-2 Privateer

Consolidated PB4Y Liberator and Privateer

and British land-based aircraft.

In August 1942 the US Navy began to receive its first Liberators, designated PB4Y-1 and equivalent to the USAAF's B-24D. Some of this initial batch went to a training squadron, but the majority equipped the US Navy's first operational PB4Y unit, based in Iceland. The proportion of deliveries to the US Navy was low at first, as production was required also for both the RAF and USAAF, but they received a substantial reinforcement of these aircraft from the USAAF in August 1943, which decided to hand over its B-24 antisubmarine squadrons to the US Navy; these Liberators were equipped with ASV (Air-to-Surface Vessel) radar, and were also designated PB4Y-I. In addition to the foregoing, and allocations from the production line to make a total of 977, the US Navy also received a number of Liberators configured for transport duties, equivalent to the USAAFs C-87 and C-87A, with the US Navy designations RY-2 and RY-1 respectively. Under the designation PB4Y-1P, a number of naval Liberators were modified subsequently for operation in a reconnaissance role, and some remained in service with Navy Squadrons VP-61 and VP-62 until 1951.

However, the near 1,000 Liberators which the US Navy received had all been configured and equipped for operation by the USAAF. Good though they were, it was decided that a variant developed specifically for US Navy use would be even better, and in May 1943 Consolidated received a contract for the evolution of such a variant, which duly became designated as the PB4Y-2 Privateer. Three prototypes were built initially, the first making its maiden flight on 20 September 1943, and these were seen to retain the wing and landing gear of the B-24. They introduced a new tail unit, with tall single fin and rudder, similar to that designed for the XB-24K, had a lengthened forward fuselage and changes in the armament installation to provide a maximum of 12 0.50-in (12.7-mm) machine-guns. Pratt & Whitney R-1830-94 Twin Wasp engines were installed, without turbochargers since high-altitude operation was not required, which meant that the distinguishing oval engine cowlings were replaced by others in which the longer axis was vertical rather than horizontal. Thus the two main recognition characteristics of the Liberator disappeared at once from this new US Navy aircraft.

Only a small number of PB4Y-2s were used operationally before the end of World War II; 46 of a transport variant designated RY-3 were also to be built for the US Navy, and 27 similar aircraft were delivered in early 1945 for use by RAF Transport Command.

Specification

Type: land-based maritime patrol aircraft
Powerplant: four 1,350-hp (1 007-kW) Pratt & Whitney R-1830-94 Twin Wasp radial piston engines
Performance: maximum speed 237 mph (381 km/h) at 13,750 ft (4 190 m); cruising speed 140 mph (225 km/h); service ceiling 20,700 ft (6 310 m); range 2,800 miles (4 506 km)
Weights: empty 37,485 lb (17 003 kg); maximum take-off 65,000 lb (29 48 kg)
Dimensions: span 110 ft 0 in (33.53 m); length 74 ft 7 in (22.73 m); height 30 ft 1 in (9.17 m); wing area 1,048 sq ft (97.36 m²)
Armament: 12 0.50-in (12.7-mm) machine-guns in turrets and waist positions, plus up to 12,800 lb (5 806 kg) of other weapons
Operators: RAF, USN

Derived from the B-24 bomber, the Consolidated PB4Y Privateer patrol bomber proved itself a great maritime aircraft. Seen here is an example of the PB4Y-2 version from NAS Patuxent River, Maryland.

Consolidated XP4Y-1

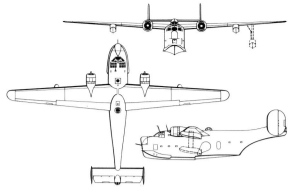

History and Notes

During 1938 Consolidated originated the design of a new flying boat which the company identified as the Model 31. A feature of this aircraft was the inclusion of a new high aspect ratio wing, of constant taper from root to wingtip, designed by David R. Davis, and which offered reduced drag at both low and high speeds. The union of wide span, narrow chord and a special aerofoil section offered a very low drag coefficient, which meant that any powerplant installation in combination with the Davis wing should offer greater range than the same powerplant combined with conventional lifting surfaces. Consolidated had chosen to use this wing for the Model 31 as long range was regarded as an essential requirement of both the civil and military versions which were planned.

A prototype was completed in May 1939, flying shortly after that, but by the time flight testing was completed war had broken out in Europe. This suggested to Consolidated that development emphasis should lead towards a military, rather than a civil version, and eventually the company were awarded a contract for a single prototype for the US Navy. Designated XP4Y-1 and unofficially named Corregidor, almost three years of redesign and test were to result before this prototype (27852) became available for service testing. Construction was all-metal with smooth flush-riveted skins, but all control surfaces had a basic structure of metal with fabric covering. The cantilever monoplane wing was mounted high on the hull, and the tailplane, also high set on an upswept aft fuselage, had large endplate fins and rudders. Retractable stabilising floats were installed beneath each wingtip, and dummy nose, dorsal and tail turrets were provided for flight testing.

Service tests were satisfactory and large-scale construction was planned, but a shortage of the Wright R-3350 Cyclone engines which powered the XP4Y-1 led to cancellation, in the summer of 1943, of the US Navy's contract for 200 P4Y-1s, for which serial numbers 44705-44904 had been allocated.

Consolidated XP4Y-1 Corregidor (dashed lines: retractable stabilising floats)

Specification

Type: long-range maritime patrol flying boat
Powerplant: two 2,300-hp (1 715-kW) Wright R-3350-8 Cyclone 18 twin-row radial piston engines
Performance: maximum speed 247 mph (398 km/h) at 13,600 ft (4 145 m); cruising speed 136 mph (219 km/h); service ceiling 21,400 ft (6 520 m); maximum patrol range 3,280 miles (5 279 km)
Weights: empty 29,334 lb (13 306 kg); maximum take-off 48,000 lb (21 772 kg)
Dimensions: span 110 ft 0 in (33.53 m); length 74 ft 1 in (22.58 m); height 25 ft 2 in (7.67 m); wing area 1,048 sq ft (97.36 m²)
Armament (intended): one 37-mm cannon in bow turret, and two 0.50-in (12.7-mm) machine-guns in dorsal and tail turrets, plus up to 4,000 lb (1 814 kg) of weapons carried externally
Operator: USN (for evaluation only)

The Consolidated Model 31 made effective use of the so-called Davis wing, of high aspect ratio and special section to minimise drag and so offer a considerable increment in range performance. Development of the XP4Y-1 patrol flying-boat followed, but engine shortages led to cancellation.

Consolidated-Vultee XP-81

History and Notes

Early turbine-powered fighter aircraft which emerged during the course of World War II had only very limited range capability. The Bell P-59 Airacomet, for example, had a range of a mere 525 miles (845 km), which meant that such an aircraft was quite unsuitable for use as a long-range bomber escort. However, the USAAF had a requirement for such an aircraft in the Pacific theatre, and in early January 1944 Consolidated-Vultee began the design of an unusual aircraft tailored for such a role.

The airframe was quite conventional in layout, with a low-set cantilever monoplane wing, retractable tricycle type landing gear and accommodation for a pilot beneath a neat transparent canopy. The unusual feature was a mixed tandem powerplant, with a turboprop engine and tractor propeller mounted in the fuselage nose, and a turbojet engine in the aft fuselage. It was intended that both engines would be used for take-off, high-speed flight and combat, but that only the far more fuel-conscious turboprop would be used for long-range cruising flight.

When the airframe of the first prototype was complete, the turboprop powerplant was not ready for installation. In order that flight testing should proceed without protracted and uncertain delay, a Packard/Rolls-Royce V-1650-7 Merlin engine was installed, and with this and the Allison J33 turbojet the XP-81 was flown for the first time on 11 February 1945. When, much later in that year, the General Electric XT31 turboprop became available and was installed, the XP-81 was flown in its initial planned form. Unfor-

tunately, the turboprop was developing only about 60 per cent of its rated power, thus proving no more effective so far as performance was concerned than the Merlin. Only limited testing followed the prototype's first flight in this form, on 21 December 1945, and with World War II ended the project was abandoned. Pre-production YP-81s had also been ordered, but these were cancelled soon after VJ-Day.

Specification

Type: single-seat long-range escort fighter
Powerplant: one 2,300-ehp (1 715-kW) General Electric XT31-GE-1 turboprop, and one 3,750-lb (1 701-kg) thrust Allison-built J33-GE-5 turbojet
Performance (estimated with above powerplant): maximum speed at sea level 478 mph (769 km/h); cruising speed 275 mph (443 km/h); service ceiling 35,500 ft (10 820 m); range 2,500 miles (4 023 km)
Weights: empty 12,755 lb (5 786 kg); maximum take-off 24,650 lb (11 181 kg)
Dimensions: span 50 ft 6 in (15.39 m); length 44 ft 10 in (13.67 m); height 14 ft 0 in (4.27 m); wing area 425 sq ft (39.48 m²)
Armament (intended): six 0.50-in (12.7-mm) machine-guns or six 20-mm cannon
Operator: USAAF (for evaluation only)

Designed to overcome the range limitations of early jet aircraft by combining a turboprop for cruising and a jet engine for combat performance, the XP-81 was an ambitious project by Consolidated-Vultee. It foundered for lack of an adequate turboprop.

Culver A-8/PQ-8

History and Notes

Anti-aircraft guns were first developed by the Prussians in 1870, in their unsuccessful attempts to destroy the hydrogen balloons which carried VIP passengers and mail out of Paris, when it was under siege during the Franco-Prussian War. Since that time the primary problem in the use of such weapons has been to hit the target. As a result, early anti-aircraft batteries put up a frightening and noisy screen of fire, intended to force enemy aircraft to avoid the area that was being defended, or to fly at such a height that the accuracy of their scouting or bombing mission was much impaired.

It was not easy, of course, for anti-aircraft gunners to become very proficient at hitting a moving object. Few pilots would enjoy the task of flying an aeroplane to serve as a target: even when the target was a drogue, towed some distance behind a tug aircraft, there had been some narrow escapes when height and speed were misjudged and over/under-correction made to the sighting of the gun. In an attempt to resolve this situation, radio-controlled target aircraft were developed, but woe betide the gunner who was clever (or lucky) enough to hit one.

In the USA, the Culver LCA Cadet lightplane was selected by the US Army Air Corps in 1940 as being suitable for development as a radio-controlled target to satisfy this requirement. The first of these acquired for test purposes (41-18889) was allocated the designation A-8 (later XPQ-8). This was a low-wing monoplane with retractable tricycle-type landing gear, and

powered by a 100-hp (75-kW) Continental O-200 flat-four engine. Successful testing resulted in a production order for 200 similar PQ-8s, and at a later date an additional 200 were ordered as PQ-8A (later Q-8A), this version being equipped with a more powerful engine.

The US Navy had a similar training problem for the anti-aircraft gunners of its warships, and in late 1941 acquired a single example of the US Army PQ-8A for evaluation. An order for 200 similar aircraft, designated TDC-2, was placed with Culver in 1942.

Specification

Type: radio-controlled target aircraft
Powerplant (PQ-8A): one 125-hp (93-kW) Continental O-200-1 flat-four piston engine
Performance: maximum speed 116 mph (187 km/h)
Weight: empty 720 lb (327 kg); maximum take-off 1,305 lb (592 kg)
Dimensions: span 26 ft 11 in (8.20 m); length 17 ft 8 in (5.38 m); height 5 ft 6 in (1.68 m); wing area 120 sq ft (11.15 m²)
Armament: none
Operators: USAAC/USAAF, USN

Derived from the Cadet lightplane, the Culver PQ-8 radio-controlled target aircraft was a useful adjunct to US Army anti-aircraft artillery training, but was too slow adequately to simulate the newer combat aircraft likely to be encountered by the US forces in World War II.

Culver PQ-14

History and Notes

Use of the Culver PQ-8/-8A and TDC-2, by the USAAF and US Navy respectively, left little doubt that targets of this nature improved very considerably the accuracy of those anti-aircraft units which had ample opportunity to train with such devices. The real problem associated with use of the PQ-8 was that its maximum speed of 116 mph (187 km/h) was completely unrealistic in 1941-2, when attacking fighters and bombers could demonstrate speeds respectively three and two times that of the PQ-8.

To meet this requirement, Culver developed during 1942 an aircraft designed specifically to serve as a target aircraft. With a general resemblance to the PQ-8, it was of low-wing monoplane configuration and with a similar tail unit; the control surfaces of the aerofoils were, however, larger in area to ensure much improved response and manoeuvrability when under radio control. The landing gear was of the retractable tricycle type. The powerplant comprised a Franklin O-300 flat-six engine, of greater power to provide increased performance.

In early 1943 the USAAF acquired a single prototype for evaluation under the designation XPQ-14. This proving satisfactory from the maintenance, launch and flight aspect, a batch of 75 YPQ-14As were ordered for service trials in its target aircraft role. Proving successful the PQ-14 was ordered into large-scale production, and the USAAF was to acquire 1,348 PQ-14As (later Q-14A), of which 1,201 were transferred to the US Navy, being designated TD2C-1 in that service. A heavier version was built subsequently, the USAAF acquiring an initial batch of 25 for service trials as YPQ-14Bs, and following these up with the procurement of 594 PQ-14Bs for training units. A single

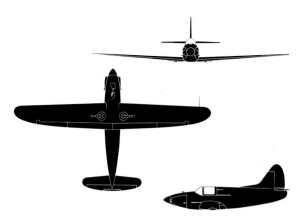

Culver PQ-14A

example with a Franklin O-300-9 engine was designated PQ-14C.

Specification

Type: radio-controlled target aircraft
Powerplant (PQ-14A): one 150-hp (112-kW) Franklin O-300-11 flat-six piston engine
Performance: maximum speed 180 mph (290 km/h)
Weight: maximum take-off 1,820 lb (826 kg)
Dimensions: span 30 ft 0 in (9.14 m); length 19 ft 6 in (5.94 m); height 7 ft 11 in (2.41 m)
Armament: none
Operators: USAAF, USN

The US Navy received from the USAAF some 1,201 examples of the improved PQ-14A drone, which were redesignated TD2C-1s in naval service.

Curtiss AT-9

History and Notes

In 1940, and with Europe already at war, the US Army Air Corps knew that it was essential to begin preparations for the very real possibility that in the not too distant future, and despite a national isolationist policy which was still very active, the United States of America would become involved in that war. The only positive action which could be taken at that time was to intensify the training programme so that a maximum number of active and reserve flying crews would be available if the need arose.

As a part of this general thinking, the US Army had already begun evaluation of the Cessna T-50 as an 'off-the-shelf' twin-engined trainer which would prove suitable for the transition of a pilot qualified on single-engined aircraft to a twin-engined aircraft and its very different handling technique. Procured as the AT-8, Cessna's T-50 was to be built in large numbers.

For the more specific transition to a 'high-performance' twin-engined bomber it was considered that something rather less stable than the T-50 was needed. However, Curtiss-Wright had anticipated this requirement in the design of their Model 25, a twin-engined pilot transition trainer which had the take-off and landing characteristics of a light bomber aircraft. The Model 25 was of low-wing cantilever monoplane configuration, provided with retractable tailwheel-type landing gear, and powered by two Lycoming R-680-9 radial engines, with Hamilton-Standard two-blade constant-speed metal propellers. The single prototype acquired for evaluation had a welded steel-tube fuselage structure, and the wings, fuselage and tail units were fabric-covered.

Evaluation proving satisfactory, the type was ordered into production under the designation AT-9. These production examples differed from the prototype by being of all-metal construction. A total of 491 AT-9s was produced, and these aircraft were followed into

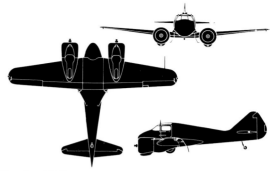

Curtiss AT-9

service by 300 generally similar AT-9As. They remained in use for a comparatively short time, for the USA's involvement in World War II in late 1941 resulted in the early development of far more effective training craft.

Specification

Type: twin-engined advanced trainer
Powerplant: two 295-hp (220-kW) Lycoming R-680-9 radial piston engines
Performance: maximum speed 197 mph (317 km/h); cruising speed 175 mph (282 km/h); range 750 miles (1 207 km)
Weights: empty 4,600 lb (2 087 kg); maximum take-off 6,000 lb (2 722 kg)
Dimensions: span 40 ft 4 in (12.29 m); length 31 ft 8 in (9.65 m); height 9 ft 10 in (2.99 m); wing area 233 sq ft (21.65 m²)
Armament: none
Operator: USAAC

The Curtiss AT-9 transition trainer saw only limited use before the arrival of more versatile types.

Curtiss C-46 Commando

History and Notes

In common with its more prolific contemporary, the Douglas C-47, the Curtiss C-46 Commando was derived from a design initially developed for the civil market. Work on the Curtiss CW-20 began in 1937 when Chief Engineer George A. Page was instructed to develop a 24-34 passenger commercial airliner with a gross weight of 36,000 lb (16 329 kg) and powered by two 1,600-hp (1 193-kW) Pratt & Whitney R-2800 Double Wasp engines. The cabin was to be pressurised for above-the-weather operations and was to be capable of accommodating 20 sleeping berths. The latter, arranged across the width of the aircraft, dictated a wide cabin while, to overcome the drag that would have been induced by a completely circular fuselage of adequate diameter, the fuselage was designed as two segments intersecting at a common chord which effectively became the floor line. As a result the cabin was particularly capacious and, in addition to the 2,300 cu ft (65.13 m³) of space in the upper segment, there was a usable volume of 455 cu ft (12.88 m³) below the floor.

The prototype was built at St Louis, Missouri and, powered by two 1,600-hp (1 193-kW) Wright Cyclone 586-C14-BA2 engines rather than the planned Double Wasps, was first flown by Eddie Allen on 26 March 1940. Fairing plates were introduced to smooth out the join between the two fuselage segments and a twin-finned tail unit was fitted. The latter was replaced subsequently by a large single fin, presumably to correct low-speed asymmetric handling problems, and the machine was redesignated CW-20A. It was later purchased by the US Army to become the sole C-55, and was sold to BOAC in November 1941; in 24-seat

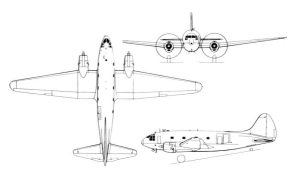

Curtiss C-46D Commando

configuration with long-range tanks it was used on long-haul routes and to link Malta with Gibraltar in 1942. It was broken up at Filton, Bristol in October 1943.

The deterioration of the situation in Europe resulted in increasing awareness of the inevitability of US entry into the war, and in September 1940 large orders were placed for fighters, bombers and transports, including 200 of a military version of the CW-20, which was designated C-46. The first 25 aircraft, built at Buffalo, New York, differed in detail from the CW-20, having fewer cabin windows and dispensing with the fuselage-join fairing plates: these had offered little aerodynamic advantage, while conferring a weight penalty of 275 lb (125 kg) and adding to the manufacturing process.

The C-46A was the main production variant of the successful Curtiss Commando transport series.

Curtiss C-46 Commando

Cabin pressurisation was not included and the engines were replaced by 2,000-hp (1 491-kW) Pratt & Whitney R-2800-43s.

This initial production version was followed by the C-46A which had double cargo doors, a strengthened floor and a hydraulic cargo-loading winch; 40 folding seats were fitted. The engines were R-2800-51s, early aircraft having three-bladed Hamilton propellers which were superseded by four-bladed Curtiss electrically-operated propellers. Production reached 1,041 at Buffalo and 10 at St Louis before production was transferred to a government-owned factory at Louisville, Kentucky where 439 were manufactured before the programme was returned to St Louis. Higgins Industries Inc. of New Orleans completed two of a contract for 500 C-46As before the order was cancelled.

The first aircraft built after the return to St Louis was the sole XC-46B with water-injected 2,100-hp (1 566-kW) R-2800-34W engines and a more conventional 'stepped' windscreen. Meanwhile, at Buffalo, Curtiss had begun production of 1,410 C-46Ds, equivalent to C-46As but with a revised nose and doors for paratroop operations, and equipped to carry 50 men. Also produced at Buffalo were 234 C-46Fs, with R-2800-75 engines and blunt wing-tips, and a single C-46G with R-2800-34W engines and which was later to become the XC-113, a test-bed for the Curtiss TG-100 turboprop engine. The C-46G had a single cargo door, as did the St Louis-built C-46E which also featured the stepped windscreen of the XC-46B, and R-2800-75 engines driving three-blade Hamilton propellers. Only 17 were built of a contract for 550 which was cancelled after VE-Day.

The CW-20 would have been only a marginal commercial proposition in cargo configuration as, although by comparison with the Douglas DC-3 it offered twice the cabin volume, a 25 per cent increase in fuel capacity and a 45 per cent increase in gross weight, these favourable features were offset by 50 per cent greater fuel consumption and the fact that, at a gross weight of 40,000 lb (18 144 kg) the cabin could not be filled unless cargo density was less than 4.5 lb/cu ft (16.02 kg/m³), a relatively low figure.

In service use, however, the C-46 was cleared to operate at a military overload weight of 50,675 lb (22 986 kg), allowing almost 6,000 lb (2 722 kg) more payload to be carried, and the cabin capacity became a considerable asset. The first of the US Army's C-46s was rolled out at Buffalo in May 1942 and delivered on 12 July. Some of the earliest deliveries were to Air Transport Command's Caribbean Wing (Eastern Air Lines' Military Transport Division, formed on 1 September 1942). This used some of the first aircraft from the line to build up operational expertise, flying a military service from Miami to Middleton, Pennsylvania, commencing on 1 October, and then from Miami to Natal, Brazil, from February 1943. As the transfer of men and equipment to North Africa built up, C-46s were introduced to the South Atlantic ferry route, from Natal to Accra, Gold Coast, via Ascension Island, until sufficient Douglas C-54s became available in 1944.

Some of the early training was also carried out by the airlines, notably Northwest and Western, and USAAF schools included No. 2 OTU at Homestead, Florida and No. 3 OTU, initially at Rosecrans, St Joseph, Missouri, soon transferred to Reno, Nevada, where the local terrain was more similar to that which very many C-46 crews were to experience with the India-China Wing, flying over the 'Hump'. These operations took place in arduous conditions, involving take-offs from primitive airfields, climbing in the vicinity of the base to more than 20,000 ft (6 095 m) often on instruments in icing and turbulence, to cross mountain ranges that lay across the track while carrying hazardous cargoes, including fuel and ammunition.

Specification

Type: 54-seat troop carrier and cargo transport
Powerplant (C-46A): two 2,000-hp (1 491-kW) Pratt & Whitney R-2800-51 Double Wasp radial piston engines
Performance: maximum speed 269 mph (433 km/h) at 15,000 ft (4 570 m); cruising speed 183 mph (295 km/h); ceiling 27,600 ft (8 410 m); range 1,200 miles (1 931 km)
Weights: empty 32,400 lb (14 696 kg); maximum take-off 56,000 lb (25 401 kg)
Dimensions: span 108 ft 1 in (32.94 m); length 76 ft 4 in (23.27 m); height 21 ft 9 in (6.63 m); wing area 1,360 sq ft (126.34 m²)
Armament: none
Operators: USAAF, USMC, USN

Curtiss C-76 Caravan

History and Notes

During 1941 Curtiss received a contract from the US Army covering the design and construction of an all-wooden military transport. Like a number of aircraft built in the USA during the period, it was one of a series constructed in prototype form to develop the manufacturing techniques for new generation all-wood aircraft. Because of the volume of contracts that were piling up, involving the construction of then-current all-metal designs in enormous numbers, this programme for all-wood aircraft was regarded as an insurance policy against future shortages of light alloys, a circumstance which, fortunately, did not arise.

The initial contract called for 11 YC-76 pre-production aircraft to be built to the Curtiss design, and this medium-size twin-engined transport had some resemblance to the earlier, larger C-46 Commando. The most noticeable change concerned the monoplane

Curtiss C-76 Caravan

wing, which in the Commando was of low-wing and in the Caravan of high-wing configuration. Landing gear was of retractable tricycle type, the nose unit carrying twin wheels. The powerplant comprised two 1,200-hp (895-kW) Pratt & Whitney R-1830-92 Twin Wasp engines in wing-mounted nacelles. Accommodation was provided for a total of 23 persons, including the flight crew.

The first YC-76 made its initial flight on 1 January 1943, and in addition to the original contract, follow up orders were received for five C-76s and nine YC-76As, all of which were delivered during 1943. Production was terminated when it was clear that a serious shortage of light alloy was unlikely, 175 C-76As then on order being cancelled.

Specification

Type: medium military transport
Powerplant: two 1,200-hp (895-kW) Pratt & Whitney R-1830-92 Twin Wasp radial piston engines
Performance: maximum speed 200 mph (322 km/h)
Weight: maximum take-off 28,000 lb (12 700 kg)
Dimensions: span 108 ft 2 in (32.97 m); length 68 ft 4 in (20.83 m)
Armament: none
Operator: USAAF

Designed to lessen the USA's reliance on highly vulnerable sources for light alloy production, the all-wood Curtiss C-76 Caravan was cancelled when the alloy shortage failed to materialise.

Curtiss P-36

History and Notes

The Curtiss Aeroplane and Motor Company had established a continuing relationship with the US Army, which had got off to a good start in 1914 with the then large-scale procurement (94) of the first of the classic JN ('Jenny') series. Twenty years later, when the Curtiss Aeroplane Company had become a division of the Curtiss-Wright Corporation, it was decided to design and develop as a private venture a new monoplane pursuit (fighter) aircraft. Known as the Curtiss Model 75, it had such advanced features as retractable landing gear and an enclosed cockpit for the pilot, and the company believed that the US Army would be prepared to consider it as a replacement for the lower-performance Boeing P-26.

Contemporary with the famous wartime triad of the Messerschmitt Bf 109, Hawker Hurricane and Super-

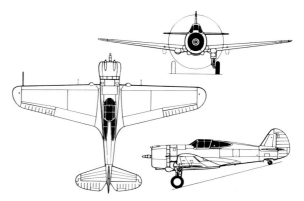

Curtiss P-36A (dashed lines: underwing gondolas of P-36C)

Curtiss P-36

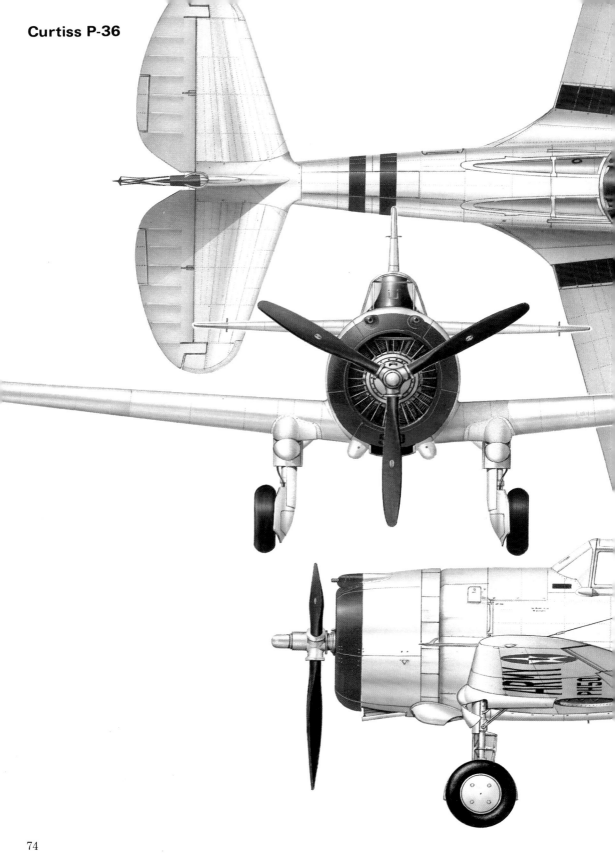

**Curtiss P-36A attached to the 35th Pursuit
Squadron (Fighter), during 1939-40 while that unit
was at Langley Field, Virginia in transition to new
quarters at Mitchell Field, New York.**

Curtiss P-36

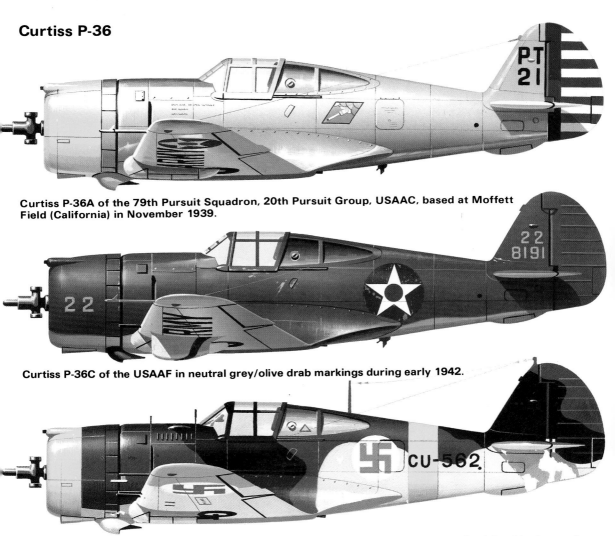

Curtiss P-36A of the 79th Pursuit Squadron, 20th Pursuit Group, USAAC, based at Moffett Field (California) in November 1939.

Curtiss P-36C of the USAAF in neutral grey/olive drab markings during early 1942.

Curtiss Hawk 75A-3 of LeLv 32, Suomen Ilmavoimat (Finnish air force), based at Suulajärvi in September 1941.

marine Spitfire, it failed to emulate their performance, for at that era of fighter development designers had not yet succeeded in realising the full potential of bluff, radial air-cooled engines. At a later date an attempt was made to upgrade the Model 75 by the installation of an Allison V-1710 turbocharged inline engine, but the prototype and a batch of service test YP-37s with the V-1710-21 engine failed to achieve a sufficiently high standard of performance and were, in any event, superseded by the P-40.

The Model 75 prototype, powered by a 900-hp (671-kW) Wright XR-1670-5 radial engine, was submitted to the US Army Air Corps in May 1935 for evaluation in a design competition for a single-seat pursuit aircraft. This failed to materialise because no competing designs were ready, and it was not until April 1936 that the twice-postponed contest began. By then the Model 75 had been re-engined with an 850-hp (634-kW) Wright XR-1820-39 Cyclone radial, and was identified as the Model 75-B with this powerplant.

The Seversky Aircraft Corporation won the USAAC's competition with a not-too-dissimilar aircraft, which was ordered into production as the P-35. However, Curtiss were awarded a contract for just three examples of their design, to be powered by a derated version of the 1,050-hp (783-kW) Pratt & Whitney R-1830-13 Twin Wasp radial engine, and to be used for test and evaluation under the designation Y1P-36. By comparison with the original Model 75 prototype, these had cockpit modifications to improve fore and aft view, and introduced a retractable tailwheel. A feature of the main landing gear units was that they swivelled through 90° as they retracted aft, so that the main wheels could lie flush in the undersurface of the wing.

Service testing of the Y1P-36s was considered so successful that a contract for 210 production P-36As was awarded on 7 July 1937, then the US Army's peacetime contract for pursuit aircraft. Delivery of these began in April 1938, but by late 1941 when the United States became involved in World War II, they

Curtiss P-36

Curtiss Hawk 75A-5 of the Chinese Nationalist air force, based at Kunming in 1942.

Curtiss Hawk 75A-7 flown by Colonel Boxman of 1. Vliegtuigafdeling, KNIL Luchtvaartafdeling (Dutch East Indies air force), based at Madioen (Dutch East Indies) in December 1941.

Curtiss Hawk 75A-8 of the Norwegian flying training centre, based at Island Airport, Toronto (Canada) during 1941.

were already considered obsolete. Circumstances compelled limited use of P-36As in the opening stage of hostilities with Japan, but they were very soon relegated for use in a training role.

Variants included a single XP-36B with a 1,000-hp (746-kW) Pratt & Whitney R-1820-25 engine, and the last 31 of the original production run were completed as P-36Cs with a more powerful Twin Wasp engine. The designations XP-36D/-36E/-36F were applied to experimental examples with differing armament.

Export Hawk 75As were supplied to the French Armée de l'Air as Hawk 75A-1/A-2/A-3/A-4 but the majority were transferred to the UK after the fall of France, being designated respectively Mohawk I/II/III/IV. The type was supplied also to Norway and Persia and, indirectly, served also with the air forces of Finland, India, Netherlands East Indies, Peru, Portugal, South Africa and Vichy France.

Specification

Type: single-seat fighter

Powerplant (P-36C): one 1,200-hp (895-kW) Pratt & Whitney R-1830-17 Twin Wasp radial piston engine

Performance: maximum speed 311 mph (501 km/h) at 10,000 ft (3 050 m); cruising speed 270 mph (435 km/h); service ceiling 33,700 ft (10 270 m); range 820 miles (1 320 km)

Weights: empty 4,620 lb (2 096 kg); maximum take-off 6,010 lb (2 726 kg)

Dimensions: span 37 ft 4 in (11.38 m); length 28 ft 6 in (8.69 m); height 9 ft 6 in (2.90 m); wing area 236 sq ft (21.92 m²)

Armament: one 0.50-in (12.7-mm) and three 0.30-in (7.62-mm) machine-guns

Operators: China, FAF, FVAF, Finland, (Luftwaffe), Netherlands East Indies, Norway, Peru, Portugal, RAF, SAAF, USAAC/USAAF

Curtiss Hawk 75

History and Notes

The belief that there would be a market for a less sophisticated export version of the Curtiss Model 75 led to the development in 1937 of such a version, known as the Curtiss Hawk 75. Generally similar in construction to the Y1P-36, which had been ordered in July 1936 for evaluation by the US Army Air Corps, this export Hawk had fixed tailwheel type landing gear, with the main units contained within streamlined fairings, and a lower-powered Wright radial engine was installed.

The original demonstration model was bought by the Chinese government during 1937, and this was followed by an order in 1938 for an additional 112 examples under the designation Hawk 75M. These were used to equip the newly reorganised Chinese air force, but as a result of the limited ability of their pilots they were not especially successful in combat with Japanese aircraft involved in the Sino-Japanese War.

In 1939 a similar version was supplied to the Thai air force under the designation Hawk 75N. This model had two additional 0.30-in (7.62-mm) machine-guns, mounted beneath the wings outboard of the main landing gear, and provision for the carriage of 10 25-lb (11-kg) or 30-lb (14-kg) bombs, or six 50-lb (23-kg)

bombs. Twenty-five of these Hawk 75Ns were supplied and were used in combat during 1941.

Production of the final version, the Hawk 75O for the Argentine, totalled 30 examples. This nation also negotiated with Curtiss a licence to build the type, and 200 were produced from 1940 in a government aircraft factory in Cordoba.

Specification

Type: single-seat fighter/fighter-bomber
Powerplant: one 875-hp (652-kW) Wright GR-1820-G3 radial piston engine
Performance: maximum speed 280 mph (451 km/h) at 10,700 ft (3 260 m); service ceiling 31,800 ft (9 690 m) range 545 miles (877 km)
Weights: empty 3,975 lb (1 803 kg); maximum take-off 6,418 lb (2 911 kg)
Dimensions: span 37 ft 4 in (11.38 m); length 28 ft 7 in (8.71 m); height 9 ft 4 in (2.84 m); wing area 236 sq ft (21.92 m²)
Armament: one 0.50-in (12.7-mm) machine-gun and one or three 0.30-in (7.62-mm) machine-guns, plus 10 25-lb (11-kg) or 30-lb (14-kg) bombs, or six 50-lb (23-kg) bombs
Operators: Argentina, China, Thailand

Curtiss P-40 Warhawk

History and Notes

The last of the Curtiss Hawks from the P-1 Hawk of 1925, the P-40 Warhawk has always been something of an enigma. By no stretch of the imagination could it be numbered among the 'great' fighter aircraft of World War II. Yet, with the exceptions of the Republic P-47 and North American P-51, it was the most extensively built US fighter, with almost 14,000 delivered before production ended in December 1944.

Construction of what the company designated as the Hawk 81 began in 1938, when the tenth P-36A production aircraft was withdrawn from the line for an experimental conversion from radial to inline engine. At that period in the development of fighter aircraft the latter type of powerplant was much in favour: the Schneider Trophy contests, which terminated in the early 1930s, were assumed to have demonstrated the superiority of the inline engine, and early examples of the British Hawker Hurricane and Supermarine Spitfire, and German Messerschmitt Bf 109, appeared to leave little doubt that this was the case. There were, of course, problems with the liquid cooling system of high-powered inline engines: they added another vital system that was vulnerable to combat damage, and they were heavier. But it was clear that inline engines did develop considerably more power per unit of frontal area. It remained for Germany's Kurt Tank to prove that the radial engine was inferior to nothing, when his superb Focke-Wulf Fw 190 became operational at the beginning of 1941.

Therefore the tenth P-36A became powered by a

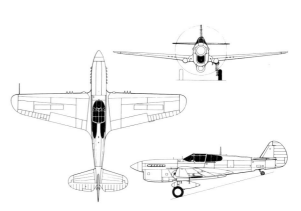

Curtiss P-40C (Tomahawk Mk IIB)

1,160-hp (865-kW) Allison V-1710-19 engine, instead of the 1,050-hp (783-kW) Pratt & Whitney R-1830-13 radial which was standard. In other respects it varied little from the P-36A.

When the Hawk 81 was first flown, on 14 October 1938, the coolant radiator was mounted beneath the aft fuselage; soon after this it was resited beneath the nose, associated with the oil cooler in a common cowling.

In May 1939 the Hawk 81, by then designated XP-40, was flown in competition against other pursuit prototypes and was selected for production as most closely meeting US Army Air Corps requirements. A total of 524 P-40s was ordered into production on 27 April 1939, this then representing the largest single

Curtiss P-40 Warhawk

Hawk 81A-2 flown by Charles Older of the 3rd Squadron ('Hell's Angels'), American Volunteer Group, based at Kunming (China) in spring 1942.

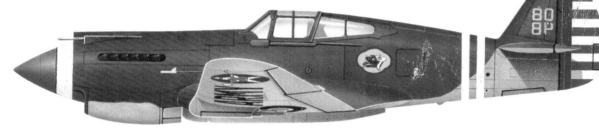

Curtiss P-40 of the 36th Squadron, 8th Pursuit Group, USAAF, during 1941.

Curtiss Tomahawk Mk I of No. 400 (Canadian) Squadron, RAF, based at Odiham (UK) in spring 1942.

Curtiss Tomahawk Mk IIA of No. 349 (Belgian) Squadron, RAF, based at Ikeja (West Africa) in February 1943.

Curtiss P-40C of the 39th Squadron, 31st Pursuit Group, USAAF, based at Selfridge Field (Michigan) in 1941.

Curtiss P-40 Warhawk

order for fighters to originate from the US Army. Just over a year later, in May 1940, the first P-40s began to come off the production line, the first three being used for service trials, as no YP-40s had been ordered for this purpose. These differed considerably from the initial XP-40, with a less powerful supercharged Allison V-1710-33 engine installed, flush-headed rivets replacing the drag-inducing domed type, and with two 0.50-in (12.7-mm) machine-guns mounted in the fuselage nose. By September 1940 a total of 200 of these aircraft had been delivered to the USAAC.

Before that, in April 1940, production priority had been accorded to delivery of 140 generally similar Hawk 81A-1 fighters ordered by France. None, however, left the assembly line before that nation's collapse and, instead, these aircraft were diverted to the UK where they were designated Tomahawk I. Deliveries to England, and to Takoradi in West Africa, began in late 1940, but the Tomahawk was unsuited for operational employment in Europe. Initial deliveries were made to No. 2 Squadron in August 1941, to supersede its Westland Lysanders, but although limited use was made of the Tomahawks in a low-level tactical reconnaissance role, the majority were relegated to training duties. The first of the RAF's overseas squadrons to receive these aircraft was No.112, which operated them with some success in a low-level ground-attack role in the Western Desert.

The next version for the RAF (Hawk 81A-2) was designated Tomahawk IIA and this, basically the same as the USAAC's P-40B, had self-sealing fuel tanks, armour, and was armed with two wing-mounted 0.303-in (7.7-mm) machine-guns. Unfortunately, the increased weight resulting from these improvements reduced performance; while improved self-sealing tanks and the addition of two more wing guns on the P-40C brought a further erosion of performance. A total of 930 of this version was built for the RAF (Hawk 81A-3), those which entered service becoming designated Tomahawk IIB. These had US radio equipment but were armed with six 0.303-in (7.7-mm) machine guns. Tomahawk IIs were operated by Nos. 2, 26, 73, 112, 136, 168, 239, 241, 250, 403, 414, 430 and 616 Squadrons of the RAF, No. 3 Squadron of the Royal Australian Air Force and Nos. 2 and 4 Squadrons of the South African Air Force. A total of 100 of the allocation of Hawk 81A-3s for the RAF was diverted to China, 90 of these reaching the American Volunteer Group (AVG) operating from Kunming and Mingaladon; 49 shipped direct from the USA, plus 146 reshipped from Great Britain, were supplied to the Soviet Union, and a small number went to the Turkish air force.

Some American P-40s were modified in 1941 to serve in a reconnaissance role, under the designation RP-40, but Curtiss had already begun redesign of the Hawk 81A in an attempt to improve its performance and effectiveness. The changes included installation of the 1,150-hp (858-kW) Allison V-1710-39, which could maintain this power output to an altitude of 11,700 ft (3 565 m), and airframe adjustments associated with installation of this engine, including a reduction of 6 in (0.15 m) in length, the addition of armour, provision of four 0.50-in (12.7-mm) wing-

The Curtiss P-40F Warhawk introduced the Rolls-Royce Merlin engine to the series. These examples, British Kittyhawk Mk IIs repossessed by the USAF, were flown off carriers for Operation 'Torch' in November 1941.

Curtiss P-40 Warhawk

Curtiss Tomahawk Mk IIB of the 154th Fighter Aviation Regiment, Red Banner Baltic Fleet Air Force, based in the Leningrad area during 1942.

Curtiss Tomahawk Mk IIB of No. 112 Squadron, RAF, based at Sidi Haneish (North Africa) in autumn 1941.

Curtiss Tomahawk Mk IIB of No. 112 Squadron, RAF, based at Sidi Haneish (North Africa) in October 1941.

Curtiss Tomahawk Mk IIB of the Turkish air force during 1942.

Curtiss P-40E Warhawk of the 11th Squadron, 343rd Fighter Group, USAAF, based in the Aleutian Islands during 1942.

Curtiss P-40 Warhawk

mounted machine-guns, and an underfuselage hard-point for the carriage of a 500-lb (227-kg) bomb or 52-US gallon (197-litre) fuel drop tank. First flown on 22 May 1941 as the Kittyhawk I, having been ordered as such by Britain, it was identified as Hawk 87A-1 by Curtiss, and as the P-40D by USAAC, which had also ordered this version in September 1940. Only the first 22 aircraft delivered to the USAAF had the armament of four wing-mounted guns, subsequent deliveries having six guns with the designation P-40E. A total of 1,500 of this version, identified as the P-40E-1 (Hawk 87A-3), was procured by the USAAF for supply to Britain under Lend-Lease, the model being designated Kittyhawk IA in RAF service; in addition, large numbers of this version were supplied to British Commonwealth air forces. In North Africa, where the RAF and supporting Commonwealth squadrons began to deploy Kittyhawk I/IAs in mid-1942, they proved most successful in a ground-attack role, often working in conjunction with Hurricane units. Other P-40Es were supplied to Brigadier General Claire Chennault's AVG in China where, deployed with considerable skill, the P-40 in several variants achieved notable success against Japanese aircraft, both fighter and bomber. A few P-40Es were converted as two-seat trainers, losing their fuselage fuel tank to provide space for the second cockpit.

Despite the attempts which had been made to upgrade the P-40 as a more effective fighter aircraft, its performance at altitude was totally inadequate for it to fulfil such a role, and it was decided to test the effect of using a Rolls-Royce Merlin 28 as powerplant. Accordingly, one of these engines was installed in a P-40D airframe, the resulting amalgamation of US and British technology being designated XP-40F and first flown on 25 November 1941. As had been anticipated, there was an improvement in performance at high altitude, resulting in production of the P-40F Warhawk with a 1,300-hp (969-kW) Packard-built V-1650-1 Merlin engine. The majority of the P-40Fs had an increase in overall length of 2 ft 2 in (0.66 m) as the fin and rudder were resited aft of the tailplane to improve directional stability. More than 1,300 of this version were built, of which 250 were intended for the RAF as Kittyhawk IIs. In fact none of these served with the British air force, for they were distributed instead to Free French and Russian units, some retained by the USAAF, and the balance lost at sea during convoy transit. Before the next major production version appeared, a single prototype with armament and fuel tank changes was built as the XP-40G.

Although performance of the P-40F was a considerable improvement over earlier Warhawks, there was a serious shortage of licence-built Merlin engines in the US. One project to overcome this had the designation P-40J, covering the addition of a turbo-charger to the Allison engine to provide comparable performance, but nothing came of the project. Instead, production continued with the P-40K, which introduced the 1,325-hp (988-kW) Allison V-1710-73 engine, and P-40Ks began to enter service in August 1942. Initial production batches had a fuselage similar to that of the P-40E, but later examples had a dorsal fin to overcome a tendency to swing on take-off as a result of the more

Seen before a mission in the 1944 Burma campaign is a Curtiss P-40N Warhawk carrying a centreline tank and two underwing bombs.

Curtiss P-40 Warhawk

Curtiss Kittyhawk Mk I flown by Flying Officer Neville Duke of No. 112 Squadron, RAF, and based at Landing Ground 91 (south of Alexandria in Egypt) in September 1942.

Curtiss Kittyhawk Mk III of No. 112 Squadron, No. 239 Wing, RAF, based at Cutella (Italy) in 1944.

Curtiss Kittyhawk Mk III of No. 15 Squadron, Royal New Zealand Air Force, based on Guadalcanal island (Solomon group) during summer 1943.

Curtiss Kittyhawk Mk III of No. 250 'Sudan' Squadron, No. 239 Wing, No. 211 Group, RAF, based at El Assa (Tunisia) in March 1943.

Curtiss P-40 Warhawk

For an aircraft which was basically undistinguished, the Curtiss P-40 series was remarkable in being built in more than substantial numbers: total production amounted to some 13,740, which was over 4,300 more than the total of Lockheed P-38 Lightnings, and only some 2,000 less than the totals for the great North American P-51 Mustang and Republic P-47 Thunderbolt. Designed as an interceptor fighter, the P-40 proved to have disappointing high-altitude performance, but proved to have good capabilities as a ground-attack aircraft. The type was widely used by the USAAF, but was also extensively dispersed as Lend-Lease equipment. The example illustrated is a Kittyhawk Mk I, built under the company designation Model A87A-2 and roughly comparable with the P-40C. It was on the strength of No. 112 Squadron, No. 239 Wing, No. 211 Group, Desert Air Force during May 1942, when the squadron was located at Gambut in Libya. The pilot was Sergeant Pilot H.G. Burney, and the aircraft was lost near Bir Hakeim on 30 May 1942.

Curtiss P-40 Warhawk

Curtiss Kittyhawk Mk IV flown by Group Captain G.C. Atherton, No. 80 Squadron, No. 78 Wing, 1st Tactical Air Force, Royal Australian Air Force, based on Morotai island (Moluccas group) in February 1942.

Curtiss Kittyhawk Mk IV flown by Sergeant Pilot G.F. Davis of No. 112 Squadron, RAF, based at Cutella/San Angelo in May 1944.

powerful engine. A total of 1,300 of this version was built, the majority supplied to the RCAF and USAAF, but others went to the AVG and the RAF received 21, designated Kittyhawk III.

In an attempt, presumably born of desperation, to provide the Warhawk with that little extra punch which it needed to convert it from marginal to worthwhile performance as a fighter, the P-40L which followed became the subject of weight pruning. This was achieved by restricting the armament to four 0.50-in (12.7-mm) guns, a reduction of armour, and a small loss of fuel capacity, but the resulting weight saving of about 250 lb (113 kg) did so little for performance that it seemed hardly worthwhile. A total of 700 P-40Ls was built for the USAAF, followed by 600 generally similar P-40Ms, which differed by having 1,200-hp (895-kW) Allison V-1710-81 engines. All but five of this version went to Britain under Lend-Lease, being designated Kittyhawk III like the small number of P-40Ks which had preceded them. They were redistributed for service with the RAAF, RAF, RNZAF and the 5th Squadron of the SAAF. Nos. 7 and 11 Squadrons of this latter air force were also equipped with Kittyhawks.

By late 1943 and early 1944 it was clear to all, manufacturer and users alike, that unless something was done, once and for all, to boost the performance of the Warhawk, its days were numbered. This is not surprising, for by that time much improved fighters were in service with the Axis air forces, in whatever theatre they were deployed. Significant design changes

were not possible without causing at least temporary chaos to the production lines, leading to the decision to build a new type. Thus the P-40N came into being, the last production version and also the most extensively built, with more than 5,000 manufactured in several variants.

Higher performance was achieved by a combination of weight saving and retention of the V-1710-81 engine, or subsequent installation on the production line of V-1710-99 or -115 engines of equivalent power. Initial weight saving was achieved by a reduction from six to four wing-mounted machine-guns, a reduction in fuel capacity, the introduction of light alloy oil coolers and engine liquid-cooling radiators, and lighter wheels. As a result the first production batch of P-40Ns had an empty weight some seven per cent below that of the P-40K, and the maximum speed of these proved to be the best of all basic production aircraft, the type reaching 378 mph (608 km/h) at 10,500 ft (3 200 m). Subsequent P-40Ns, according to production batches, introduced the alternative engines mentioned above, as well as a modified cockpit canopy; provision for an external bomb load of up to 1,500 lb (680 kg) and the re-introduction of six-gun armament made the aircraft effective fighter-bombers. Other improvements included the introduction of self-sealing fuel tanks of synthetic material, oxygen equipment, and exhaust flame dampers.

Many of these P-40Ns served with the USAAF in the Middle East and the Pacific, but more than half were allocated under Lend-Lease to Allied nations,

Curtiss P-40 Warhawk

including Australia, China, South Africa, the UK and the Soviet Union. Those supplied to the UK were designated Kittyhawk IV, equipping RAAF and RNZAF squadrons, and also the RAF's Nos. 250 and 450 Squadrons which were then based in Italy.

The designations P-40R-1 and P-40R-2 were allocated respectively to P-40F and P-40L aircraft, totalling some 300 combined, which were re-engined with Allison V-1710-81 engines and employed by the USAAF as advanced trainers.

The final designation to be applied to the Warhawk was XP-40Q, identifying three prototypes derived from two P-40Ks and a P-40N, which were used in continuing efforts to enhance the aircraft's performance. The powerplant used for these was the 1,425-hp (1 063-kW) Allison V-1710-121, and a variety of improvements/experiments were carried out. These included, not necessarily collectively, the installation of wing leading-edge radiators, a reduction in wing span, use of a bubble canopy and a change in fuselage lines to achieve a far better streamlined profile, and the use of water injection for the attainment of higher engine output. The result of the optimum combination of improvements was a better rate of climb, higher service ceiling, and a speed of 422 mph (679 km/h) at 20,500 ft (6 250 m). But even this considerable improvement left the performance of the Warhawk below that of contemporary Allied and Axis fighters, and production ended in December 1944.

Specification
Type: single-seat fighter-bomber
Powerplant (P-40N): one 1,200-hp (895-kW) Allison V-1710-81 inline piston engine
Performance: maximum speed 343 mph (552 km/h) at 15,000 ft (4 570 m); service ceiling 31,000 ft (9 450 m); range with auxiliary fuel at 10,000 ft (3 050 m) 1,080 miles (1 738 km)
Weights: empty 6,200 lb (2 812 kg); maximum take-off 8,850 lb (4 014 kg)
Dimensions: span 37 ft 4 in (11.38 m); length 33 ft 4 in (10.16 m); height 12 ft 4 in (3.76 m); wing area 236 sq ft (21.92 m²)
Armament: six 0.50-in (12.7-mm) machine-guns, plus up to 1,500 lb (680 kg) of bombs
Operators: China, Egypt, FFAF, Netherlands, RAAF, RAF, RCAF, RNZAF, SAAF, Soviet Union, Turkey, USAAC/USAAF

Curtiss O-52 Owl

History and Notes
In the late 1930s the US Army notified CurtissWright of its requirements for a two-seat observation aircraft. The resulting configuration suggests that rather more than an observation role was envisaged by the specification, for it was very different to any other Curtiss design, and efforts had clearly been made to confer good low-speed manoeuvrability and landing characteristics.

Identified as the Model 85 by Curtiss, the design was for a high-wing monoplane with one streamlined bracing strut on each side. Construction was all-metal, except that ailerons and tail control surfaces were fabric-covered. Good low-speed handling came from the provision of full-span automatic leading-edge slots interconnected with wide-span trailing-edge flaps: when the slots were extended the flaps were lowered automatically. The landing gear was of the retractable tailwheel type, with the main units retracting into wells in the lower sides of the fuselage. Dual controls were standard, and inward folding doors in the floor of the observer's cockpit were provided to facilitate the use of a camera. The retractable turtle back, which the company had developed for the SOC Seagull, was also incorporated, to ensure that the observer had a maximum field of fire for his flexibly-mounted machine-gun. Powerplant consisted of a Pratt & Whitney Wasp radial engine.

Ordered into production in 1939, some 203 Owls were built for the US Army under the designation O-52 with deliveries beginning in 1940. None, however, were used in first-line service, all being directed for use

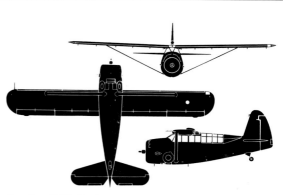
Curtiss O-52 Owl

in a training role.

Specification
Type: two-seat observation aircraft
Powerplant: one 600-hp (447-kW) Pratt & Whitney R-1340-51 Wasp radial piston engine
Performance: maximum speed 220 mph (354 km/h); cruising speed 192 mph (309 km/h); service ceiling 21,000 ft (6 400 m); range 700 miles (1 127 km)
Weights: empty 4,231 lb (1 919 kg); maximum take-off 5,364 lb (2 433 kg)
Dimensions: span 40 ft 9½ in (12.43 m); length 26 ft 4 in (8.03 m); height 9 ft 3¼ in (2.83 m); wing area 210.4 sq ft (19.55 m²)
Armament: one 0.30-in (7.62-mm) synchronised forward-firing machine-gun and one 0.30-in (7.62-mm) machine-gun on flexible mount
Operator: USAAC

Curtiss SBC Helldiver

History and Notes

Requiring a new two-seat fighter, the US Navy ordered a prototype from Curtiss in 1932 under the designation XF12C-1. This flew for the first time during 1933, in the form of a two-seat parasol-wing monoplane with retractable landing gear, and powered by 625-hp (466-kW) Wright R-1510-92 Whirlwind 14 engine. When, at the end of the year, it was considered desirable to utilise this aircraft in a scout capacity, its designation was changed to XS4C-1. Following yet another change of heart, its role became that of a scout-bomber in January 1934 and a Wright R-1820 Cyclone engine was installed. Extensive trials followed, and during a dive test in September 1934 there was structural failure of the wing and the XSBC-1, as it had by now been designated, was damaged extensively.

The parasol wing was clearly unsuitable for the dive-bombing requirement, and a new prototype was ordered as the XSBC-2, this having biplane wings and a 700-hp (522-kW) Wright R-1510-12 Whirlwind 14 engine. When, in March 1936, this engine was replaced by an 825-hp (615-kW) Pratt & Whitney R-1535-82 Twin Wasp Junior engine, the designation changed yet again to XSBC-3. Production SBC-3s, of which the US Navy ordered 83 on 29 August 1936, were generally similar, were armed with two 0.30-in (7.62-mm) machine-guns and could carry a single 500-lb (227-kg) bomb beneath the fuselage for use in a dive-bombing attack. First deliveries, to Navy Squadron VS-5, were made on 17 July 1937, and these aircraft were also to equip Squadrons VS-3 and VS-6.

The last SBC-3 on the production line was used as the prototype of an improved version. With a more powerful Wright R-1820-22 engine, and the ability to carry a 1,000-lb (454-kg) bomb, this was designated XSBC-4. Following an initial contract of 5 January 1938, the first of 174 production SBC-4s for the US Navy was delivered in March 1939. Because of the desperate situation in Europe in early 1940, the US Navy diverted 50 of its SBC-4s to France but these were received too late to be used in combat. Five of them were recovered for use by the RAF, being assembled in Britain in August 1940, and these were issued to RAF Little Rissington for allocation as ground trainers under the designation Cleveland. The US Navy's deficiency of 50 aircraft was made good by delivery of 50 out of the 90 aircraft which had been in production for France. Retaining the SBC-4 designation, these differed from standard in having self-sealing fuel tanks.

By the time the USA became involved in World War II, the SBC-3s had become obsolescent, but SBC-4s were then in service with US Navy Squadrons VB-8 and VS-8 on board the USS *Hornet* and with US Marine Squadron VMO-151.

Specification

Type: two-seat carrier-based scout-bomber
Powerplant (SBC-4): one 950-hp (708-kW) Wright R-1820-34 Cyclone 9 radial piston engine
Performance: maximum speed 237 mph (381 km/h) at 15,200 ft (4 635 m); cruising speed 127 mph (204 km/h); service ceiling 27,300 ft (8 320 m); range with 500-lb (227-kg) bomb 590 miles (950 km)
Weights: empty 4,841 lb (2 196 kg); maximum take-off 7,632 lb (3 462 kg)
Dimensions: span 34 ft 0 in (10.36 m); length 28 ft 4 in (8.64 m); height 12 ft 7 in (3.84 m); wing area 317 sq ft (29.45 m²)
Armament: one 0.30-in (7.62-mm) forward-firing and one 0.30-in (7.62-mm) machine-gun on flexible mounting, plus one 500-lb (227-kg) or one 1,000-lb (454-kg) bomb
Operators: RAF (ground training), USAAC

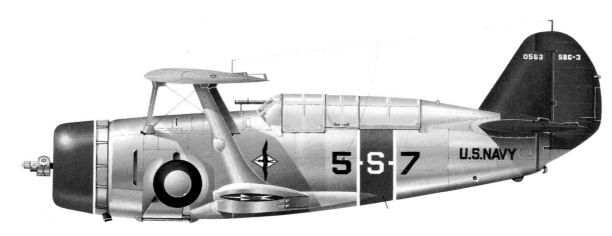

Curtiss SBC-3 of Scouting Squadron VS-5, US Navy, based aboard USS *Yorktown* in 1937.

Curtiss SBC Helldiver

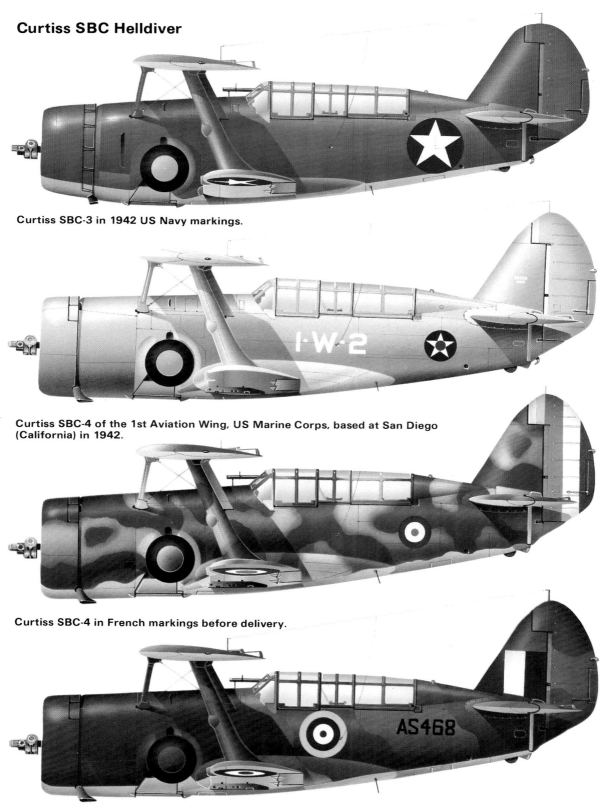

Curtiss SBC-3 in 1942 US Navy markings.

Curtiss SBC-4 of the 1st Aviation Wing, US Marine Corps, based at San Diego (California) in 1942.

Curtiss SBC-4 in French markings before delivery.

Curtiss Cleveland Mk I in RAF markings during 1940.

Curtiss SB2C Helldiver

History and Notes

The first Curtiss-built US Navy aircraft to bear the name Helldiver was the F8C/O2C biplane of the early 1930s. Second in line was the SBC Helldiver of the late 1930s, the last combat biplane to be built for the US services. The last and most famous of the line was the SB2C Helldiver of the early 1940s: this was the final combat aeroplane built by Curtiss for the US Marine Corps/US Navy, and the most extensively built of all US Navy dive-bombers.

In 1938 the US Navy began the process of procuring a new scout-bomber, to replace the SBC Helldiver which was then still in production. From the proposals received, Brewster and Curtiss were awarded contracts for prototypes of their contenders, on 4 April 1939 and 15 May 1939 respectively. Brewster's prototype was designated XSB2A-1, and duly entered production as the SB2A Buccaneer. The Curtiss prototype, designated XSB2C-1 (1758), flew for the first time on 18 December 1940, but was destroyed in a flying accident in early January 1941. Fortunately, the US Navy had great faith in this design (to the extent that large scale production had been authorised on 29 November 1940), but it was not until 18 months later, in June 1942, that the first production SB2C-1 was flown. There were a number of valid reasons for this delay: the construction and equipping of a new factory at Columbus, Ohio, in which to establish the production line; and a US Army Air Corps order of April 1941 for 900 A-25As, similar to the SB2C-1, but which caused great delay in achieving compatibility of design and equipment to satisfy both US Navy and US Army. In the final analysis only a few of this number entered US Army service; the majority were re-assigned to the US Marine Corps under the designation SB2C-1A.

Production SB2C-1s began to enter service with US Navy Squadron VS-9 in December 1942, but further protracted delays in finalising details of the best combat configuration prevented their initial operational employment until late 1943.

In configuration the SB2C was a low-wing cantilever monoplane of all-metal construction, except for fabric-covered rudder and elevators, and ailerons with metal upper skins and fabric-covered undersurfaces. The outer panels of the wings folded upwards for carrier stowage. The trailing-edge flaps were perforated and of split type so that they could be deployed as dive-brakes, opening both upward and downward. When the flaps were required to function conventionally for landing, hydraulic locks ensured that only the split lower surface was lowered. Wingtip leading-edge slats, of approximately the same span as the ailerons, were deployed automatically as the landing gear was lowered, thus ensuring that at even low landing speeds the ailerons remained fully effective. Wide-track landing gear was of the hydraulically-retractable tailwheel type: the main units retracted inward to lie flush in the undersurface of the wing centre-section, and the steerable tailwheel was only semi-retractable. Arrester gear and catapult launching spools were

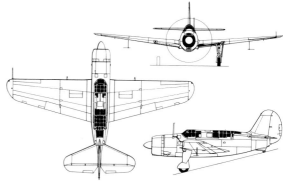

Curtiss SB2C-1C/-3 Helldiver

standard, but this latter equipment and wing-folding capability was deleted from the A-25 version produced for the US Army. Powerplant of the prototype and SB2C-1s consisted of a 1,700-hp (1 268-kW) Wright R-2600-8 Cyclone 14 twin-row radial engine. As a result of lessons learnt from early wartime experience, all fuel tanks were of the self-sealing variety; many fuel, oil and hydraulic lines had some protection against combat damage; both crew positions were protected by armour; and an inflatable dinghy was stowed in a container located beneath the cockpit canopy over the area separating the two cockpits. To simplify the pilot's task as much as possible, an autopilot was installed. Armament comprised four wing-mounted 0.50-in (12.7-mm) machine-guns, two 0.30-in (7.62-mm) guns in the rear cockpit, and up to 1,000 lb (454 kg) of bombs carried in an underfuselage bomb bay.

It is not surprising that with production totalling more than 7,000 examples, there were several variants of the basic design. Construction of the SB2C-1 totalled 978, and there were also 900 generally similar aircraft for the US Army, as noted above. One of the first batch of production SB2C-1s was taken off the line, and modified by the addition of twin floats and other specialised equipment to serve as an experimental long-range reconnaissance-bomber prototype. It was delivered to the US Navy for test in 1943 under the designation XSB2C-2.

The next production version, the SB2C-3 which was first delivered during 1944, had the more powerful R-2600-20 engine driving a Curtiss four-blade propeller, and a change in armament substituted two 20-mm cannon for the four wing-mounted machine-guns. Production of this version totalled 1,112. The SB2C-4s which followed (2,045 built) differed by having underwing racks for the carriage of four 5-in (127-mm) rocket projectiles or a 500-lb (227-kg) bomb beneath each wing. A number of SB2C-4Es carried radar equipment. Final production version was the SB2C-5 (970 built) with increased fuel capacity. Anticipating continued production, two XSB2C-6 prototypes were built, and these were powered by Pratt & Whitney R-2800-28 Double Wasp radial engines: with the war's end these prototypes were not developed as production aircraft.

Curtiss SB2C Helldiver

SB2C-1 of VB-17 aboard USS *Bunker Hill* (CV-17) for strikes against Rabaul, November 1943.

SB2C-1 of VB-8, which replaced VB-17 aboard USS *Bunker Hill.*

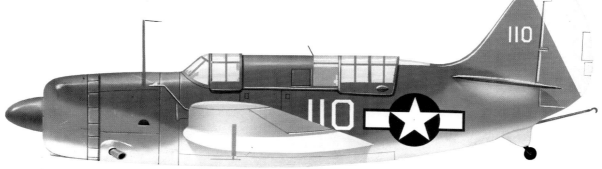

SB2C-3 of VB-3 aboard USS *Yorktown* (CV-10), February 1945, covering Iwo Jima landings.

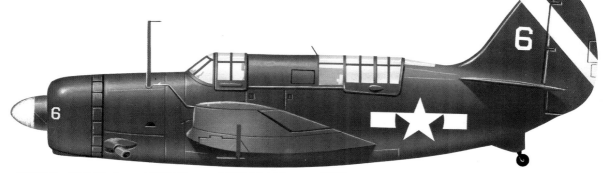

SB2C-3 of VB-80 aboard USS *Hancock* (CV-19), February 1945.

Curtiss SB2C Helldiver

To provide added productive capacity, manufacture of Helldivers was initiated at two Canadian factories, those of the Canadian Car & Foundry Co. Ltd and Fairchild Aircraft Ltd. The former built 894 aircraft, equivalent to Curtiss production models detailed above, with the respective designations SBW-1, SBW-3, SBW-4, SBW-4E and SBW-5; Fairchild built 300, not including equivalents of the SB2C-4 and SB2C-5, and these carried the respective designations SBF-1, SBF-3, and SBF-4E.

Under Lend-Lease 26 Canadian-built aircraft were supplied to the UK under the designation SWB-1B. Nine of these were used to form No. 1820 Squadron of the Fleet Air Arm at Squantum, in the USA. All were eventually shipped to Great Britain, but the type was not used operationally. These 26 aircraft were the only Helldivers to serve with any other nation during World War II, for the type was of such great value in the Pacific theatre that the US Navy absorbed the entire production. Many continued in service with the US Navy in early postwar years, and some were eventually sold to other nations.

Specification

Type: two-seat carrier-based scout-bomber
Powerplant (SB2C-4): one 1,900-hp (1 417-kW) Wright R-2600-20 Cyclone 14 twin-row radial piston engine
Performance: maximum speed 295 mph (475 km/h) at 16,700 ft (5 090 m); cruising speed 158 mph (254 km/h); service ceiling 29,100 ft (8 870 m); range 1,165 miles (1 875 km)
Weights: empty 10,547 lb (4 784 kg); maximum take-off 16,616 lb (7 537 kg)
Dimensions: span 49 ft 9 in (15.16 m); length 36 ft 8 in (11.18 m); height 13 ft 2 in (4.01 m); wing area 422 sq ft (39.20 m²)
Armament: two wing-mounted 20-mm cannon and two 0.30-in (7.62-mm) machine-guns in rear cockpit, plus up to 2,000 lb (907 kg) of bombs accommodated in bomb bay and underwing
Operators: RN, USAAF, USMC, USN

Curtiss SB2C Helldivers return to the carriers of Task Force 58.1 after a June 1944 Pacific mission.

Curtiss SB2C Helldiver

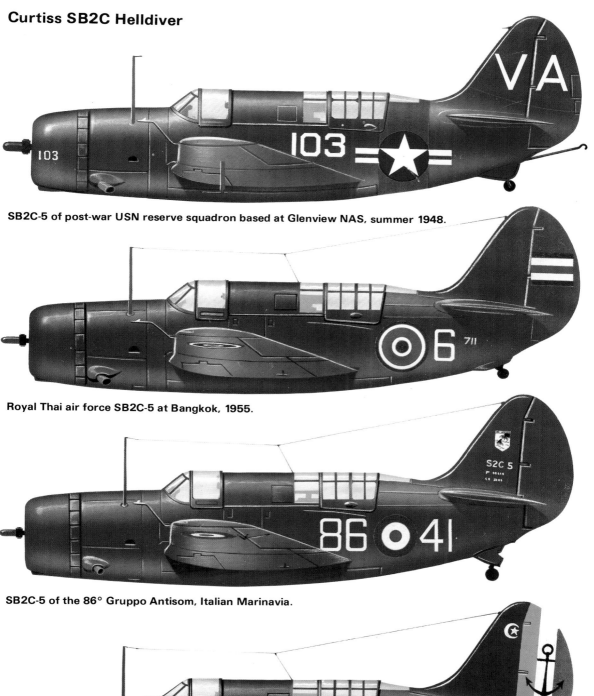

SB2C-5 of post-war USN reserve squadron based at Glenview NAS, summer 1948.

Royal Thai air force SB2C-5 at Bangkok, 1955.

SB2C-5 of the 86° Gruppo Antisom, Italian Marinavia.

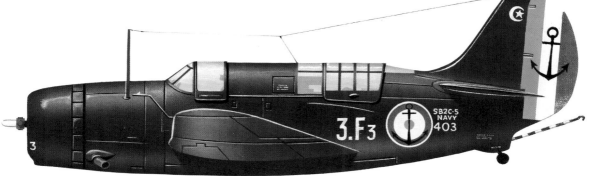

SB2C-5 of the French Aéronavale's Flottille 3F.

Curtiss SC Seahawk

History and Notes

Development of the Curtiss SC Seahawk began in June 1942, when the US Navy requested the company to submit proposals for an advanced wheel/float scout aircraft. The easily convertible landing gear configuration was required so that the aircraft could be operated from aircraft carriers and land bases, or be catapulted from battleships, and they were required to replace the rather similar Curtiss Seamew and Vought Kingfisher which stemmed from 1937 procurements to satisfy a similar role. The Curtiss design proposal was submitted on 1 August 1942 but it was not until 31 March 1943 that a contract for two XSC-1 prototypes was issued.

The XSC-1 was low-wing monoplane configuration and of all-metal construction. The wings had considerable dihedral on their outer panels, were provided with strut-mounted wingtip stabiliser floats, and were foldable for shipboard stowage. The main units of the tailwheel-type landing gear shared common attachment points with the larger single-step central float. If desired, auxiliary fuel could be accommodated in the central float. The powerplant comprised a 1,350-hp (1 007-kW) Wright R-1820-62 Cyclone 9 radial engine.

The first prototype made its maiden flight on 16 February 1944, and was followed by 500 production SC-1 Seahawks which had been contracted in June 1943. All production aircraft were delivered as landplanes, the stabiliser floats and Edo central float being purchased separately and installed as and when required by the US Navy. Delivery of production aircraft began in October 1944, the first of these equipping units aboard the USS *Guam*. A second batch of 450 SC-1s was contracted, but of these only 66 had been delivered before contract cancellation at VJ-Day.

Meanwhile, an improved version had been developed using one of the original prototypes for modification. Changes included the installation of a 1,425-hp (1 063-kW) R-1820-76 engine, provision of a clearview cockpit canopy and jump seat behind the pilot, and changes in rudder and fin profile. The modified prototype, designated XSC-1A, led to receipt of a contract for similar SC-2s, but only 10 had been delivered by the war's end.

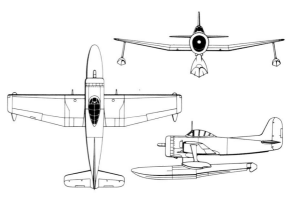

Curtiss SC-1 Seahawk

Specification

Type: single-seat scout or ASW aircraft
Powerplant (SC-1): one 1,350-hp (1 007-kW) Wright R-1820-62 Cyclone 9 radial piston engine
Performance: maximum speed 313 mph (504 km/h) at 28,600 ft (8 715 m); cruising speed 125 mph (210 km/h); service ceiling 37,300 ft (11 370 m); range 625 miles (1 006 km)
Weights: empty 6,320 lb (2 867 kg); maximum take-off 9,000 lb (4 082 kg)
Dimensions: span 41 ft 0 in (12.50 m); length 36 ft 4½ in (11.09 m); height 12 ft 9 in (3.89 m); wing area 280 sq ft (26.01 m²)
Armament: two 0.50-in (12.7-mm) machine-guns, plus two underwing bomb racks with a combined capacity of 650 lb (295 kg)
Operator: USN

Curtiss SNC

History and Notes

Under the designation SNC, the US Navy ordered in 1940 a variant of the Curtiss-Wright CW-21, the SNC indicating that it was required for deployment in the Scout Trainer category. By comparison with the CW-21 it had a very different fuselage to provide accommodation for a crew of two, and since the training role did not require such high performance as the CW-21 fighter, a very much lower-powered engine was installed.

Of all-metal construction, except for fabric-covered ailerons, the SNC had retractable landing gear of the tailwheel type; this was identical to that of the early manufacture CW-21s, with the main units retracting aft to be housed in split fairings beneath the wings. The major emphasis, however, had been to ensure that both accommodation and equipment should be suitable for an advanced training role. The tandem cockpits were enclosed by one extended canopy, the forward cockpit being intended for the pupil; dual flight controls were standard; and full instrumentation was duplicated so that the Falcon, as the SNC was unofficially named, could be used for instrument flight training. In addition, machine-guns, and light bombs on underwing racks, could be installed when needed for gunnery or bombing training, and there were provisions for a radio transmitter and receiver, and for an oxygen supply.

An initial contract for 150 of these trainers was placed in November 1940, and subsequent contracts for 150 and 5 aircraft brought numbers built to 305 before production was terminated.

Curtiss SNC

Specification
Type: two-seat advanced multi-role trainer
Powerplant: one 420-hp (313-kW) Wright R-975-E3 Whirlwind 9 radial piston engine
Performance: maximum speed 201 mph (323 km/h); cruising speed 195 mph (314 km/h) at 2,500 ft (760 m); service ceiling 21,900 ft (6 675 m); range 515 miles (829 km)
Weights: empty 2,610 lb (1 184 kg); maximum take-off 3,626 lb (1 645 kg)
Dimensions: span 35 ft 0 in (10.67 m); length 26 ft 6 in (8.08 m); height 7 ft 6 in (2.29 m); wing area 174.3 sq ft (16.19 m²)
Armament: one forward-firing 0.30-in (7.62-mm) machine-gun and light bombs on underwing racks as required for training purposes
Operator: USN

The Curtiss SNC-1 was an interim trainer developed from the CW-21 Demon to provide the US Navy with training reinforcements during 1940-1.

Curtiss SO3C Seamew

History and Notes
In 1937 the US Navy invited proposals for the design of a scout monoplane which would offer improved performance over the Curtiss SOC Seagull then in operational service. In accordance with what at that time had become fairly conventional practice, it was required for operation from either ships at sea or land bases, which meant that easily interchangeable float/wheel landing gear was essential. From the proposals received both Curtiss and Vought were awarded prototype contracts in May 1938 under the respective designations XSO3C-1 and XSO2U-1. The latter prototype (1440), powered by a 550-hp (410-kW) Ranger XV-770-4 engine, was duly flown in competition, but it was the Curtiss design which was ordered into production.

One must assume that the performance of the Vought prototype left much to be desired, for the Curtiss XSO3C-1, first flown on 6 October 1939, was found to have serious instability problems. These were resolved finally by the introduction of upturned wingtips and increased tail surfaces, but the resulting aircraft in its landplane form was almost certainly the ugliest aircraft to be produced by the Curtiss company. Of all-metal construction, except for fabric-covered control surfaces, the aircraft had wide-span split trailing-edge flaps, and the crew of two was accommodated in tandem enclosed cockpits. The floatplane landing gear comprised a large single-step central float and strut-mounted wingtip stabiliser floats. The wheeled landing gear was conspicuous by having large streamlined fairings, and picking up on to the central float attachment points the main units were located so far aft that an undignified nose-up position on the ground and difficult ground handling resulted. An unusual feature was attachment of the front section of the dorsal fin to the top of the aft cockpit sliding hatch: when the hatch was slid forward the fin lost its foremost section. The prototype and production SO3C-

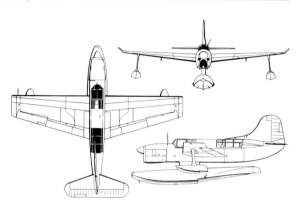

Curtiss SO3C-1 Seagull/Seamew

1s, initially named Seagull, were powered by a 520-hp (388-kW) Ranger V-770-6 engine.

SO3C-1 production aircraft began to enter service on board the USS *Cleveland* in July 1942, and 300 were built before production was switched to the SO3C-2. This differed in having equipment for carrier operations, including an arrester hook, plus an underfuselage rack on the landplane version to mount a 500-lb (227-kg) bomb. Production of this model totalled 456, of which 250 were allocated to Britain under Lend-Lease, although British records would seem to suggest that only 100 were received. The designation of the version originally intended for the Royal Navy was SO3C-1B, but those actually delivered were SO3C-2Cs, with a more powerful engine, 24-volt electrical system and improved radio. Wheeled aircraft were equipped with hydraulic brakes. In British service these aircraft were designated Seamew, a name subsequently adopted by the US Navy, but none were used operationally in Great Britain. Instead, they equipped Nos. 744 and 745 Training Squadrons, based at Yarmouth, Canada, and Worthy Down, Hampshire, respectively, for the instruction of air gunners/wireless operators.

The unsatisfactory performance of the SO3C-1 in the

Curtiss SO3C Seamew

US Navy led to its withdrawal from first-line service. Many were converted for use as radio-controlled targets, 30 being assigned to Britain, where they were designated Queen Seamews and used to supplement the fleet of de Havilland Queen Bee target aircraft.

In an attempt to retrieve the situation, Curtiss introduced in late 1943 a lighter-weight variant equipped with the more powerful SGV-770-8 engine; designated SO3C-3, only 39 were built before production ended in January 1944. Plans to introduce an SO3C-3 variant with arrester gear, and production by the Ryan Aeronautical Corporation of SO3C-1s under the designation SOR-1, were cancelled.

Specification

Type: two-seat scout/observation aircraft

Powerplant (SO3C-2C): one 600-hp (447-kW) Ranger SGV-770-8 inline piston engine
Performance (floatplane): maximum speed 172 mph (277 km/h) at 8,100 ft (2 470 m); cruising speed 125 mph (201 km/h); service ceiling 15,800 ft (4 815 m); range 1,150 miles (1 851 km)
Weights (floatplane): empty 4,284 lb (1 943 kg); maximum take-off 5,729 lb (2 599 kg)
Dimensions (floatplane): span 38 ft 0 in (11.58 m); length 36 ft 10 in (11.23 m); height 15 ft 0 in (4.57 m); wing area 290 sq ft (26.94 m²)
Armament (floatplane): one 0.30-in (7.62-mm) forward-firing machine-gun and one 0.50-in (12.7-mm) machine-gun on flexible mount, plus two 100-lb (45-kg) bombs or 325-lb (147-kg) depth charges beneath wings
Operators: RN, USN

Curtiss SOC Seagull

History and Notes

Last of the Curtiss biplanes to be used operationally by the US Navy, the SOC Seagull had a service history which very nearly duplicates that of the Royal Navy's Fairey Swordfish torpedo-bomber. Both originated in 1933, both should have become obsolescent during the early stages of World War II, both remained operational until the end of the war surviving, superbly, later designs intended to replace them.

The US Navy's requirement for a new scouting/observation aircraft was circulated to US manufacturers in early 1933, resulting in proposals from Curtiss, Douglas and Vought. A competing prototype was ordered from each under the respective designations of XO3C-1, XO2D-1 and XO5U-1, but it was the XO3C-1, ordered on 19 June 1933 and first flown in April 1934, which was ordered into production as the SOC-1. This changed designation reflected the combination of scout and observation roles.

When first flown, the prototype was equipped with an amphibious landing gear, the twin main wheels being incorporated in the central float. However, standard production aircraft were built as floatplanes, with non-retractable tailwheel type landing gear optional; but in any event they were easily convertible from one configuration to the other. Construction was mixed, with wings and tail unit of light alloy, a welded steel-tube fuselage structure, and a mixture of light alloy and fabric covering. Other features of the design included braced biplane wings which could be folded for shipboard stowage: the upper was the 'business' wing, carrying full-span Handley Page leading-edge slots, trailing-edge flaps, and ailerons. The pilot and gunner/observer were accommodated in tandem cockpits, enclosed by a continuous transparent canopy with sliding sections for access. To provide a maximum field of fire for the flexibly-mounted gun in the rear cockpit, the turtleback could be retracted.

Deliveries of the first SOC-1 production aircraft

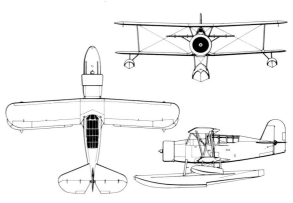

Curtiss SOC Seagull

began on 12 November 1935. These were powered by 600-hp (447-kW) Pratt & Whitney R-1340-18 Wasp engines, and the first squadrons to become fully equipped with the type comprised Scouting Squadrons VS-5B/-6B/-9S/-10S/-11S/-12S. Production of 135 SOC-1s was followed by 40 SOC-2s, supplied to contract with wheeled landing gear, and minor improvements in detail, and with R-1340-22 Wasp engines. A total of 83 SOC-3s was built, the aircraft being generally similar to the SOC-1. SOC-2s and -3s, after modification to install arrester gear during 1942, were redesignated SOC-2A and SOC-3A respectively. Curtiss also built three aircraft virtually the same as the SOC-3 for service with the US Coast Guard: these were acquired by the US Navy in 1942 and equipped with arrester gear to bring them up to SOC-3A standard. In addition to the SOC Seagulls built by Curtiss, 44 were produced by the Naval Aircraft Factory at Philadelphia, Pennsylvania. Basically the same as the Curtiss-built SOC-3, these were designated SON-1 or, if fitted with arrester gear, SON-1A.

Following termination of SOC production in early 1938, Curtiss became involved in the development and

Curtiss SOC Seagull

manufacture of a successor, designated SO3C Seamew. However, when the operational performance of the Seamew proved unsatisfactory it was withdrawn from first-line service, and all available SOCs were brought back into operational service, continuing to fulfil their appointed role until the end of the war.

Specification

Type: two-seat scout/observation aircraft
Powerplant (SOC-1): one 600-hp (447-kW) Pratt & Whitney R-1340-18 Wasp radial piston engine
Performance (floatplane): maximum speed 165 mph (266 km/h) at 5,000 ft (1 525 m); cruising speed 133 mph (214 km/h); service ceiling 14,900 ft (4 540 m); range 675 miles (1 086 km)
Weights (floatplane): empty 3,788 lb (1 718 kg); maximum take-off 5,437 lb (2 466 kg)
Dimensions: span 36 ft 0 in (10.97 m); length 26 ft 6 in (8.08 m); height (floatplane) 14 ft 9 in (4.50 m); wing area 342 sq ft (31.77 m²)
Armament: one 0.30-in (7.62-mm) forward-firing machine-gun and one 0.30-in (7.62-mm) gun on flexible mount, plus external racks for up to 650 lb (295 kg) of bombs
Operator: USN

Curtiss-Wright CW-21 Demon

History and Notes

Most World War II aircraft bearing the Curtiss name were products of the Curtiss Airplane Division of the Curtiss-Wright Corporation; very few carried a Curtiss-Wright label. One of the exceptions was the CW-21 Demon, a lightweight fighter based on the CW-19R general-purpose monoplane. A prototype (NX19431) of this new design was built as a private venture and flown for the first time in January 1939. Of low-wing monoplane configuration, the CW-21 possessed an almost prophetic wing planform with pronounced sweepback of the leading edges. Construction was all metal, and the main units of the tailwheel type landing gear retracted aft to be housed in split fairings beneath the wings. The powerplant consisted of a Wright R-1820 Cyclone radial engine.

The CW-21 Demon, as the prototype was designated by the company, was intended primarily for export, as were many other successful designs emanating from Curtiss. The first foreign order was received from the Chinese government, this covering the supply of three complete aircraft, and 32 sets of components for assembly in China. The order was completed during

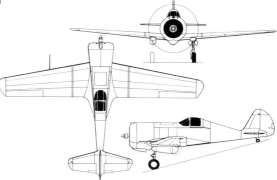

Curtiss-Wright CW-21 Demon (dashed line: main landing gear when retracted)

1939, and was followed in 1940 by an order from the Netherlands East Indies covering the supply of 24 improved aircraft. These differed by having a semi-

The Netherlands East Indies Army Air Corps was the only operator of the CW-21B Demon in World War II.

Curtiss-Wright CW-21 Demon

retractable tailwheel and main landing gear units which retracted inward to lie flush in the undersurface of the wing centre-section, slightly increased fuel tankage, and double the armament of the CW-21. In this revised form the Demon was designated CW-21B.

Delivery of the Dutch aircraft began in June 1940, but only 17 had been received by 8 December 1941 when the Netherlands declared war on Japan. Inadequately armed and with no worthwhile protection for the pilot, they were fair game for the Japanese Mitsubishi A6M Zeros when they clashed during the Japanese assault on Java in early 1942: within the brief space of two days all but five were eliminated.

Three other CW-21Bs were built for China, for supply to the American Volunteer Group (AVG) which, under the leadership of Brigadier General Claire Chennault, was doing its utmost to defend the Burma Road. Flown by AVG pilots, all three were lost on 23 December 1941 during a flight from Rangoon to Kunming, when in conditions of bad visibility they crashed into a mountain.

Specification
Type: single-seat fighter
Powerplant (CW-21B): one 1,000-hp (746-kW) Wright R-1820-G5 radial piston engine
Performance: maximum speed 315 mph (507 km/h) at 17,000 ft (5 180 m); cruising speed 282 mph (454 km/h); service ceiling 34,300 ft (10 455 m); range 630 miles (1 014 km)
Weights: empty 3,382 lb (1 534 kg); maximum take-off 4,500 lb (2 041 kg)
Dimensions: span 35 ft 0 in (10.67 m); length 27 ft 2 in (8.28 m); height 8 ft 11 in (2.72 m); wing area 174.3 sq ft (16.19 m²)
Armament: two 0.30-in (7.62-mm) and two 0.50-in (12.7-mm) fixed machine-guns
Operators: China, Netherlands East Indies

Curtiss-Wright P-60

History and Notes
Throughout its extended period of production, the performance of the Curtiss P-40 Warhawk was progressively improved: for example, the P-40C had a maximum speed of 345 mph (555 km/h), while that of the XP-40Q prototype was 422 mph (679 km/h) at 20,500 ft (6 250 m). However, as early as 1940 when the first P-40s had become established on the production line, the company had sought to develop an improved version which would offer superior performance. The company's proposal which first aroused the interest of the US Army Air Corps involved a variation of the P-40 design to incorporate a wing with a low-drag laminar-flow section, eight-gun armament, and powerplant comprising a Continental XIV-1430-3 inline inverted-Vee engine.

Awarded a contract on 1 October 1940 for two prototypes under the designation XP-53, Curtiss-Wright had made little progress before the USAAC decided that it would like to evaluate one of these airframes with a Rolls-Royce Merlin engine installed. The second XP-53 was cancelled, and a new contract awarded for the Merlin-engined version as the XP-60. Priority was given to this latter aircraft and, with a Merlin 28 engine installed, this flew for the first time on 18 September 1941. The original XP-53 prototype was subsequently converted to serve as a static test airframe for the production P-60 version.

Meanwhile, flight testing of the XP-60 was not proceeding too happily, and with the realisation that Packard-built Merlins were likely to be in short supply a decision was made to use instead an Allison V-1710-75 with turbocharger. With this as the chosen powerplant, Curtiss-Wright received a USAAF contract for 1,950 P-60As on 31 October 1941. It was but the beginning of a fruitless saga which extended into 1944.

Curtiss-Wright estimated that the P-60A with the V-1710-75 engine would not meet the USAAF's performance specification, leading to suspension of the P-60A contract. Instead, instructions were given on 2 January 1942 to built one example of each of three different prototypes designated XP-60A, XP-60B and XP-60C, to be powered respectively by a Allison V-1710-75 with General Electric turbocharger, a V-1710-75 with a Wright turbocharger, and a Chrysler XIV-2220 inline engine. The company pointed out that there were major problems in relation to this last powerplant and its installation, suggesting as an alternative the Pratt & Whitney R-2800 Double Wasp with contra-rotating propellers. This was approved by the USAAF, and official interest centred on this version. However, because it was not possible at the time to obtain an engine with suitable reduction gear and contra-rotating propellers, an R-2800-10 was installed with a single four-blade propeller, the aircraft being designated XP-60E. The XP-60D designation had been applied to the original XP-60 following installation of a Merlin 61 engine and the introduction of vertical tail surfaces with increased area.

The XP-60C with an R-2800-53 engine and contra-rotating propellers was eventually flown on 27 January 1943, and at that time the XP-60E, delayed by engine installation modifications, had not flown at all. There was some consternation therefore when, towards the end of April 1943, the USAAF demanded that this latter aircraft must be available within four days for service trials. It still had not made its first flight, so the XP-60C was hurriedly prepared and flown to Patterson Field where its performance was so disappointing that this virtually brought to an end all US Army interest. The only hope remaining lay in one YP-60E powered by a 2,100-hp (1 566-kW) Pratt & Whitney R-2800-18 engine, first flown on 13 July 1944, but this was flown only twice before being abandoned.

Curtiss-Wright P-60

Specification
Type: single-seat interceptor fighter
Powerplant (XP-60C): one 2,000-hp (1 491-kW) Pratt & Whitney R-2800-53 Double Wasp radial piston engine
Performance: maximum speed 414 mph (666 km/h) at 20,350 ft (6 205 m); service ceiling 37,900 ft (11 550 m); range 315 miles (507 km)
Weights: empty 8,698 lb (3 945 kg); maximum take-off 10,785 lb (4 892 kg)
Dimensions: span 41 ft 3¾ in (12.59 m); length 33 ft 11 in (10.34 m); height 12 ft 4¼ in (3.77 m); wing area 275.15 sq ft (25.56 m²)
Armament: four 0.50-in (12.7-mm) machine-guns in wings
Operator: USAAF (for evaluation only)

Begun as the XP-60B with an inline engine, this aircraft was completed as the XP-60E with radial engine.

Curtiss-Wright XP-55

History and Notes
On 27 November 1939 the US Army Air Corps issued to interested manufacturers its specification for a single-seat interceptor fighter: this was to be powered by a newly-developed Pratt & Whitney engine which had the designation X-1800-A3G; the USAAC also intimated that unconventional designs could be offered if they achieved the three desirable characteristics of low drag, powerful armament and exceptional visibility for the pilot.

Three manufacturers submitted design proposals in early 1940 and from evaluation of these Vultee was placed first, Curtiss-Wright second and Northrop third, with prototypes being ordered from each under the respective designations XP-54, XP-55 and XP-56. Vultee's design was not too unconventional, but it would be difficult to suggest if it was the Curtiss-Wright or Northrop design which was the more radical. The USAAC apparently believed it to be the

former, for it insisted on a powered model for wind tunnel testing, and when these figures seemed inconclusive the company produced at their own cost a full-scale flying model which they identified as the Model 24-B.

This aircraft had an all-wood wing, fabric-covered welded steel-tube fuselage structure, and was powered by a 275-hp (205-kW) Menasco C68 inline engine. Flight testing indicated some instability problems, but in the course of more than 160 flights between November 1941 and May 1942, these seemed to have been resolved by a number of modifications. On 10 July 1942 Curtiss-Wright were awarded a USAAF contract for

Produced in response to a USAAC requirement for an unorthodox fighter, the Curtiss-Wright XP-55 Ascender was a swept wing canard pusher that failed as a result of inherent stability problems.

Curtiss-Wright XP-55

the construction of three XP-55 prototypes, to be powered by Allison V-1710s because the Pratt & Whitney X-1800 engine had failed to materialise.

The XP-55 was of tailless configuration, with an aft-mounted low-set cantilever monoplane wing. Of thin laminar-flow section, this all-metal wing was sharply swept, carried ailerons and trailing-edge flaps and, near to the wingtips, vertical fins and rudders which extended above and below the wing. The oval-section fuselage was of all-metal construction, and a small fixed horizontal surface with elevators hinged to its trailing edge was mounted at the fuselage nose. Landing gear was of the retractable tricycle type, the main units retracting into the undersurface of the wing, nosewheel unit aft. Final choice for powerplant was the Allison V-1710-95, mounted in the rear fuselage and driving a three-blade pusher propeller. A vertical fin was incorporated in both the upper and lower engine cowling. Armament comprised four 0.50-in (12.7-mm) machine-guns in the fuselage nose.

The first prototype made its maiden flight in July 1943, but was destroyed in an accident four months later, on 15 November, while stall tests were being carried out. The second prototype made its first flight on 9 January 1944, and its test programme was kept well clear of the stall area until the third prototype, incorporating modifications to overcome this short-

coming, was completed and flying. This event occurred on 25 April 1944, and with the second prototype similarly modified both were handed over to the USAAF for evaluation in September 1944. This showed that low-speed handling characteristics were poor, and while these aircraft were satisfactory in level flight their performance fell below that of contemporary conventional fighters and their development was abandoned.

Specification

Type: single-seat interceptor fighter
Powerplant: one 1,275-hp (951-kW) Allison V-1710-95 inline piston engine
Performance: maximum speed 390 mph (628 km/h) at 19,300 ft (5 885 m); cruising speed 296 mph (476 km/h); service ceiling 34,600 ft (10 545 m); range 635 miles (1 022 km)
Weights: empty 6,354 lb (2 882 kg); maximum take-off 7,929 lb (3 597 kg)
Dimensions: span 44 ft 0½ in (13.42 m); length 29 ft 7 in (9.02 m); height 10 ft 0¾ in (3.07 m); wing area 235 sq ft (21.83 m²)
Armament: four nose-mounted 0.50-in (12.7-mm) machine-guns
Operator: USAAF (for evaluation only)

Douglas A-20 Havoc

History and Notes

It would perhaps be unkind to describe the Douglas A-20 (company identification DB-7) as undistinguished, especially when it was one of the most extensively built light bombers of World War II. It was, in fact, an ubiquitous aeroplane, used in a variety of roles, and performing well in that chosen role no matter where it was deployed.

The basic design originated as early as 1936, when the Douglas Aircraft Company began to consider the creation of an attack aircraft which, although un-specified by the US Army Air Corps, would serve as a more effective replacement for the single-engined Model 8A, itself derived from the earlier Northrop Model 2-C. By discussion with engineering staff of the USAAC, it became possible to outline a fairly advanced specification, leading to the company project identified as the Model 7A but this, with its twin-engined powerplant, was certainly breaking new ground, for all previous attack aircraft procured by the US Army had been of single-engined layout. There was, however, then no alternative to the twin-engined layout if the suggested performance and gun/weapon-carrying capability was to be achieved: even this was to need revision very shortly, as information began to filter through of the aircraft involved, and their advantages or shortcomings, when large-scale civil war erupted in Spain.

Redesign in 1938 produced the Model 7B, also of

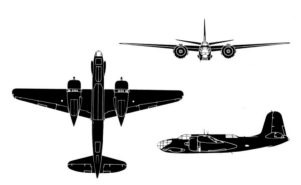

Douglas A-20C (Boston III)

twin-engined configuration, but with the then-proposed 450-hp (336-kW) engines replaced by two 1,100-hp (820-kW) Pratt & Whitney R-1830 Twin Wasps. Of cantilever shoulder-wing configuration, the Model 7B had an upswept aft fuselage, mounting a conventional tail unit. Landing gear was of the retractable tricycle type, but a most unusual feature was the introduction of interchangeable fuselage nose sections that would make for easy production of either attack or bomber versions: for deployment in the former role a solid nose housed four 0.30-in (7.62-mm) machine-guns, to supplement the standard six 0.30-in (7.62-mm) guns of which two were mounted in a blister on each side of the fuselage, plus one each in retractable dorsal and

Douglas A-20 Havoc

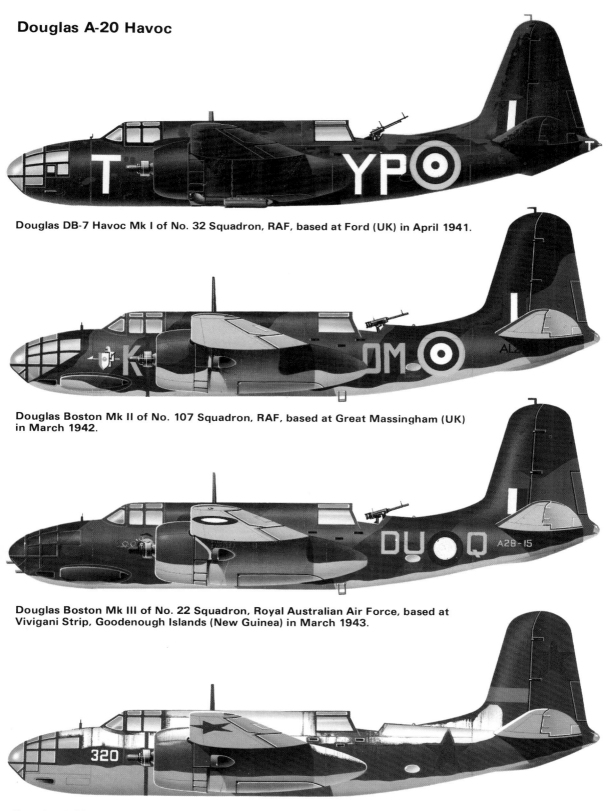

Douglas DB-7 Havoc Mk I of No. 32 Squadron, RAF, based at Ford (UK) in April 1941.

Douglas Boston Mk II of No. 107 Squadron, RAF, based at Great Massingham (UK) in March 1942.

Douglas Boston Mk III of No. 22 Squadron, Royal Australian Air Force, based at Vivigani Strip, Goodenough Islands (New Guinea) in March 1943.

Douglas A-20B of the Northern Fleet Air Force (VVS SF) based in the Arctic Front during 1943-4.

Douglas A-20 Havoc

ventral turrets; and for deployment in the latter role the bombardment nose was conventionally glazed, but had an obliquely-mounted optically-flat bomb-aiming panel. First flown in this form on 26 October 1938, the Model 7B evinced the characteristics of a thoroughbred: it was fast, highly manoeuvrable and, in fact, could be regarded as a 'pilot's aeroplane'.

Immediately the company realised its potential it offered the type for export, as the USAAC then had no requirement for such a machine. The first order, for 100 aircraft, came from a French purchasing mission in February 1939. However, although impressed by the performance of the Model 7B, many modifications were demanded to render the aircraft more suitable for what were considered to be essential requirements for its deployment in Europe, where advanced aircraft in service with the Luftwaffe had demonstrated their potential in the recently ended Spanish Civil War.

So extensive were the modifications that even the basic configuration of the Model 7B was changed, with the fuselage deepened to increase internal bomb capacity and fuel tankage, and its cross-section reduced, thus preventing both navigator/bomb aimer and gunner moving from their operational stations; the wing was lowered from the shoulder- to mid-wing position; a longer oleo-strut for the nosewheel was introduced; provision of armour protection for the crew and fuel tanks was made; and uprated Twin Wasp engines developing 1,200 hp (895 kW) each were installed. In view of the foregoing changes, the resulting aircraft was redesignated DB-7 (Douglas Bomber), and the production prototype was flown for the first time on 17 August 1939. Despite a miraculous effort made by Douglas to complete the manufacture of the initial 100 DB-7s by the end of 1939, the French had only managed to get just over 60 into service at the time of the German attack on 10 May 1940. Of these, only 12 aircraft of *Groupement* 2 were used operationally, on 31 May 1940, in low-level attacks on German armoured columns.

During the period when Douglas was developing the DB-7, a new French order for an improved version was received. Required to operate at a gross weight about 24 per cent higher than that of the DB-7, as a result of additional equipment, this necessitated the provision of 1,600-hp (1 193-kW) R-2600 Wright Cyclone 14 radial engines in revised nacelles, with associated changes in the engine installation, and the revised aircraft was designated DB-7A. Moreover, because the DB-7 had shown that directional stability was bordering on the marginal, even with 1,200-hp (895-kW) engines, increased fin and rudder area was provided to cater for the higher power engines.

When it was clear that the collapse of France was imminent, steps were taken to arrange for the UK to take over the balance of the French orders, plus a small quantity which had been ordered by Belgium. Thus some 15 to 20 DB-7s entered service with the RAF. These were allocated the name Boston I and used as conversion trainers in Operational Training Units,

including No.13 OTU at RAF Bicester. Most of the Gallic oddities had been ironed out before the aircraft were delivered, but instruments with metric calibration caused a few eyebrows to be raised in horror. The next batch to be received, about 125 DB-7 aircraft, were originally allocated the designation Boston II. However, their load-carrying capability and high speed confirmed a suitability for conversion to desperately needed night fighters, and in the winter of 1940 these were provided with AI (Airborne Interception) radar, additional armour, eight 0.303-in (7.7-mm) machine-guns in the nose, flame-damping exhaust systems, and overall matt black finish, under the new designation Havoc I. One very unusual addition was the provision of basic dual flying controls in the gunner's position: as no crew member could get to the pilot's aid in emergency, there was thus at least a long-odds chance that the gunner might achieve a non-calamitous landing. First delivered to the RAF in December 1940, these aircraft became operational with No. 85 Squadron on 7 April 1941. A second batch of about 100 DB-7As was converted similarly, but were each provided with 12 nose-mounted machine-guns, and were designated Havoc II. About 40 DB-7s were modified to serve as night intruders, retaining the bomb aimer's nose and able to accommodate up to 2,400 lb (1 089 kg) of bombs; an armament of four 0.303-in (7.7-mm) machine-guns was mounted beneath the nose. Officially named Havoc I (Intruder), the type also acquired such unofficial names as Moonfighter, Ranger and Havoc IV. Whatever the name, the aircraft were operated with considerable success by No. 23 Squadron. In order to enhance the somewhat limited capability of the AI radar installed in the Havoc Is, 31 were each equipped with a Hellmore/GEC searchlight of some 2,700 million candlepower intensity. Known as Havoc II Turbinlites, the aircraft were used, with little success, to illuminate German aircraft after stalking to within contact distance, when escorting Hawker Hurricane fighters would be able to attack and destroy the well-lit target. The name Havoc was adopted subsequently by the USAAF as the general name for its A-20s of all versions.

A few DB-7As were retained for use in a light bomber role under the designation Boston III, but the UK had ordered an improved version, the DB-7B, with changed electric and hydraulic systems, and instrumentation which conformed to RAF requirements and layout. These also were designated Boston III, and carried four 0.303-in (7.7-mm) guns in the nose, another two on a high-speed mounting in the aft cockpit, and a seventh gun firing through a ventral tunnel, plus a bomb load of up 2,000 lb (907 kg). These Boston IIIs were used extensively by squadrons of No. 2 Group, incuding Nos. 88, 107, 226 and 342. They served also with Nos. 13, 14, 18, 55 and 114 Squadrons in North Africa from early 1942, replacing Bristol Blenheims.

Initial USAAC contracts for the DB-7, placed in May 1939, produced 63 A-20s with turbocharged Wright

Douglas A-20 Havoc

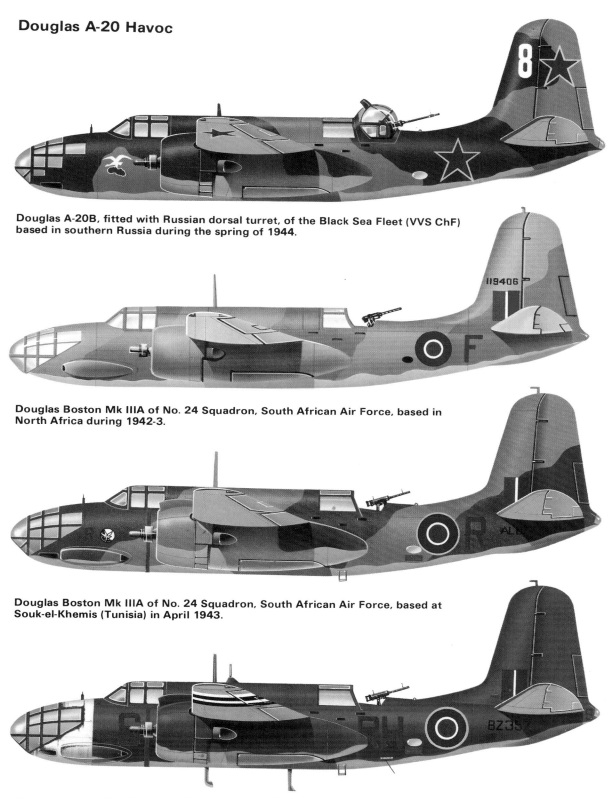

Douglas A-20B, fitted with Russian dorsal turret, of the Black Sea Fleet (VVS ChF) based in southern Russia during the spring of 1944.

Douglas Boston Mk IIIA of No. 24 Squadron, South African Air Force, based in North Africa during 1942-3.

Douglas Boston Mk IIIA of No. 24 Squadron, South African Air Force, based at Souk-el-Khemis (Tunisia) in April 1943.

Douglas Boston Mk IIIA of No. 88 (Hong Kong) Squadron, RAF, based at Hartford Bridge (UK) in June 1944.

Douglas A-20 Havoc

The slimness of the Douglas A-20 series' fuselage is evident in this shot of a BD-2, a version of the A-20B used in small numbers (eight) by the US Marine Corps as target tugs and utility aircraft.

R-2600-7 Cyclone 14 engines: of these three were converted to serve in a photo-reconnaissance role with the designation F-3; the remainder became the XP-70 prototype and 59 P-70 production night fighters, the prototype with unsupercharged R-2600-11 engines, and all with British-built AI radar and an armament of four 20-mm cannon mounted beneath the fuselage. These night fighters were used primarily in a training role, so that USAAC crews could become conversant with the newly developed technique of radar interception.

The first bomber version to serve with the USAAC was the A-20A, generally similar to the A-20, but powered by unsupercharged R-2600-3 engines and with armament as for the DB-7B except that the machine-guns were of 0.30-in (7.62-mm) calibre. In addition, two remotely-controlled aft-firing guns were mounted in the rear of each engine nacelle, and the bomb load was 1,100 lb (499 kg). One XA-20B prototype was modified from a production A-20A, and had a changed armament. This was not adopted for production A-20Bs, which had two 0.50-in (12.7-mm) nose-mounted guns, and which were in most respects similar to the DB-7A.

Large-scale production dictated more standardisation, so that the RAF's Boston III and USAAC A-20C were one and the same, equipped with R-2600-23 engines. To boost production, Douglas granted a

licence to Boeing and this latter company produced 140 A-20Cs for supply to the RAF under Lend-Lease as Boston IIIAs: they differed in their electrical system, and in some changes to the ancillary equipment of the engines. DB-7s of this version were supplied also to the USSR under Lend-Lease during 1942.

The next major production variant was the A-20G, of which almost 3,000 were built by Douglas at Santa Monica. These also had R-2600-23 engines, and were some 8 in (0.20 m) longer to provide a nose armament comprising two 0.50-in (12.7-mm) machine-guns and four 20-mm cannon, and either two 0.50-in (12.7-mm) guns or one 0.50-in (12.7-mm) and one 0.30-in (7.62-mm) gun in the rear cockpit. Most of the early production A-20Gs in this configuration were supplied to the USSR; the next A-20G variant had the 20-mm cannon replaced by 0.50-in (12.7-mm) machine-guns; and the final variant introduced a rear fuselage 6 in (0.15 m) wider to accommodate an electrically-operated dorsal turret with two 0.50-in (12.7-mm) guns, underwing bomb racks to accept an additional 2,000 lb (907 kg) of bombs, extra fuel tanks in the bomb bay, and provision for an underfuselage drop tank to provide a ferry range of more than 2,000 miles (3 219 km). This was, of

Douglas A-20 Havoc

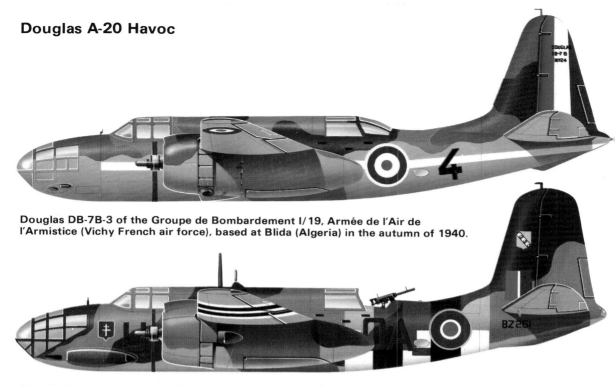

Douglas DB-7B-3 of the Groupe de Bombardement I/19, Armée de l'Air de l'Armistice (Vichy French air force), based at Blida (Algeria) in the autumn of 1940.

Douglas Boston Mk IIIA of No. 342 'Lorraine' Squadron, RAF, based at Hartford Bridge (UK) in June 1944.

course, vital for the type's deployment in the Pacific theatre where their arrival in 1942 came as something of a mixed blessing to Major General George C. Kenney's 5th Air Force, struggling to defeat the Japanese threat to New Guinea. As delivered, the aircraft were considered to be too lightly armed, so the basic armament was supplemented by four 0.50-in (12.7-mm) machine-guns, and as there were no bombs available as required for their employment in a close-support role, Kenney suggested the provision of 23-lb (10-kg) fragmentation bombs with small parachutes attached. With the A-20s each able to carry 40 of these 'parafrag' bombs, the aircraft played a vital role in dislodging the enemy from Burma.

Other improvements introduced gradually to A-20G Havocs included improved armour, navigation equipment and bomb-aiming controls, and winterisation accessories for aircraft to be operated in the low-temperature zones. Also produced were 412 A-20Hs, with little change from the A-20Gs except for the installation of 1,700-hp (1 268-kW) R-2600-29 engines. Neither the G nor H version served with the RAF, but the A-20J and A-20K, bomb-leader versions of the A-20G and A-20H respectively, were built for both the USAAF and RAF, with the respective designations Boston IV and Boston V in service with the latter air force. They differed only by having a frameless transparent nose to enhance the bomb aimer's view.

When production ended, on 20 September 1944, Douglas had built 7,385 DB-7s of all versions, and these had been used by the USAAF and its Allies in the widest imaginable number of roles. They had been

supplied also to Brazil, the Netherlands and the USSR, and small numbers from those received by the UK had been diverted to serve with the Royal Australian Air Force, Royal Canadian Air Force, Royal New Zealand Air Force and South African Air Force. In addition one A-20A had been supplied to the US Navy, under the designation BD-1, and used for evaluation. In 1942 eight A-20Bs were procured for use as target tugs under the designation BD-2.

Specification

Type: three-seat light bomber
Powerplant (A-20G): two 1,600-hp (1 193-kW) Wright R-2600-23 Cyclone 14 radial piston engines
Performance: maximum speed 317 mph (510 km/h) at 10,000 ft (3 050 m); cruising speed 230 mph (370 km/h); service ceiling 25,000 ft (7 620 m); range with 725 US gallons (2 744 litres) of fuel and 2,000 lb (907 kg) of bombs 1,025 miles (1 650 km)
Weights: empty 15,984 lb (7 250 kg); maximum take-off 27,200 lb (12 338 kg)
Dimensions: span 61 ft 4 in (18.69 m); length 48 ft 0 in (14.63 m); height 17 ft 7 in (5.36 m); wing area 465 sq ft (43.20 m²)
Armament: six 0.50-in (12.7-mm) forward-firing machine-guns, two 0.50-in (12.7-mm) guns in power-operated dorsal turret and one 0.50-in (12.7-mm) gun firing through ventral tunnel, plus up to 4,000 lb (1 814 kg) of bombs
Operators: Brazil, FAF, FFAF, FVAF, Netherlands, RAAF, RAF, RCAF, SAAF, Soviet Union, USAAC/USAAF, USMC, USN

Douglas A-26/B-26 Invader

History and Notes

A US Army Air Corps requirement of 1940 for a multi-role light bomber called for fast low-level capability, with an alternative deployment from medium altitude for precision bombing attack, plus heavy defensive armament. To meet this specification Douglas proposed what was, in effect, a developed and enlarged version of the three-seat A-20 with Pratt & Whitney R-2800 Double Wasp engines. Three prototypes were ordered in June 1941, the first XA-26 making its initial flight on 10 July 1942. Like the A-20 it was of shoulder-wing configuration, the twin-engined powerplant was mounted in large underwing nacelles, and electrically-operated single-slotted trailing-edge flaps were provided. The hydraulically retractable landing gear was of the tricycle type: the main units were housed in the aft portion of the engine nacelles, but the nose unit rotated through 90° during retraction for the wheel to lie flush in the undersurface of the fuselage nose. Considerable thought had been devoted to satisfy the requirement for heavy armament, and the XA-26 had remotely-controlled dorsal and ventral turrets, each with two 0.50-in (12.7-mm) machine-guns, two similar guns in the fuselage nose, which had transparent panels and a bomb aimer's position, and up to 3,000 lb (1 361 kg) of bombs could be accommodated in the bomb bay. To widen the assessment of suitable armament, the second (XA-26A) prototype was equipped as a night fighter, with AI (Airborne Interception) radar in the nose, four 0.50-in (12.7-mm) guns in an upper turret and four 20-mm cannon carried in a ventral fairing; the third (XA-26B) prototype had only a single 75-mm cannon mounted in the fuselage nose.

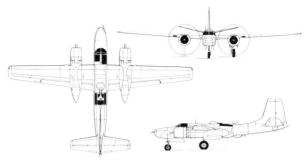

Douglas A-26B

Extensive service trials showed that Douglas had provided an aircraft more than suitable for the intended role, and in so doing had exceeded every performance specification. Not only was the aircraft some 700 lb (318 kg) below the design weight, but it could carry almost double the load of bombs which the US Army had considered necessary. The only area of delay in completing the trials and ordering the new attack aircraft into production came from the searching investigation to ensure the selection of the most effective armament. This was finalised for the first production attack version, designated A-26B, as six (later eight) 0.50-in (12.7-mm) machine-guns mounted in a solid nose; dorsal and ventral electrically operated and remotely controlled turrets each with two 0.50-in

The JD-1 was the US Navy target-tug version of the USAAF's Douglas A-26C Invader, of which 140 conversions were made from 1945.

Douglas A-26/B-26 Invader

Douglas A-26B Invader of the 552nd Bomb Squadron, 386th Bomber Group, 9th Air Force, based at Beaumont-sur-Oise (France) in April 1945.

(12.7-mm) guns; and the bomb load increased to 4,000 lb (1 814 kg). The powerplant of these production aircraft comprised two R-2800-27 or -71 engines, and it is an interesting sidelight on the scale of the engine nacelles to appreciate that not only did the aft nacelle house the landing gear when retracted, but all of the engine accessory equipment; moreover, a door was provided in the fireproof bulkhead so that a mechanic could enter this area for maintenance of this equipment as required. Fast engine changes were made possible by the provision of quick release attachment for all pipes and wires connected to each engine.

Soon after entering service, the desirability of enhancing firepower for the ground-attack role was achieved by the provision of eight more 0.50-in (12.7-mm) guns mounted in underwing packs, and modification of the upper turret so that it could be locked in position to fire forward, under control of the pilot, thus providing a total of 18 forward-firing guns. Not surprisingly, this combination of firepower and high speed (the Invader being one of the fastest bombers used by the USAAF during World War II) made the A-26 a highly effective aircraft.

Operational service of the Invader began on 19 November 1944, the type being deployed initially with 9th Air Force units operating in Europe: later it entered service in the Pacific, where its combination of firepower, speed and fairly useful range made it a valuable close-support aircraft during the last of the bitterly contested landings as the Allies hopped from one island to another en route to Japan's home islands.

In early 1945 the A-26Bs operating in the Pacific were joined by a new version, designated A-26C, in which some fairly extensive changes had been made. These included reinstatement of the bomb-aimer's position in the fuselage nose, with the necessary transparent panels, two of the nose guns being retained, plus the four turret guns. The fuselage was widened to accommodate pilot and co-pilot/bomb aimer side-by-side, and dual controls were provided as standard.

Other variants included a small number of FA-26Cs, equipped with cameras for use in the reconnaissance role; one XA-26D, developed from an A-26B, and armed with eight nose-mounted machine-guns, four guns in the two turrets, and six guns mounted beneath the wings to provide a maximum of 16 forward-firing and two aft-facing defensive machine-guns. With this armament 750 A-26Ds were ordered, but none were

produced as such, for VJ-Day contract cancellation came before any were built. Subsequently, however, in-service A-26Bs were modified retrospectively to this weapon standard. Other nose armament experiments which had been carried out with the A-26B included a 75-mm cannon in lieu of the six 0.50-in (12.7-mm) guns; a 75-mm cannon plus two 0.50-in (12.7-mm) guns; and two 37-mm cannon. The A-26E, a development of the A-26C, was also cancelled at VJ-Day and the one other wartime variant, the XA-26F, was used to flight test a General Electric J31 turbojet engine, this being mounted in the aft fuselage.

The performance and potential of the A-26 Invader was such that it continued in USAAF/USAF service for many years after the end of the conflict for which it had been designed, the designation B-26 being adopted in June 1948, when the USAF discontinued the classification of aircraft in an Attack category. B-26s became early reinforcements for NATO forces in Europe, were involved in the Korean War from 27 June 1950, and were operational in Vietnam in 1962. In this latter conflict, a number of B-26s were converted to a new COIN (counter insurgency) configuration by On Mark Engineering, under the designation B-26K (later A-26A).

Following conversion of a USAAF A-26B Invader to target tug configuration in 1945 (designation XJD-1), the US Navy acquired 140 JD-1s converted from A-26Cs, for operation by US Navy Squadrons VU-3/4/ -7 and -10. Some were converted subsequently for the launch and control of target drones.

Specification

Type: three-seat attack/light bomber aircraft

Powerplant (A-26C): two 2,000-hp (1 491-kW) Pratt & Whitney R-2800-79 Double Wasp radial piston engines

Performance: maximum speed 373 mph (600 km/h) at 10,000 ft (3 050 m); cruising speed 284 mph (457 km/h); service ceiling 22,100 ft (6 735 m); range 1,400 miles (2 253 km)

Weights: empty 22,850 lb (10 365 m); maximum take-off 35,000 lb (15 876 kg)

Dimensions: span 70 ft 0 in (21.34 m); length 51 ft 3 in (15.62 m); height 18 ft 3 in (5.56 m); wing area 540 sq ft (50.17 m²)

Armament: six 0.50-in (12.7-mm) machine-guns (two in each of nose position, dorsal turret and ventral turret), plus up to 4,000 lb (1 814 kg) of bombs

Operator: USAAF

Douglas A-33

History and Notes

In 1932 Jack Northrop, formerly a Douglas engineer, established the Northrop Aircraft Company at El Segundo, California. Then, in 1939, Douglas and Northrop merged as one company, with Douglas as the major stockholder, which explains the apparent illogicality of the Douglas Company being responsible for the development of a Northrop design.

The US Army Air Corps' requirement for an attack aircraft in the early 1930s led to design and development of the Northrop A-17, of which delivery began in 1935, with the production of these and later versions continuing into 1937. In 1939, with an increasing state of unrest around the world, Douglas decided to develop this Northrop design for export to foreign nations, identifying the resulting aircraft as the Model 8A.

The Model 8A was generally similar to the Northrop A-17 which had been built for the USAAC; it differed primarily by having the much more powerful Wright R-1820 Cyclone 9 engine and retractable tailwheel type landing gear. Of all-metal light alloy construction, except for fabric-covered control surfaces, the Model 8A had a monoplane wing which was low-set, and had wide-span perforated trailing-edge flaps of the split type which, in addition to their normal function, provided stability and control during the dive. An unusual feature of the fuselage was its construction in upper and lower halves, the wing centre-section being built integrally with the latter. The powerplant comprised the Wright R-1820 engine with a three-blade constant-speed metal propeller, and accommodation was provided for a pilot and gunner in tandem cockpits.

Among the Model 8As produced by Douglas for export was a batch of 34 8A-5s under construction for Peru, and 31 of these were commandeered by the US Army in early 1942 for deployment in an attack role. These were provided with four 0.30-in (7.62-mm) machine-guns in the wing leading edge, two more on a flexible mounting in the gunner's cockpit, and provision was made to carry up to 1,800 lb (816 kg) bombs internally and externally. All of these aircraft, which were given the designation A-33, were used for training purposes.

Specification

Type: two-seat attack aircraft
Powerplant: one 1,200-hp (895-kW) Wright GR-1820-G205A Cyclone 9 radial piston engine
Performance: maximum speed 265 mph (426 km/h) at 9,000 ft (2 745 m); cruising speed 200 mph (322 km/h) at 10,000 ft (3 050 m); service ceiling 32,000 ft (9 755 m); range 910 miles (1 464 km)
Weights: empty 5,370 lb (2 436 kg); maximum take-off 8,949 lb (4 059 kg)
Dimensions: span 47 ft 8¾ in (14.55 m); length 32 ft 5 in (9.88 m); height 9 ft 9 in (2.97 m); wing area 363 sq ft (33.72 m²)
Armament: four 0.30-in (7.62-mm) machine-guns in wing leading edges and two 0.30-in (7.62-mm) guns on flexible mount, plus up to 1,800 lb (816 kg) of bombs stowed internally and on external racks
Operator: USAAF

Essentially an upengined and refined version of the Northrop A-17, the Douglas Model 8A was produced in small numbers and taken over for USAAF service.

Douglas AD-1 Skyraider

History and Notes

The importance of the naval torpedo/dive-bomber in the Pacific theatre had led to development of the single-seat Douglas BTD-1 Destroyer, derived from the XSB2D-1 prototype first flown on April 1943. The Destroyer had proved a disappointment, with only a small number built before cancellation of the contract, but the US Navy had little doubt that the basic concept was right and initiated a competition to procure a far more competent aircraft in this category. Two production aircraft resulted from this: the Martin AM-1 Mauler and the Douglas AD-1. Two other submissions, the Curtiss XBTC-1 and Kaiser-Fleetwings XBTK-1, were built only in prototype form.

Martin's AM-1, which looked very similar to the Douglas design, did not enjoy the same success. Contract cancellation came after 151 had been built because, in what had then become the postwar years, the US Navy desired to limit its procurement to just one aircraft in this category, and the chosen vehicle was the Douglas AD-1. This had been ordered in prototype form under the designation XBT2D-1 on 6 July 1944, and selection of the big Wright R-3350 Cyclone 18 as its powerplant not only ensured that the all-important attribute of great load-carrying capability would be achieved, but dictated a large airframe structure. A decision not to include internal accommodation for weapons resulted in a lightweight, strong airframe, its large size providing 15 external hardpoints from which a wide variety of weapons could be dispensed. This was to be the secret of the type's success, permitting postwar deployment in many roles, carrying such weapons as bombs, depth charges, mines, napalm, rockets and torpedoes, armed with 20-mm cannon, able to launch nuclear weapons, and provided with add-on kits which allowed for deployment in ambulance, freighter and target-tug roles.

The AD-1 was of conventional low-wing all-metal monoplane construction, the outer wing panels folding hydraulically for carrier stowage. The ailerons were confined to the outer wing panels, and the trailing-edge flaps to the wide centre-section. The flaps were of the area-increasing Fowler type to give optimum low-speed performance. Three hydraulically extended dive-brakes were a part of the fuselage structure; and the tail unit included an electrically actuated variable-incidence tailplane. Retractable tailwheel type landing gear was provided, the main units retracting to lie flat within the wing, the tailwheel forward, and a deck arrester hook was mounted aft of the tailwheel.

During the period of prototype construction an order for 25 pre-production aircraft was placed, and although the designation remained XBT2D, the name Destroyer II was bestowed. These proved to be only temporary, however, for when 548 production examples were ordered in April 1945 the designation became AD-1 in the 'Attack' category and the name Skyraider was allocated. It was not until the closing months of World War II that the prototype was flown for the first time, on 18 March 1945, but although delivery of production aircraft began in June, the type was not used operationally before VJ-Day.

Forged during the bitter conflict in the Pacific theatre, a demanding background which created an outstanding aeroplane, the Skyraider was to become

Though it was too late for service in World War II, the Douglas AD-1 was the progenitor of the classic series of Skyraider attack aircraft.

Douglas AD-1 Skyraider

one of the most valuable weapons available to both US Navy and USAF during the Korean and Vietnam wars. In Korea the Skyraider's contribution was sufficient to win a US Navy accolade of 'the best and most effective close support airplane in the world', and Skyraiders remained in production until February 1957, by which time 3,160 had been built.

Specification
Type: single-seat carrier-based attack aircraft
Powerplant (AD-1): one 2,400-hp (1 790-kW) Wright R-3350-24 Cyclone 18 radial piston engine

Performance: maximum speed 366 mph (589 km/h) at 13,500 ft (4 115 m); cruising speed 204 mph (328 km/h); service ceiling 33,000 ft (10 060 m); range 1,900 miles (3 058 km)
Weights: empty 10,264 lb (4 656 kg); maximum take-off 18,030 lb (8 178 kg)
Dimensions: span 50 ft 0 in (15.24 m); length 38 ft 2 in (11.63 m); height 15 ft 5 in (4.70 m); wing area 400 sq ft (37.16 m²)
Armament: two 20-mm cannon, plus up to 6,000 lb (2 722 kg) of mixed weapons on underfuselage and underwing hardpoints
Operator: USN

Douglas B-18

History and Notes
Faced with a US Army Air Corps requirement of early 1934 for a bomber with virtually double the bomb load and range capability of the Martin B-10, which was then the USAAC's standard bomber, Douglas had little doubt that it could draw upon engineering experience and design technology of the DC-2 commercial transport which was then on the point of making its first flight. Designed to be at least comparable with, and possibly better than, Boeing's Model 247 which had first flown 12 months earlier and introduced new performance standards for twin-engined commercial transports, the DC-2 and later DC-3 marked the beginning of a new era for airlines all over the world.

Private venture prototypes to meet the US Army's requirements were evaluated at Wright Field, Ohio, in August 1935, these including the Boeing Model 299, Douglas DB-1 and Martin 146. The first was to be built as the B-17 Flying Fortress, the last was produced as an export variant of the Martin B-10/B-12 series, and

the Douglas DB-1 (Douglas Bomber 1) was ordered into immediate production under the designation B-18 in January 1936. Derived from the commercial DC-2, the DB-1 prototype retained a basically similar wing, tail unit and powerplant. There were, however, two differences in the wing: while retaining the same basic planform of the DC-2, that of the DB-1 had a 5 ft 6 in (1.68 m) reduction in span and was mounted in a mid-wing instead of low-wing position. The powerplant comprised two 930-hp (694-kW) Wright R-1820-45 Cyclone 9 engines, each driving a three-blade constant-speed metal propeller. The entirely new fuselage was considerably deeper than that of the commercial transport, to provide adequate accommodation for a crew of six, and to include nose and dorsal turrets, a bomb aimer's position, and an internal bomb bay; there

Developed from the DC-2 civil transport, the Douglas DB-1 was the losing contender in the USAAC contest won by the Boeing B-17 prototype, but it entered production as the B-18 and, with the revised nose illustrated, as the B-18A.

Douglas B-18

was, in addition, a third gunner's position, with a ventral gun discharging via a tunnel in the under-fuselage structure.

A total of 133 B-18s was covered by the first contract, this number including the single DB-1 which had served as a prototype. True production aircraft however, had a number of equipment changes, producing an increase in the normal loaded weight. The last B-18 to come off the production line differed by having a power-operated nose turret, and carried the company identification DB-2, but this feature did not become standard on subsequent production aircraft.

The next contracts, covering 217 B-18As, were placed in June 1937 (177) and mid-1938 (40). This version differed by having the bomb aimer's position extended forward and over the nose gunner's station, and the installation of more powerful Wright R-1820-53 engines. Most of the USAAC's bomber squadrons were equipped by B-18s or B-18As in 1940, and the majority of the 33 B-18As which equipped the USAAC's 5th and 11th Bomb Groups, based on Hawaiian airfields, were destroyed when the Japanese launched their attack on Pearl Harbor.

When in 1942 B-18s were replaced in first-line service by B-17s, some 122 were equipped with search radar and magnetic anomaly detection (MAD) equipment for deployment in the Caribbean on anti-submarine patrol. The Royal Canadian Air Force also acquired 20 B-18As which, under the designation Digby I, were employed on maritime patrol. The designation B-18C applied to two other aircraft reconfigured for ASW patrol. Another two aircraft were converted for use in a transport role as C-58s, but many others were used similarly without conversion or redesignation.

Specification

Type: medium bomber and ASW aircraft
Powerplant (B-18A); two 1,000-hp (746-kW) Wright R-1820-53 Cyclone 9 radial piston engines
Performance: maximum speed 215 mph (346 km/h) at 10,000 ft (3 050 m); cruising speed 167 mph (269 km/h); service ceiling 23,900 ft (7 285 m); range 1,200 miles (1 931 km)
Weights: empty 16,321 lb (7 403 kg); maximum take-off 27,673 lb (12 552 kg)
Dimensions: span 89 ft 6 in (27.28 m); length 57 ft 10 in (17.63 m); height 15 ft 2 in (4.62 m); wing area 965 sq ft (89.65 m²)
Armament: three 0.30-in (7.62-mm) machine-guns (in nose, ventral and dorsal positions), plus up to 6,500 lb (2 948 kg) of bombs
Operators: RCAF, USAAC/USAAF

Douglas B-23 Dragon

History and Notes

The Douglas B-18, which had been designed to meet a US Army Air Corps requirement of 1934 for a high-performance medium bomber, was clearly not in the same league as the Boeing B-17 Flying Fortress, which was built to the same specification. Figures highlight the facts: 350 B-18s were procured in total, by comparison with almost 13,000 B-17s. In an attempt to rectify the shortcomings of their DB-1 design, Douglas developed in 1938 an improved version and the proposal seemed sufficiently attractive for the US Army to award a contract for 38 of these aircraft under the designation B-23 and with the name Dragon.

Although the overall configuration was similar to the earlier aircraft, when examined in detail it was revealed as virtually a new aircraft. Wing span was increased, the fuselage entirely different and of much improved aerodynamic form, and the tail unit had a much higher vertical fin and rudder. Landing gear was the same retractable tailwheel type, but the engine nacelles had been extended so that when the main units were raised in flight they were faired by the nacelle extensions and created far less drag. Greatly improved performance was expected from these refinements, plus the provision of 60 per cent more power by the use of two Wright R-2600-3 Cyclone 14 engines, each with a three-blade constant-speed propeller. An innovation was the provision of a tail gun position, this being the first US bomber to introduce

A refined version of the B-18, the Douglas B-23 was the first US bomber to have a tail gun position, fitted with a single 0.50-in (12.7-mm) machine-gun.

such a feature.

First flown on 27 July 1939, the B-23s were all delivered to the US Army during that year. Early evaluation had shown that performance and flight characteristics were disappointing. Furthermore, information received from the European theatre during 1940 made it clear that development would be unlikely to result in range, bomb load and armament capabilities to compare with the bomber aircraft then in service with the combatant nations, or already beginning to emerge in the USA. As a result these aircraft saw only limited service in a patrol capacity along the US Pacific coastline before being relegated to training duties. During 1942 12 of these aircraft were converted to serve as utility transports under the designation VC-67, and were used also as glider tugs.

Douglas B-23 Dragon

Specification
Type: four/five-seat medium bomber
Powerplant: two 1,600-hp (1 193-kW) Wright R-2600-3 Cyclone 14 radial piston engines
Performance: maximum speed 282 mph (454 km/h) at 12,000 ft (3 660 m); service ceiling 31,600 ft (9 630 m); range 1,455 miles (2 342 km)
Weights: empty 19,059 lb (8 645 kg); maximum take-off 30,475 lb (13 823 kg)
Dimensions: span 92 ft 0 in (28.04 m); length 58 ft 4 in (17.78 m); height 18 ft 6 in (5.64 m); wing area 993 sq ft (92.25 m²)
Armament: one 0.50-in (12.7-mm) machine-gun in tail position and three 0.30-in (7.62-mm) guns in nose, dorsal and ventral positions
Operator: USAAF

Douglas BTD Destroyer

History and Notes
Early service use of the Douglas SBD Dauntless had convinced the US Navy of its capability as a dive-bomber: its later wartime record, in such actions as the Battle of the Coral Sea (May 1942) and the Battle of Midway (June 1942), merely provided confirmation. Long before that date, however, the US Navy had initiated the procurement of a more advanced dive-bomber, leading to the development by Douglas of a two-seat aircraft in this category, of which two prototypes were ordered by the US Navy in June 1941.

Designated XSB2D-1 Destroyer, the first prototype (03551) made its initial flight on 8 April 1943, but instead of being ordered into production it was used as the basis of a new aircraft which the cut and thrust of war in the Pacific had shown to be more essential. As the XSB2D-1, the prototype appeared as a clean and purposeful-looking two-seat dive-bomber, introducing an internal bomb bay and, for the first time for an aircraft to operate from an aircraft carrier, retractable tricycle type landing gear. The US Navy's new requirement was for a single-seat torpedo/dive-bomber, and the XSB2D-1 was modified for this new role by conversion to a single-seat cockpit, the addition of two wing-mounted 20-mm cannon, and enlargement of the bomb bay to accommodate a torpedo or up to 3,200 lb (1 451 kg) of bombs. Air-brakes were installed in each side of the fuselage, and a big Wright Cyclone 18 engine was provided to give the requisite high performance.

A contract on 31 August 1943 increased earlier orders for this aircraft, designated BTD-1 and named Destroyer, to 358. Deliveries of production aircraft began in June 1944, but only 28 had been delivered before contract cancellation was initiated soon after

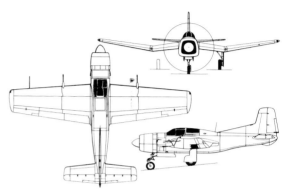

Douglas BTD-1 Destroyer

VJ-Day. The Destroyer's performance was disappointing and, so far as is known, the type was not used operationally.

Specification
Type: single-seat torpedo/dive-bomber
Powerplant: one 2,300-hp (1 715-kW) Wright R-3350-14 Cyclone 18 twin-row radial piston engine
Performance: maximum speed 344 mph (554 km/h) at 16,100 ft (4 905 m); service ceiling 23,600 ft (7 195 m); range 1,480 miles (2 382 km)
Weights: empty 11,561 lb (5 244 kg); maximum take-off 19,000 lb (8 618 kg)
Dimensions: span 45 ft 0 in (13.72 m); length 38 ft 7 in (11.76 m); height 13 ft 7 in (4.14 m); wing area 373 sq ft (34.65 m²)
Armament: two 20-mm cannon in wing leading edges, plus one torpedo or up to 3,200 lb (1 451 kg) of bombs in internal bay
Operator: USN

Douglas DC-2

History and Notes
When TWA, faced with an urgent need to replace its Fokker airliners, found itself behind United Air Lines in the queue for Boeing's Model 247, the airline's vice president of operations, Jack Frye, drew up a specification for an all-metal three-engined airliner with seats for at least 12 passengers and able to cruise at 146 mph (235 km/h) for more than 1,000 miles (1 609 km). Gross weight was to be a maximum of 14,200 lb (6 441 kg), rate of climb 1,200 feet (366 m) per minute and service ceiling 21,000 ft (6 400 m).

Donald W. Douglas responded within a fortnight to Frye's specification, issued to the US industry on 2 August 1932, and a contract was signed on 20 September, Douglas having convinced TWA technical adviser Charles Lindbergh that the required performance could be achieved safely on only two engines. The prototype, identified as the DC-1, was rolled out

Douglas DC-2

on 22 June 1933 and, powered by two Wright R-1820 Cyclones, made its maiden flight at 12.36 on 1 July, piloted by Carl Cover with project engineer Fred Herman as co-pilot.

Despite initial carburettor problems, the test programme was completed successfully and the aircraft handed over to TWA at Los Angeles Municipal Airport in December. The DC-1 never entered service, being used for promotional purposes by TWA. These included a coast-to-coast record flight of 13 hours 4 minutes through the night of 18/19 February 1934, when Frye and Eddie Rickenbacker of Eastern Airlines flew the aircraft on the last Los Angeles-Newark airline mail flight. This was in protest against the Roosevelt administration's decision to cancel all existing mail contracts with effect from midnight on 19 February 1934.

An initial contract was signed for 25 production aircraft, with some structural changes which resulted in redesignation to DC-2, and the first example was delivered to TWA on 14 May 1934, entering service four days later. The DC-2 was quickly adopted by other US airlines, including American and Eastern, and attracted export orders which included aircraft to KLM and Swissair.

The US Navy's use of the DC-2 was limited to a single R2D transport procured in 1934, later supplemented by four R2D-1s. The US Army Air Corps, however, opened its purchases for Fiscal Year 1936 with a 16-seat DC-2, which was evaluated as the XC-32 and which led to orders for two externally similar YC-34s and 18 C-33s, the latter type having enlarged vertical tail surfaces and a cargo door. In 1937 a C-33 was fitted with a DC-3 tail unit and redesignated C-38; from it was developed the C-39, with other DC-3 components, which included the wing centre-section and landing gear, and 975-hp (727-kW) R-1820-55 engines. Thirty-five were ordered for the army's transport groups, entering service in 1939.

The fourth and fifth C-39s were converted while still on the production line to C-41 and C-42 standard respectively. The first was fitted with 1,200-hp (895-kW) Pratt & Whitney R-1830-21 Twin Wasps and cleared to operate at a gross weight of 25,000 lb (11 340 kg), while the second was powered by 1,200-hp (895-kW) Wright R-1820-53 Cyclones and cleared at 23,624 lb (10 716 kg). Two more C-39s were later converted to C-42s, while 24 civil DC-2s impressed in 1942 received the designation C-32A. These aircraft lacked cargo doors.

The DC-2s in military service were used extensively, especially in the early years of World War II, and are remembered especially for their role in carrying US survivors from the Philippines to Australia in December 1941. A number of DC-2s, impressed for wartime service by the RAF, were used by No. 31 Squadron in India.

Specification

Type: 18-seat cargo and passenger transport
Powerplant (C-39): two 975-hp (727-kW) Wright R-1820-55 radial piston engines
Performance: maximum speed 210 mph (338 km/h) at 5,000 ft (1 525 m); cruising speed 155 mph (249 km/h); service ceiling 20,600 ft (6 280 m); range 900 miles (1 448 km)
Weights: empty 14,729 lb (6 681 kg); maximum take-off 21,000 lb (9 525 kg)
Dimensions: span 85 ft 0 in (25.91 m); length 61 ft 6 in (18.75 m); height 18 ft 8 in (5.69 m); wing area 939 sq ft (87.23 m²)
Armament: none
Operators: Finland, Germany, RAAF, RAF, USAAC/USAAF, USMC, USN

In mint factory condition, the third R2D-1 for the US Navy awaits delivery at the Douglas facility at Santa Monica on 23 December 1934.

Douglas C-47 Skytrain

A Douglas Dakota III of No. 24 Squadron, Royal Air Force.

History and Notes

The word ubiquitous has been associated with a number of aircraft in wide-scale use during World War II, but the most ubiquitous of all has to be the Douglas C-47/C-53/R4D/Skytrain/Skytrooper Dakota/'Gooney Bird'. Use any name you like for this superlative wartime transport aircraft, produced in greater numbers than any other in this category, with almost 11,000 manufactured by the time production ended in 1945: but whatever name you choose, it can be spelled 'dependable', for this was the secret of the type's greatness and enduring service life.

Its design originated from the DC-2/DST/DC-3 family of commercial transports that followed in the wake of the DC-1 prototype which flew for the first time on 1 July 1933. The US Army had gained early experience of the basic aircraft after the acquisition of production DC-2s in 1936, followed by more specialised conversions for use as cargo and personnel transports. In August 1936 the improved DC-3 began to enter service with US domestic airlines, its larger capacity and enhanced performance making it an even more attractive proposition to the US Army, which very soon advised Douglas of the changes in configuration which were considered desirable to make it suited for

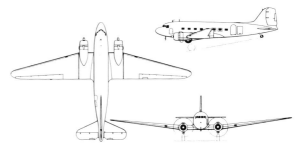

Douglas C-47 Skytrain (Dakota)

operation in a variety of military roles. These included the provision of more powerful engines, a strengthened rear fuselage to cater for the inclusion of large cargo doors, and reinforcement of the cabin floor to make it suitable for heavy cargo loads. Much of the basic design work had already been completed by Douglas,

Douglas C-47A Skytrains line up to receive stores before the D-Day assult. The C-47A differed from the C-47 only in its 24-volt instead of 12-volt electrics.

Douglas C-47 Skytrain

Very early C-47-DL aircraft towing Waco Hadrian troop-carrying gliders on a training flight in the United States.

for a C-41 cargo prototype had been developed by the installation of 1,200-hp (895-kW) Pratt & Whitney Twin Wasp engines in a C-39 (DC-2) fuselage. Thus, when in 1940 the US Army began to issue contracts for the supply of these new transport aircraft under the designation C-47, the company was well prepared to meet the requirements and to get production under way. The only serious problem was lack of productive capacity at Santa Monica, where European demands for the DB-7 light bomber had already filled the factory floor, resulting in the C-47 being built in a new plant at Long Beach, California.

Initial production version was the C-47, of which 953 were built at Long Beach, and since the basic structural design remained virtually unchanged throughout the entire production run, this version will serve for a description of the structure and powerplant. Of all-metal light alloy construction, the cantilever monoplane wing was set low on the fuselage, and provided with hydraulically operated split type trailing-edge flaps; the ailerons comprised light alloy frames with fabric covering. The fuselage was almost circular in cross-section. The tail unit was conventional but, like the ailerons, the rudder and elevators were fabric-covered. Pneumatic de-icing boots were provided on the leading edges of wings, fin and tailplane. Landing gear comprised a semi-retractable main units which were raised forward and upward to be housed in the lower half of the engine nacelles, with almost half of the main wheels exposed. The powerplant of the C-47 comprised two Pratt & Whitney R-1830-92 Twin Wasp engines, supercharged to provide an output of 1,050 hp (783 kW) at 7,500 ft (2 285 m), and each driving a three-blade constant-speed metal propeller. The crew consisted of a pilot and co-pilot/navigator situated in

a forward compartment with the third member, the radio operator, in a separate compartment.

The all-important cabin could be equipped for a variety of roles. For the basic cargo configuration, with a maximum load of 6,000 lb (2 722 kg), pulley blocks were provided for cargo handling and tie-down rings to secure it in flight. Alternative layouts could provide for the transport of 28 fully-armed paratroops, accommodated in folding bucket type seats along the sides of the cabin; or for 18 stretchers and a medical team of three. Racks and release mechanism for up to six parachute pack containers could be mounted beneath the fuselage, and there were also under-fuselage mountings for the transport of two three-blade propellers.

The first C-47s began to equip the USAAF in 1941, but initially these were received only slowly and in small numbers, as a result of the establishment of the new production line at Long Beach which, like any other, needed time to settle down to routine manufacture. With US involvement in World War II in December 1941, attempts were made to boost production, but in order to increase the number of aircraft in service as quickly as possible DC-3s already operating with US airlines, or well advanced in construction for delivery to operators, were impressed for service with the USAAF.

As Douglas began to accumulate contracts calling for production of C-47s in thousands, it was soon obvious that the production line at Long Beach would be quite incapable of meeting requirements on such a

A Douglas C-47A-65-DL, built by the Douglas plant at Long Beach, California. It is marked with the livery of the 81st Troop Carrier Squadron, 436th Troop Carrier Group at the time of the airborne assault on Normandy in June 1944. The group also took part in the airborne operations in southern France, the Netherlands and Germany later in the war.

Douglas C-47 Skytrain

large scale, so a second production line was established at Tulsa, Oklahoma. The first model to be built at Tulsa was the second production version, the C-47A, which differed from the C-47 primarily by the provision of a 24-volt, in place of a 12-volt, electrical system. Tulsa was to build 2,099 and Long Beach 2,832 of the type, 962 of them being delivered to the RAF which designated them Dakota IIIs. Last of the major production variants was the C-47B, which was provided with R-1830-90 or -90B engines that had two-stage superchargers to offer high altitude military ratings of 1,050 hp (783 kW) at 13,100 ft (3 990 m) or 900 hp (671 kW) at 17,400 ft (5 305 m) respectively. These were required for operation in the China-Burma-India (CBI) theatre, in particular for the 'Hump' operations over the 16,500-ft (5 030-m) high Himalayan peaks, carrying desperately needed supplies from bases in India to China. Long Beach built only 300 of the model, but Tulsa provided 2,808 C-47Bs plus 133 TC-47Bs which were equipped for service as navigational trainers. The UK was to receive a total of 896 C-47Bs, which in RAF service were designated Dakota IV.

The availability of such large numbers, in both US and British service, meant that it was possible to begin to utilise the C-47s on a far more extensive basis. The formation in mid-1942 of the USAAF's Air Transport Command saw the C-47s' wide-scale deployment as cargo transports carrying an almost unbelievable variety of supplies into airfields and airstrips which would have been complimented by the description 'primitive'. Not only were the C-47s carrying in men and materials, but were soon involved in a two-way traffic, serving in a casualty-evacuation role as they returned to their bases. These were the three primary missions for which these aircraft had been intended when first procured: cargo, casualty evacuation, and personnel transports. However, their employment by the USAAF's Troop Carrier Command from mid-1942, and the RAF's Transport Command, was to provide two new roles, arguably the most important of their

deployment in World War II, as carriers of airborne troops. The first major usage in this capacity came with the invasion of Sicily in July 1943, when C-47s dropped something approaching 4,000 paratroops. RAF Dakotas of Nos. 31 and 194 Squadrons were highly active in the support of Brigadier Orde Wingate's Chindits, who infiltrated the Japanese lines in Burma in an effort to halt their advance during the winter of 1942-3, their only means of supply being from the air. Ironically, Wingate (by then a major general) died on 24 March 1944 when a Dakota in which he was a passenger crashed into cloud-camouflaged jungle-clad mountains.

The other important role originated with the C-53 Skytrooper version, built in comparatively small numbers as the C-53B/-53C/-53D; seven C-53s supplied to the RAF were redesignated Dakota II. These were more nearly akin to the original DC-3 civil transport, without a reinforced floor or double door for cargo, and the majority had fixed metal seats to accommodate 28 fully-equipped paratroops. More importantly, they were provided with a towing cleat so that they could serve as a glider tug, a feature soon to become standard with all C-47s, and it is in this capacity that they served conspicuously in both USAAF and RAF service during such operations as the first airborne invasion of Burma on 5 March 1944 and the D-Day invasion of Normandy some three months later. In this latter operation more than 1,000 Allied C-47s were involved, carrying paratroops and towing gliders laden with paratroops and supplies. In the initial stage of this invasion 17,262 US paratroops of the 82nd and 101st Airborne Divisions and 7,162 men of the British 6th Airborne Division were carried across the English Channel in the greatest airlift of assault forces up to

Douglas C-47 Skytrain

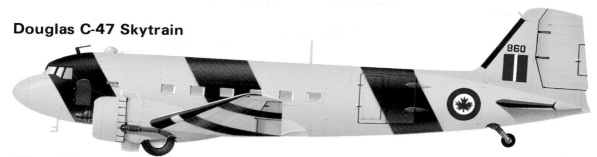

Dakota of the Royal Canadian Air Force used for pilot training duties during World War II.

that time. Not all, of course, were carried in or towed by C-47s, but these aircraft played a most significant role in helping to secure this first vital foothold on European soil: in less than 60 hours C-47s alone airlifted more than 60,000 paratroops and their equipment to Normandy.

Other C-47 variants of World War II included the XC-47C, prototype of a projected version to be equipped as a floatplane or, as was the prototype, with convertible amphibious floats. Of all-metal construction, these single-step Edo floats each had two retractable wheels, and housed a 300-US gallon (1 136-litre) fuel tank. While this version was not built as such by Douglas, a small number of similar conversions were made by USAAF maintenance units for service in the Pacific. Douglas were also contracted to build 131 staff transports under the designation C-117, these having the airline-standard cabin equipment of a commercial DC-3, plus the improvements which were current on the C-47. Their numbers, however, had reached only 17 (one C-117B built at Long Beach and 16 C-117As from Tulsa) when VJ-Day brought contract cancellation. The requirement for a large-capacity high-speed transport glider, to be towed by a C-54, resulted in experimental conversion of a C-47 to serve in this role under the designation XCG-17. Early tests had been conducted with a C-47 making unpowered approaches and landings to confirm the feasibility of the project, followed by a series of flights in which one C-47 was towed by another: for take-off the towed aircraft used some power, but shut down its engines when airborne. Conversion of a C-47 to XCG-17 configuration began after completion of these tests, with engines, propellers and all unnecessary equipment removed, and the forward end of the engine nacelles faired over. This was undoubtedly aerodynamically inefficient, and contributed to a reduction in performance of the XCG-17, but it was a USAAF requirement that any production aircraft should be capable of easy reconversion to powered C-47s. Despite any inefficiency the embryo cargo glider had a successful test programme, demonstrating a towed speed of 290 mph (467 km/h), stalling speed of only 35 mph (56 km/h) and a glide ratio of 14:1. Payload was 14,000 lb (6 350 kg), permitting the transport of 40 armed paratroops. No production aircraft were built, however, as a result of changing requirements.

In addition to the C-47s which served with the USAAF and RAF, approximately 600 were used by the US Navy. These comprised the R4D-1 (C-47), R4D-3 (C-53), R4D-4 (C-53C), R4D-5 (C-47A), R4D-6 (C-47B) and R4D-7 (TC-47B). US Navy and US Marine Corps requirements resulted in several conversions with designations which include the R4D-5E/-6E with special-purpose electronic equipment; the winterised and usually ski-equipped R4D-5L/-6L; the R4D-4Q/-5Q/-6Q for radar countermeasures; cargo versions re-equipped for passenger carrying as the R4D-5R/-6R; the air-sea warfare training R4D-5S/-6S; the navigational training R4D-5T/-6T; and the VIP-carrying R4D-5Z/-6Z. R4Ds were used initially by the Naval Air Transport Service that was established within five days of the attack on Pearl Harbor, equipping its VR-1, VR-2 and VR-3 squadrons, and soon after this by the South Pacific Combat Air Transport Service which provided essential supplies to US Marine Corps units as they forced the Japanese to vacate islands which stretched across the seas that led like stepping stones to that nation's home islands.

In addition to US production, the type was built in the USSR as the Lisunov Li-2 (2,000 examples or more) and in Japan as the Showa (Nakajima) L2D (485 examples).

C-47s had been involved from the beginning to the end of World War II, and that is but a small portion of their history in both military and civil service. Since VJ-Day military C-47s have supported the Berlin Airlift, Korean and Vietnam wars, to mention only major operations. It would not be too far from the truth to suggest that in the 42 years to 1982 there have not been many military actions or major civil disasters in which the enduring C-47 has not played some part.

Specification

Type: military transport and glider tug
Powerplant (C-47A): two 1,200-hp (895-kW) Pratt & Whitney R-1830-93 Twin Wasp radial piston engines
Performance: maximum speed 229 mph (369 km/h) at 7,500 ft (2 285 m); cruising speed 185 mph (298 km/h) at 10,000 ft (3 050 m); service ceiling 23,200 ft (7 070 m); range 1,500 miles (2 414 km)
Weights: empty 16,970 lb (7 698 kg); maximum take-off 26,000 lb (11 793 kg)
Dimensions: span 95 ft 0 in (28.96 m); length 64 ft 2½ in (19.57 m); height 16 ft 11 in (5.16 m); wing area 987 sq ft (91.69 m²)
Armament: none
Operators: India, Japan, Germany, RAAF, RAF, RCAF, Romania, Soviet Union, USAAF, USMC, USN

Douglas C-54

History and Notes

Like its stable mate the C-47, the Douglas C-54 was derived from the prototype of a civil airliner, the DC-4E, which was built to a specification initially drawn up by United Air Lines. United was joined quickly by American Airlines, Eastern Airlines, Pan American Airways and TWA, all five carriers agreeing to share the cost with Douglas in return for early delivery positions.

Powered by four of the newly developed 1,400-hp (1044-kW) Pratt & Whitney R-2180 Twin Hornet 14-cylinder radial engines, the DC-4E flew for the first time on 7 June 1938, with Carl Cover in command. The design was innovative, incorporating such relatively untried features as cabin pressurisation, power-boosted controls, a 115-volt AC electrical system and an auxiliary power unit. This very complexity, problems with the engines, and the growing realisation that the aircraft was too large for its payload-carrying capacity led to increasing lack of interest on the part of the sponsoring airlines. Although certification was achieved on 5 May 1939 and United carried out some route-proving trials, Douglas terminated development and the DC-4E was later sold to Dai Nippon Koku Kabushiki Kaisha in Japan.

Work started immediately on a redesigned DC-4 which, at 50,000 lb (22 680 kg) gross weight and with seats for 42 passengers, was slightly smaller than the prototype. The cabin pressurisation system, power-boosted controls and auxiliary power unit were deleted and the engines were replaced by four Pratt & Whitney Twin Wasps. The landing gear was also modified, the mainwheels retracting forward in to the engine nacelles rather than sideways into the wings. Orders for 61 production standard DC-4As were placed by American, Eastern, Pan American and United on 26 January 1940, but war in Europe and the apparent inevitability of increased United States' involvement led the Roosevelt administration to boost the US Army Air Corps' transport capability with orders which included nine C-54s and 62 C-54As. The nine C-54s were part of the initial batch of 24 civil DC-4As laid down at Santa Monica and the remaining 15 were diverted to USAAC order early in 1942. The first production C-54 made its maiden flight on 26 March 1942 and all 24 were in service with the Air Transport Command's Atlantic Wing by October. They were slightly modified for military service by the addition of four 464-US gallon (1 756-litre) fuel tanks in the fuselage, to bring the total capacity to 3,700 US gallons (14 006 litres), conferring a range of 2,500 miles (4 023 km) with a 9,600-lb (4 354-kg) payload which included 26 passengers.

The C-54A, 77 of which were built in Santa Monica and 117 at a new factory at Orange Place, Chicago, appeared in January 1943, featuring 33 bucket seats for troops, a large cargo door, stronger floor, cargo boom hoist and slightly larger wing tanks. The last increased total fuel capacity to 3 734 US gallons (14 134 litres) and gross weight rose to 68,000 lb (30 844 kg),

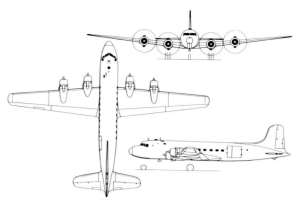

Douglas C-54D Skymaster

allowing a payload of 9,000 lb (4 082 kg) to be carried for over 3,000 miles (4 828 km).

A need to carry larger loads over shorter sectors led to the development of the C-54B, in which two of the fuselage tanks were deleted in favour of a 499-US gallon (1 889-litre) integral tank in each outer wing panel. Gross weight rose again, to 73,000 lb (33 112 kg), and up to 49 troops or 36 casualty litters could be accommodated. C-54B production totalled 89 at Santa Monica and 100 at Chicago.

A single VC-54C, with the C-54A's fuselage tanks and C-54B outer wing panels, with tanks, to give it a range of 5,500 miles (8 851 km), was delivered in June 1944 for the use of President Roosevelt. In the same month, a C-54B was delivered to the Royal Air Force for the use of Winston Churchill, fitted out with a 10-seat VIP cabin by Armstrong Whitworth Aircraft and operated by No. 24 Squadron until November 1945 when it was returned to the United States.

Replacement of the C-54B's Pratt & Whitney R-2000-7 engines by -11 models produced the C-54D, 304 of which were built for the USAAF at Chicago. Ten were supplied to the Royal Air Force under Lend-Lease, serving with No. 47 Group's Nos. 232 and 246 Squadrons on routes to the Far East.

Between January and June 1945, the Santa Monica plant manufactured 105 USAAF C-54Es, a long-range version with 3,600 US gallons (13 627 litres) of fuel wholly contained in wing tanks. The final production version was the C-54G, with 1,450-hp (1 081-kW) R-2000-9 engines, which was developed for the India-China Wing's operations over the 'Hump', the 10,000-16,500 ft (3 050-5 030 m) mountain ranges between the US bases on the plains of Assam and the Chinese air force bases, principally that at Kunming. USAAF procurement of C-54Gs totalled 162, all built at Santa Monica. In August 1945, USAAF Air Transport Command was operating 839 C-54s of all models on a world-wide route network.

The US Navy, with a particular responsibility for air transport in the Pacific, acquired 183 aircraft from the USAAF production lines between 1943 and 1945. This total comprised 19 R5D-1s (C-54A), 11 R5D-2s (C-54B)

Douglas C-54

and 20 R5D-4s (C-54E) built at Santa Monica; and 38 R5D-1s, 19 R5D-2s and 76 R5D-3s (C-54D) built at Chicago. The aircraft equipped squadrons of the Naval Air Transport Service until that formation was disbanded in July 1948, some units then becoming part of the combined Military Air Transport Service.

The R5D-5 was a planned US Navy version of the C-54G but none was delivered as such, although some R5D-2s and -3s were later converted to that standard, with R-2000-9 engines. Cargo aircraft converted to passenger configuration were given an -R suffix, as in the case of the R5D-4R and R5D-5R, and the -Z suffix denoted aircraft with much improved interior furnishing. Redesignation of the R5D fleet in 1962 resulted in the R5D-1, R5D-2, R5D-3, R5D-4R, R5D-5 and R5D-5R becoming the C-54N, C-54P, C-54Q, C-54R, C-54S and C-54T respectively. Two single-example conversions for the US Coast Guard were the EC-54U, with special electronic equipment, and the RC-54V photographic surveillance variant.

Postwar USAF versions included an experimental XC-54K with 1,425-hp (1 063-kW) Wright R-1820-HD engines, a C-54L with a modified fuel system, 38 C-54Ms stripped to increase payload by 2,500 lb (1 134 kg) for coal-carrying operations during the Berlin Airlift, and 30 C-54Es converted in 1951 as MC-54M ambulance

A Douglas C-54G Skymaster, one of the USAAF's long-range transport mainstays in World War II, comes in to land on an Indian airfield in September 1945.

aircraft for the evacuation of wounded personnel from Korea. Other specialised-role conversions included 38 SC-54Ds modified for the USAF Air Rescue Service in 1955, nine JC-54 range support aircraft used for missile nose-cone recovery, and a small number of AC-54Ds for navaid calibration.

Specification
Type: cargo and passenger transport
Powerplant (C-54B): four 1,350-hp (1 007-kW) Pratt & Whitney R-2000-7 radial piston engines
Performance: maximum speed 274 mph (441 km/h) at 14,000 ft (4 265 m); cruising speed 239 mph (385 km/h) at 15,200 ft (4 635 m); service ceiling 22,000 ft (6 705 m); range 3,900 miles (6 276 km)
Weights: empty 38,000 lb (17 237 kg); maximum take-off 73,000 lb (33 112 kg)
Dimensions: span 117 ft 6 in (35.81 m); length 93 ft 11 in (28.63 m); height 27 ft 6¼ in (8.39 m); wing area 1,463 sq ft (135.91 m²)
Armament: none
Operators: RAF, USAAC/USAAF, USCG, USN

Douglas DC-5

History and Notes
Designed at Douglas Aircraft Company's El Segundo facility, the DC-5 was developed as a 16/22-passenger commercial transport for local service operations out of the smaller airports. Interestingly, at a time when the low-wing configuration was in the ascendancy, it was a high-wing monoplane, although it also featured

the then relatively novel tricycle type landing gear. With a design gross weight of 18,500 lb (8 391 kg), the DC-5 was offered with either Pratt & Whitney R-1690 or Wright Cyclone radial engines which, on 550 US gallons (2 082 litres) of fuel, in two wing tanks, gave the aircraft a range of 1,600 miles (2 575 km) at a cruising speed of 202 mph (325 km/h). The engine nacelles forward of the firewalls, collector rings,

Douglas DC-5

control runs, rudder pedals and the pilot's seats were among the components used in the DC-3.

The prototype, powered by two 850-hp (634-kW) Wright GR-1820-F62 Cyclones, flew for the first time on 20 February 1939, piloted by Carl Cover. Orders were placed by KLM (four aircraft), Pennsylvania Central Airways (six) and SCADTA of Colombia (two), but the programme was overtaken by the war and only the KLM aircraft were delivered. Although intended for service in Europe, two went first to the Netherlands West Indies to link Curacao and Surinam and the other two to Batavia in the Netherlands East Indies. All four were used to evacuate civilians from Java to Australia in 1942 and one, damaged at Kemajoran Airport, Batavia, on 9 February 1942, was captured by the Japanese and extensively test-flown at Tachikawa air force base. The three surviving DC-5s were operated in Australia by the Allied Directorate of Air Transport and were given the USAAF designation C-110.

The earliest DC-5 military operations, however, were by the US Navy which had ordered seven examples in 1939. Three were R3D-1 16-seat personnel transports, the first of which crashed before delivery, and four were R3D-2s for the US Marine Corps with 1,000-hp (746-kW) R-1820-44 engines, a large sliding cargo door, and bucket seats for 22 paratroops. The prototype, after certification and development flying had been completed, was sold with a 16-seat executive interior to William E. Boeing, and was later impressed for US Navy use as the sole R3D-3.

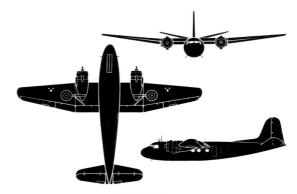

Douglas DC-5 (C-110/R3D-1)

Specification

Type: cargo and passenger/paratroop transport
Powerplant (DC-5): two 850-hp (634-kW) Wright GR-1820-F62 radial piston engines
Performance: maximum speed 221 mph (356 km/h) at 7,700 ft (2 345 m); cruising speed 202 mph (325 km/h) at 10,000 ft (3 050 m); service ceiling 23,700 ft (7 225 m); range 1,600 miles (2 575 km)
Weights: empty 13,674 lb (6 202 kg); maximum take-off 20,000 lb (9 072 kg)
Dimensions: span 78 ft 0 in (23.77 m); length 62 ft 6 in (19.05 m); height 19 ft 10 in (6.05 m); wing area 824 sq ft (76.55 m²)
Armament: none
Operators: USAAF, USN

Douglas Dolphin

History and Notes

In 1930 Douglas introduced a new twin-engined commercial amphibian flying boat which it had named Dolphin. Powered by two radial engines, strut-mounted above the high-set cantilever monoplane wing, it provided accommodation for a pilot, co-pilot and six pasengers. An unusual feature was the use of an aerofoil section structure to brace together the two engines. An attractive looking boat, it soon aroused the interest of the US armed forces, which were seeking transport amphibians to supplement the Loening observation amphibians then in service with both the US Army and US Navy.

First to acquire the Dolphin, however, was the US Coast Guard, which ordered three of the standard commercial aircraft in 1931 under the designation RD, the first of these being delivered on 9 March 1931. The next service to procure these aircraft was the US Army Air Corps which in 1932 ordered eight Y1C-21s, with 350-hp (261-kW) Wright R-975-3 Whirlwind engines and accommodation for a crew of seven, plus two eight-seat Y1C-26s with 300-hp (224-kW) Pratt & Whitney R-985-1 Wasp Juniors. Subsequent orders covered eight Y1C-26As and six C-26Bs, powered by R-985-5 and R-985-9 engines respectively, both types being rated at 350 hp (261 kW). In 1934, two C-26Bs

were converted to accommodate a crew of two and seven passengers, being redesignated C-29. In that same year the US Army changed the role of the aircraft, using them as observation amphibians with a crew of four; this brought designation changes with Y1C-21, Y1C-26, Y1C-26A and C-26B versions becoming respectively OA-3, OA-4, OA-4A and OA-4B. One OA-4 was provided experimentally with non-retractable tricycle type landing gear: this enabled the USAAC to evaluate the potential of this landing gear configuration, leading to its specification for use initially on the Douglas C-54 transport.

US Navy use of the Dolphin began with the procurement of one aircraft, comparable with the US Coast Guard RD and US Army Y1C-21, which was delivered in December 1931 and designated XRD-1. Three RD-2s followed, these differing by having strengthened mounting struts for the overwing engines, which were higher-powered Pratt & Whitney R-1340-96 Wasps, and six additional RD-3s were of generally similar configuration were next procured. From these aircraft the US Navy subsequently allocated one RD-2 and two RD-3s for service with the US Marine Corps. Final procurement covered the production of 10 RD-4s for the US Coast Guard, these

Douglas Dolphin

being virtually identical to the US Navy's RD-3s except for minor changes in detail and equipment.

One civil Dolphin, which had been used for survey purposes by an Australian petroleum company, was presented to that nation's government soon after the beginning of World War II. This duly found its way into service with the Royal Australian Air Force, which operated it in a training role.

In American service the US Army's Dolphins were used primarily for transport or patrol duties, and those of the US Navy as personnel transports, being of great value for communications between ship and shore. The US Coast Guard, however, used their small fleet in a search and rescue role. When the USA became involved in the war, all services used these aircraft extensively in the early years for transport and rescue, as well as for security patrols along the nation's coastline.

Specification
Type: general-purpose amphibian flying boat
Powerplant (RD-4): two 450-hp (336-kW) Pratt & Whitney R-1340-96 Wasp radial piston engines
Performance: maximum speed 156 mph (251 km/h); cruising speed 135 mph (217 km/h); service ceiling 17,000 ft (5 180 m); range 720 miles (1 159 km)
Weights: empty 7,000 lb (3 175 kg); maximum take-off 9,530 lb (4 323 kg)
Dimensions: span 60 ft 0 in (18.29 m); length 45 ft 1 in (13.74 m); height 14 ft 0 in (4.27 m); wing area 592 sq ft (55.00 m²)
Armament: none
Operators: RAAF, USAAC/USAAF, USCG, USMC, USN

A Douglas RD-4 of the US Coast Guard pictured at San Francisco coast guard station in June 1942.

Douglas SBD Dauntless

History and Notes
Without any doubt the Douglas SBD Dauntless is regarded as being the most successful dive-bomber to be produced by the American aviation industry during World War II. It was successful both in terms of achievement and longevity, blunting the might of the Japanese navy in actions in the Coral Sea, Midway and during the Solomons campaign, but continuing to offer a valuable contribution to US Navy/Marine Corps actions until late 1944, long after contemporary creations had disappeared from the scene.

A product of John Northrop's influence on Douglas design philosophy, the Dauntless stemmed from the Northrop BT-1 which began to enter service with the US Navy in spring 1938. One of the production BT-1s served as the prototype for a new naval dive-bomber, allocated the designation XBT-2; however, by the time that this entered production in 1940, Northrop had become a division of the Douglas Company, resulting

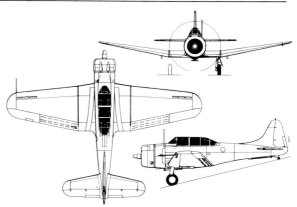

Douglas SBD Dauntless

in the SBD designation to identify Douglas as the manufacturer of the new scout/dive-bomber.

There had been structural and engine changes, and while the SBD retained a general family likeness to its

Douglas SBD Dauntless

progenitor, it was really a very different aeroplane. Of low-wing cantilever configuration, it was of all-metal construction except for fabric-covered control surfaces. Features of the wing design included slots adjacent to the leading edge forward of the ailerons, and hydraulically actuated perforated dive-brakes above and below the trailing edge of the wing outboard to the ailerons, and below the wing centre-section and beneath the fuselage. Fuselage construction included a number of watertight compartments, the tail unit was conventional, and the main units of the tailwheel type landing gear retracted inward to lie flush within wells formed in the wing centre-section. Arrester gear was provided for shipboard operation. Accommodation was provided for a crew of two in tandem cockpits, housed beneath a continuous transparent canopy, and provided with dual controls. The powerplant of the prototype was a 1,000-hp (746-kW) Wright XR-1820-32 Cyclone radial engine.

Testing of the prototype showed not only its superiority over the earlier Northrop BT-1, but performance and flight characteristics that immediately singled it out as an exceptional aircraft. Initial production orders for 57 SBD-1s and 87 SBD-2s were placed on 8 April 1939, the SBD-2s differing by having increased fuel capacity and armament revisions. SBD-1s began to enter service with the US Marine Corps in late 1940, equipping Marine Squadron VMB-2, with deliveries to VMB-1 following in early 1941. The SBD-2s went to the US Navy, and by the end of 1941 were serving aboard the USS *Enterprise* with Squadron VB-6, and with VB-2 on the USS *Lexington*.

An improved SBD-3 version began to enter service in March 1941, introducing self-sealing tanks (and with increased fuel capacity), armour protection, bulletproof windscreen, a 1,000-hp (746-kW) Wright R-1820-52 engine, and armament changes that initiated the standard of two 0.50-in (12.7-mm) and two 0.30-in (7.62-mm) machine-guns. The SBD-3 was followed into production by the SBD-4, which differed only by having a 24-volt instead of 12-volt electrical system. Production of these two versions totalled 1,364 units, making possible a wider distribution of these much-needed and important aircraft to US Navy squadrons which included VB-3, VB-5, VS-2, VS-3, VS-5 and VS-6, as well as to many US Marine Corps squadrons.

Most extensively built was the SBD-5, produced in a new Douglas factory at Tulsa, Oklahoma. This differed from earlier versions in having a 1,200-hp (895-kW) R-1820-60 engine and increased ammunition capacity, and introduced illuminated gunsights for both the fixed forward-firing and rear cockpit flexibly-mounted machine-guns. A total of 2,409 was built for the US Navy before Douglas turned to the final production variant, the SBD-6, with an even more powerful R-1820-66 engine and increased fuel tankage. Also supplied to the US Navy in small numbers were photo-reconnaissance variants of the earlier production versions, with camera installations and related equipment, under the designations SBD-1P, SBD-2P and SBD-3P. Nine examples of the SBD-5 version were supplied for service with the Royal Navy's Fleet Air Arm in January 1945, these being designated Dauntless

Though designed largely as a dive-bomber, the Douglas SBD Dauntless was also a useful scout-bomber, and these examples were deployed, with special livery, on anti-submarine patrol in the Atlantic during 1943.

Douglas SBD Dauntless

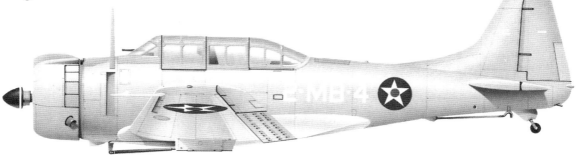

Douglas SBD-1 Dauntless of VMSB-232 (lately VMB-2) of US Marine Corps Air Group 21, based at Ewa, Oahu island (Hawaii) in December 1941.

Douglas SBD-3 Dauntless of VS-41 aboard USS *Ranger* for Operation 'Torch' in November 1942.

Douglas SBD-3 Dauntless of VSB-6 aboard USS *Enterprise* in February 1942.

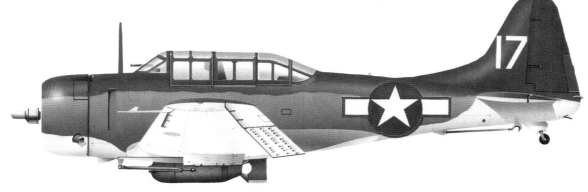

Douglas SBD-4 Dauntless of VMSB-243 of the US Marine Corps' 1st Marine Air Wing, based on Munda, New Georgia island (Solomon group) in August 1943.

Douglas SBD Dauntless

Douglas SBD-5 Dauntless of VMS-3, US Marine Corps, based in the Caribbean during May 1944.

Douglas SBD-5 Dauntless of the Royal New Zealand Air Force, based at Piva, Bougainville island (Solomon group) in April 1944.

Douglas A-24B of the 312th Bomb Group (Dive), USAAF, based on Makin island (Gilbert group) in December 1943.

Douglas A-24B of the Groupe de Chasse-Bombardement 1/18 'Vendée', Free French air force, based at Vannes (France) in November 1944.

Douglas SBD Dauntless

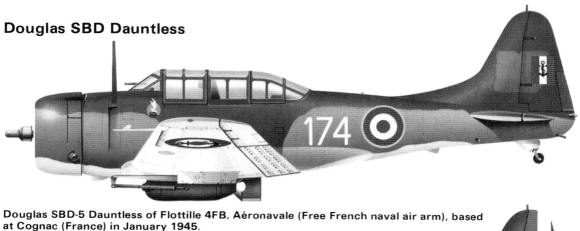

Douglas SBD-5 Dauntless of Flottille 4FB, Aéronavale (Free French naval air arm), based at Cognac (France) in January 1945.

Douglas SBD-5 Dauntless of the Escuadron Aereo de Pelea 200, Fuerza Aerea Mexicana (Mexican air force), based at Pie de la Cuesta (Mexico) in 1946.

DB Mk I, but none of them was used operationally. Another small quantity was supplied to Mexico. Although the US Navy and US Marine Corps use of the Dauntless in a first-line capacity tailed off in late 1944, many late-version aircraft remained in use for some years after the end of World War II.

The success of the German Junkers Ju 87 as a dive-bomber, when Hitler's armoured columns raced over much of Europe in 1940, made the US Army conscious of the fact that it possessed no significant aircraft within this category. Accordingly 168 of the US Navy's SBD-3 version were ordered from Douglas as a matter of some urgency, these being delivered in the summer of 1941 under the US Army designation A-24. They were virtually identical to the SBD-3, except for the deletion of the arrester hook, and the provision of an inflated tailwheel tyre instead of the solid rubber favoured by the US Navy. About a third of these aircraft were despatched to the Philippines in November 1941 for service with the USAAF's 27th Bombardment Group, but as they were still at sea when Pearl Harbor was attacked, they were diverted instead to Australia, equipping the 91st Bombardment Squadron in February 1942, and subsequently the 8th Bombardment Squadron. Both of these units found the A-24 lacking in performance and range for operational deployment in this theatre.

Despite these apparent shortcomings, the US Army continued to procure A-24s during 1942, receiving first 170 A-24As equivalent to the US Navy's SBD-4, and finally 615 A-24Bs (SBD-5). None were deployed

with significant success, confirming the experience of Ju 87 usage in Europe and Africa, that their role was strictly confined: within that limited role, of course, they were indeed the 'tool for the job'. Their failure in US Army service was due to the fact that there was no identical job for them to do. Despite this, a number remained in USAAF/USAF service for some years after the end of World War II.

Specification

Type: two-seat carrier-based scout/dive-bomber

Powerplant (SBD-6): one 1,350-hp (1 007-kW) Wright R-1820-66 Cyclone 9 radial piston engine

Performance: maximum speed 255 mph (410 km/h) at 14,000 ft (4 265 m); cruising speed 185 mph (298 km/h) at 14,000 ft (4 265 m); service ceiling 25,200 ft (7 680 m); range (scout-bomber) 773 miles (1 244 km)

Weights: empty 6,535 lb (2 964 kg); maximum take-off 9,519 lb (4 318 kg)

Dimensions: span 41 ft 6 in (12.65 m); length 33 ft 0 in (10.06 m); height 12 ft 11 in (3.94 m); wing area 325 sq ft (30.19 m²)

Armament: two forward-firing 0.50-in (12.7-mm) machine-guns and two 0.30-in (7.62-mm) machine-guns on flexible mount, plus underfuselage mountings for up to 1,600 lb (726 kg) of bombs and up to a total of 650 lb (295 kg) carried beneath the wings

Operators: France, Mexico, RN, USAAF, USMC, USN

Douglas TBD Devastator

Douglas TBD-1 Devastator of VT-6, US Navy, aboard USS *Enterprise* during early 1942.

History and Notes

In early 1934 the US Navy initiated a design competition for the development of a new torpedo-bomber for service on board US aircraft carriers and, in particular, for the USS *Ranger* which was due to be commissioned during the year. From the proposals received, prototypes were ordered from Douglas and the Great Lakes Aircraft Corporation: the XTBD-1 designed by Douglas represented the first carrier-based aircraft of monoplane configuration to be produced for the US Navy.

The prototype XTBD-1, which flew for the first time on 15 April 1935, was of fairly conventional configuration and construction. The low-set cantilever monoplane wing could be folded mechanically at approximately mid-span, with the ailerons spanning the outer panels, and split type trailing-edge flaps extending the full span of the fixed inner wing on each side of the fuselage. Construction was all-metal, except that the rudder and elevators were fabric covered. The deep fuselage housed an internal weapons bay which could accommodate a torpedo, or a large armour-piercing bomb. Only the main units of the tailwheel type landing gear were retractable, the main wheels being half exposed below the wing's lower surface when retracted. An arrester hook was mounted forward of the tailwheel. The powerplant of the prototype consisted of an 800-hp (597-kW) Pratt & Whitney XR-1830-60 radial engine, driving a three-blade controllable-pitch metal propeller. Accommodation was provided for a crew of three (pilot, bomb-aimer/navigator and gunner) housed beneath a long transparent cockpit enclosure.

Initial testing of the prototype went so well that within nine days of the first flight Douglas was able to hand it over to the US Navy for service trials, carried out over a period of nine months and resulting in a contract for 129 production TBDs awarded to Douglas on 3 February 1936. When delivery of these aircraft began, on 25 June 1937, the US Navy had in its possession what was then, unquestionably, the most advanced torpedo-bomber in the world.

The first US Navy squadron to receive its TBD-1s,

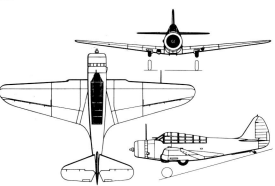

Douglas TBD-1 Devastator

on 5 October 1937, was VT-3, Squadrons VT-2, VT-5 and VT-6 being equipped during the following year. They remained in first-line service with the US Navy until after the Battle of Midway. The main clash of this battle came on 4 June 1942, when 35 TBD-1s, by then bearing the name Devastator, were shot to pieces, caught between blistering anti-aircraft fire and the guns of Mitsubishi A6M Zero naval fighters.

Specification

Type: three-seat torpedo-bomber
Powerplant: one 850-hp (634-kW) Pratt & Whitney R-1830-64 Twin Wasp radial piston engine
Performance: maximum speed 206 mph (332 km/h) at 8,000 ft (2 440 m); cruising speed 128 mph (206 km/h); service ceiling 19,700 ft (6 005 m); range with 1,000-lb (454-kg) bomb or torpedo 416 miles (669 km)
Weights: empty 6,182 lb (2 804 kg); maximum take-off 10,194 lb (4 624 kg)
Dimensions: span 50 ft 0 in (15.24 m); length 35 ft 0 in (10.67 m); height 15 ft 1 in (4.60 m); wing area 422 sq ft (39.20 m²)
Armament: one 0.30-in (7.62-mm) forward-firing machine-gun and one 0.30-in (7.62-mm) gun on flexible mounting, plus one torpedo or 1,000-lb (454 kg) armour-piercing bomb
Operator: USN

Douglas XB-19A

History and Notes

The continuing battle fought between the US Army Air Corps and the US Navy was long and bitter, and no single issue was more vigorously contested by the US Navy than the repeated argument that the USAAC, with suitable aircraft, could more economically and effectively defend the nation's enormous expanse of coastline than could the US Navy with surface vessels. Constant dripping is said to wear away a stone, and by the early 1930s the US Navy relented sufficiently to allow the USAAC to explore its beliefs.

In 1934, therefore, the USAAC's Matériel Division issued its Project 'A' specification for a long-range bomber which could be used, if necessary, to provide support to troops in Alaska, Hawaii or Panama, this requiring the carriage of a 2,000-lb (907-kg) bombload at a speed of 200 mph (322 km/h) over a range of 5,000 miles (8 047 km). It was certainly very much of a challenge at a time when aircraft manufacturers had not then succeeded in creating a civil airliner to cope with the North Atlantic.

Boeing responded with their Model 294, of which a prototype was ordered under the designation XBLR-1 (Experimental Bomber Long Range-1), later XB-15. The proposal received from Douglas was for a very much larger aircraft, and a single prototype of this was contracted as the XBLR-2, later XB-19. When completed it was then the largest aircraft built, with accommodation for a crew of 10 and maximum bombload of 36,000 lb (16 329 kg). It flew for the first time on 27 June 1941. Powered by four Wright R-3350-5 Cyclone 18 engines, each with a take-off rating of 2,000 hp (1 491 kW) and maximum continuous rating of 1,700 hp (1 268 kW) at 5,700 ft (1 740 m), it was underpowered for the duty required of it, and had to await the availability of more powerful engines. When these eventually materialised, the requirement had

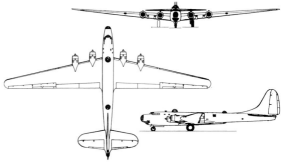

Douglas XB-19

changed, and the Douglas giant was provided instead with four Allison inline engines and operated in a transport role during World War II under the designation XB-19A.

Specification

Type: long-range heavy bomber prototype
Powerplant (XB-19A): four 2,600-hp (1 939-kW) Allison V-3420-11 inline engines
Performance (XB-19): maximum speed 209 mph (336 km/h); cruising speed 186 mph (299 km/h); service ceiling 22,000 ft (6 705 m); maximum range 7,750 miles (12 472 km)
Weights (XB-19): empty 82,253 lb (37 309 kg); maximum take-off 164,000 lb (74 389 kg)
Dimensions: span 212 ft 0 in (64.62 m); length 132 ft 0 in (40.23 m); height 42 ft 9 in (13.03 m); wing area 4,492 sq ft (417.31 m²)
Armament (XB-19): two 37-mm cannon, five 0.50-in (12.7-mm) machine-guns and six 0.30-in (7.62-mm) guns, plus up to 36,000 lb (16 329 kg) of bombs
Operator: USAAF

Fairchild 24W-41 Argus

History and Notes

When Sherman Fairchild withdrew from The Aviation Corporation in 1931, he retained control of the subsidiary Kreider-Reisner Company of Hagerstown, Maryland, renamed Fairchild Aircraft Corporation in 1935. Kreider-Reisner's Model 24C three-seat touring aircraft, first introduced in 1933, remained in production, its versions including the Models 24C8-C, 24C8-E and 24C8-F. The four-seat 24J was introduced in 1937, and was built with both Ranger and Warner engines. The Ranger-engined version was superseded by the 24K in 1938. The main production variants, however, were the 24R and 24W, respectively Ranger and Warner-powered, and produced from 1939.

The 24W-41, with a 165-hp (123-kW) Super Scarab, was developed for service with the US Army Air Corps as the UC-61 Forwarder but, of 163 built, only two were retained. The rest were supplied to the UK under Lend-Lease and were known as the Argus I. The type

Fairchild UC-61 (Model 24W-41/Argus)

was adopted as the Air Transport Auxiliary's standard transport for the carriage of ferry pilots; the ATA also received a large number of Argus IIs which were equipped with new radios and had a 24-volt electrical

Fairchild 24W-41 Argus

system rather than the 12-volt system of the Mk I. Of the RAF allocation of 364 Argus IIs from the 512 UC-61As built to USAAC order, a number were used in India and the Middle East, as were many of the RAF Argus IIIs which comprised the entire USAAC order for 306 UC-61Ks, developed from the 24R with a 175-hp (130-kW) Ranger L-440-7 engine. Civil aircraft impressed by the USAAF in 1942 were allocated designations UC-61B to UC-61J according to civil model number.

US Navy use of the Fairchild 24 was confined to two J2K-1s and two J2K-2s acquired for the US Coast Guard in 1936, all powered by 145-hp (108-kW) Ranger engines, and 13 24W-40s which were similar to the USAAC UC-61A, and taken into the inventory in 1940 and 1942 for instrument training and personnel transport with the designation GK-1.

Specification

Type: four-seat liaison and communications aircraft, or instrument trainer

Powerplant (UC-61): one 165-hp (123-kW) Warner R-500 Super Scarab radial piston engine

Performance: maximum speed 132 mph (212 km/h); cruising speed 117 mph (188 km/h); service ceiling 15,700 ft (4 785 m); range 640 miles (1 030 km)

Weights: empty 1,613 lb (732 kg); maximum take-off 2,562 lb (1 162 kg)

Dimensions: span 36 ft 4 in (11.07 m); length 23 ft 9 in (7.24 m); height 7 ft 7½ in (2.32 m); wing area 193.3 sq ft (17.96 m²)

Armament: none

Operators: Finland, RAAF, RAF, RCAF, USAAF, USCG, USMC, USN

Fairchild AT-21 Gunner

History and Notes

Heavy firepower had been a distinguishing feature of the fighter aircraft which faced each other in Europe when World War II began, and progressive development aimed to increase this as much as possible. To neutralise the advantage held by attacking fighters, power-operated multigun turrets were evolved, to provide fast aiming and ranging of their concentrated firepower. One operational British fighter, the Boulton Paul Defiant, was even provided with a gun turret in this category for attack, rather than defence.

The importance of these requirements were not at first appreciated by the US Army: the B-17B Fortress, for example, was protected by only five machine-guns in five separate mountings. When the significance was understood, as combat experience in Europe became

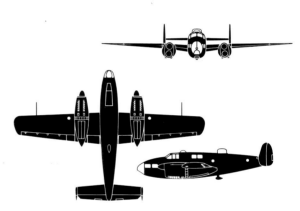

Fairchild AT-21 Gunner

This is the sole AT-21A Gunner built at Hagerstown by Fairchild. Notable are the two gun positions provided for air gunner training.

Fairchild AT-21 Gunner

available to US Army Air Corps planners, steps were taken immediately to ensure that adequate offensive and defensive firepower, including the introduction of power-operated turrets, was written into procurement specifications. These steps did not, however, cover deficiencies in the training schedule: not only were there no air gunners with the experience to use a gun turret, if and when provided, but there were also no specialised air gunnery training schools. Neither were there any suitable aircraft in which pupils could gain the essential air-to-air firing practice.

To resolve this latter shortcoming, the USAAC ordered two specialised gunnery training prototypes from the Fairchild Engine & Airplane Corporation. The first XAT-13 was intended to serve for the training of all members of a bomber's crew working as a team, and the single prototype (41-19500) was powered by two 600-hp (447-kW) Pratt & Whitney R-1340-AN-1 radial engines. The second XAT-14 prototype (41-19503) was powered by two 520-hp (388-kW) Ranger V-770-6 inline engines and was generally similar in layout, but was adapted subsequently as a more specialised trainer for bomb aimers under the designation XAT-14A, with its defensive guns removed. Testing of these aircraft served to crystallise ideas, resulting in the procurement of a special gunnery trainer under the designation AT-21 and given the name Gunner.

Of the 175 AT-21s constructed, 106 were built by Fairchild, and to speed deliveries to the USAAF, 39 were built by Bellanca Aircraft Corporation and 30 by McDonnell at St Louis. Entering service with newly established air gunnery schools they remained in service until 1944, displaced eventually by the production of training examples of the operational type in which the air gunners would eventually serve.

Specification

Type: specialised gunnery trainer
Powerplant: two 520-hp (388-kW) Ranger V-770-15 inline piston engines
Performance: maximum speed 225 mph (362 km/h) at 12,000 ft (3 660 m); cruising speed 196 mph (315 km/h) at 12,000 ft (3 660 m); service ceiling 22,150 ft (6 750 m); range 910 miles (1 464 km)
Weights: empty 8,654 lb (3 925 kg); maximum take-off 11,288 lb (5 120 kg)
Dimensions: span 52 ft 8 in (16.05 m); length 38 ft 0 in (11.58 m); height 13 ft 1½ in (4.00 m); wing area 378 sq ft (35.12 m²)
Armament: one 0.30-in (7.62-mm) machine-gun in fuselage nose and two 0.30-in (7.62-mm) guns in power-operated dorsal turret
Operator: USAAF

Fairchild JK

History and Notes

In 1936 the US Navy acquired a single example of the five-seat Fairchild Model 45, for use primarily as a staff transport for more senior officers, but also for general communication duties, and allocated the designation JK. Typical of the light transport aircraft built by so many US companies during the second half of the 1930s, it was a cantilever low-wing monoplane, its wings and conventional tail unit of light alloy construction with fabric covering. The fuselage was also fabric-covered but its basic structure was of welded steel tube. The main units of the tailwheel type landing retracted aft into the wing centre-section, the wheels remaining half exposed when

retracted. The powerplant consisted of a Wright R-760 Whirlwind 7 radial engine, driving a two-blade variable-pitch metal propeller. The five occupants of the cabin were accommodated on two forward seats, with dual controls as standard, plus a bench seat for three at the rear of the cabin.

In early 1942, when civil aircraft were being impressed to serve with the US armed forces until the US industry had geared itself up to large-scale production, the US Navy acquired two additional examples of the Model 45. Designated JK-1, these were used for general communciation and transport duties.

Specification

Type: five-seat cabin monoplane
Powerplant: one 320-hp (239-kW) Wright R-760-E2 Whirlwind 7 radial piston engine
Performance: maximum speed 170 mph (274 km/h); cruising speed 164 mph (264 km/h) at 8,000 ft (2 440 m); service ceiling 19,000 ft (5 790 m); range 840 miles (1 352 km)
Weights: empty 2,512 lb (1 139 kg); maximum take-off 4,000 lb (1 814 kg)
Dimensions: span 39 ft 6 in (12.04 m); length 30 ft 3 in (9.22 m); height 8 ft 0 in (2.44 m); wing area 248 sq ft (23.04 m²)
Armament: none
Operator: USN

The Fairchild 45 was used in very small numbers by the US Navy as the JK-1 staff and light transport.

Fairchild PT-19 Cornell

History and Notes

Stretching back over a period of many years it had become traditional to utilise light two-seat biplanes for primary flying training. While this meant that the inexperienced pilot began his tuition in what was generally accepted to be the most easily mastered of all aeroplanes, slow, stable, and forgiving of errors and hard treatment, there was a school of thought which suggested that this bred over-confidence, making the next stage of training more difficult. This latter school thought that a reasonably stable aircraft should by all means be used, but why not a monoplane with higher wing loading which needed to be thoughtfully flown for more of the time, thus ensuring that the step to be climbed from primary to advanced training was not quite so high, and less likely to cause a tumble.

This was the line of thought which led to a break with tradition so far as the US Army Air Corps was concerned, and with a need for more primary training aircraft in 1939, it carried out an evaluation of the Fairchild company's M-62 two-seat monoplane. By comparison with the US Army's most advanced biplane trainer then in service (the Stearman PT-13) maximum speed, rate of climb and service ceiling were very nearly the same. But the wing loading of the M-62 was almost 43 per cent higher, which meant that its stalling speed was also higher and its low-speed handling characteristics just that little more critical. It seemed to be exactly what was needed, and in 1940 an initial order was placed for these trainers under the designation PT-19.

Construction of this aircraft was fairly typical of its type and period, the cantilever monoplane wing mounted low on the fuselage being a conventional two-spar wooden structure with plywood skins. The ailerons comprised light alloy frames with fabric covering, and manually-operated split type trailing-edge flaps were provided. Fuselage structure was of welded steel tube, with mainly fabric covering, but the tail unit was all-wood, except for metal-frame fabric-covered rudder and elevators, and landing gear was of the non-retractable tailwheel type. The powerplant of the initial PT-19 version consisted of a 175-hp (130-kW) Ranger L-440-1 inverted inline engine, driving a two-blade fixed-pitch propeller. Two open cockpits accommodated pupil and instructor, and though dual controls were standard, instrumentation of the PT-19 was only very basic.

Delivery of PT-19s began in 1940, and the aircraft soon proved that they were not lethal instruments of destruction in the hands of embryo pilots. With the expansion of flying training in 1941 Fairchild rapidly discovered they had contracts for more aeroplanes than they could possibly build in their existing factory. Steps were taken to double the capacity of their plant, and arrangements made with Aeronca Aircraft Corporation at Middletown, Ohio, and the St Louis Aircraft Corporation at St Louis, Missouri, to initiate production on their behalf. At a later stage the Howard Aircraft Corporation of St Charles, Illinois, provided an additional source of production.

A total of 270 PT-19s was built before a new PT-19A version was introduced on the production lines of Fairchild, Aeronca and St Louis, these companies

The Fairchild PT-19 Cornell provided US Army Air Corps pilot trainees with a trainer more similar to the fighters they were to fly, rather than biplanes of a previous era.

Fairchild PT-19 Cornell

turning out 3,182, 477 and 33 respectively. The only significant change in this version was the introduction of the slightly more powerful Ranger L-440-3 engine and some refinements in detail. The PT-19A, like the original version, had only basic instrumentation and so was unsuitable for blind-flying or instrument flight training. This shortcoming was rectified in the subsequent PT-19B, which was provided with full blind-flying instrumentation and a hood to cover the pupil's front cockpit when such training was in progress. Production totalled 774 by Fairchild and 143 by Aeronca.

The combination of production contracts covering numbers far in excess of those which Fairchild had anticipated with the urgency of the US Army's requirements, resulted in 1942 in a famine of Ranger engines. To resolve the situation the company produced an XPT-23 prototype by the installation of an uncowled Continental R-670 radial engine and, after evaluation, this was put into production with the designation PT-23. A total of 869 was built by Fairchild (2), Aeronca (375), Howard (199), St. Louis (200), as well as 93 by Fleet Aircraft Ltd of Fort Erie, Ontario, for use in the Commonwealth Air Training Scheme which had been established in Canada. A version of the PT-23, with the blind-flying instrumentation and hood which had been introduced on the PT-19B, was built by Howard (150) and St. Louis (106) under the designation PT-23A. This was the last version to be built for the USAAF in America, with almost 6,000 delivered before the production lines closed down.

The PT-23s which Fleet in Canada had built for service under the Commonwealth Air Training Scheme had resulted in the request for a slightly more advanced version, and this reverted to the use of a Ranger L-440-C5 engine. Improvements included a continuous transparent canopy covering both cockpits, with all controls and blind-flight and navigation instruments duplicated in each. And since the temperature in Canada could often be somewhat lower than in the USA, cockpit heating and ventilation was provided. Fleet built 1,057 of these in Canada under the designations PT-26A and PT-26B, while Fairchild built another 670 for supply to the RCAF under Lend-Lease: designated PT-26, these had the name Cornell II in RCAF service.

Specification

Type: two-seat primary trainer
Powerplant (PT-26A): one 200-hp (149-kW) Ranger L-440-C5 inline piston engine
Performance: maximum speed 122 mph (196 km/h); cruising speed 101 mph (163 km/h); service ceiling 13,200 ft (4 025 m); range 400 miles (644 km)
Weights: empty 2,022 lb (917 kg); maximum take-off 2,736 lb (1 241 kg)
Dimensions: span 36 ft 0 in (10.97 m); length 27 ft 8½ in (8.45 m); height 7ft 7½ in (2.32 m); wing area 200 sq ft (18.58 m²)
Armament: none
Operators: RCAF, USAAC/USAAF

The Fairchild PT-19B differed from its predecessors largely in the fitting of blind-flying instruments and provision for a hood over the front cockpit. Production of the variant totalled 774 by Fairchild at Hagerstown and 143 by the Aeronca Aircraft Corporation at Middletown, Ohio.

Fleetwings BT-12

History and Notes

US involvement in World War II, following the Japanese attack on Pearl Harbor in December 1941, was to highlight the unprepared state of the USAAF. This was in no way due to the lack of foresight or enthusiasm of its higher-ranking policy-forming staff officers, but to a national administration which, traditionally, had followed a policy of isolation from the world's squabbles. Although some preparations had been made before the 'day of infamy' on 7 December, they were too little and too late. After the event manufacturers could not produce aircraft fast enough to meet the demands of the nation's armed forces, and the requirement for training aircraft was insatiable.

This helps to explain why Fleetwings Inc, of Bristol, Pennsylvania, a specialist in the fabrication of stainless steel, and a manufacturer of components and assemblies in this material for the US aviation manufacturers, came to build a basic trainer for the USAAF under the designation BT-12.

Identified by the company as the Model 23, the aircraft was in appearance a fairly conventional low-wing monoplane, with fixed tailwheel type landing gear and the powerplant consisting of one Pratt & Whitney R-985 radial engine. Accommodation was provided for instructor and pupil in separate fully-duplicated cockpits with a continuous transparent canopy covering both.

The unconventional feature of the BT-12 lay in its construction which (with the exception of 65 per cent of the wing skin, ailerons, flaps, part of the fuselage skin, rudder and elevators which were fabric-covered) was entirely of stainless-steel construction. Fabrication was almost entirely of spot or seam welding.

A single XBT-12 prototype (39-719) was flown and

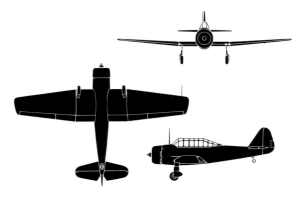

Fleetwings BT-12

evaluated, and a contract for an additional 24 BT-12 aircraft was awarded, these being produced and delivered in the period 1942-3.

Specification

Type: two-seat basic trainer
Powerplant: one 450-hp (336-kW) Pratt & Whitney R-985-25 Wasp Junior radial piston engine
Performance: maximum speed 195 mph (314 km/h); cruising speed 175 mph (282 km/h); service ceiling 23,800 ft (7 255 m); range 550 miles (885 km)
Weights: empty 3,173 lb (1 439 kg); maximum take-off 4,410 lb (2 000 kg)
Dimensions: span 40 ft 0 in (12.19 m); length 29 ft 2 in (8.89 m); height 8 ft 8 in (2.64 m); wing area 240.4 sq ft (22.33 m²)
Armament: none
Operators: USAAC/USAAF

General Motors FM Wildcat

History and Notes

With the need to concentrate on development and production of the F6F Hellcat, designed to provide the US Navy with an advanced carrier-based fighter to supersede the F4F Wildcat, Grumman had to seek an alternative production source for the earlier aircraft. After the necessary negotiations had been completed, General Motors' Eastern Aircraft Division was awarded a contract on 18 April 1942 to build 1,800 F4F-4 Wildcats under the designation FM-1, and production lines were established in the company's five factories at Baltimore, Maryland; Bloomfield, Linden and Trenton, New Jersey; and Tarrytown, New York. The first of these FM-1s, which had in fact been assembled from Grumman-manufactured components, flew for the first time on 1 September 1942, but only a small number of aircraft actually built by the company were delivered by the end of the year.

Generally similar to the Grumman F4F Wildcat, the FM-1 differed only by having changes in armament,

with the wing-mounted 0.50-in (12.7-mm) machine-guns reduced from six to four, but with ammunition capacity increased by some 20 per cent. During 1943 General Motors built 1,127 FM-1s, and was at the same time working upon an improved version of the Wildcat which was to become designated FM-2. This was the production version of two Grumman-built XF4F-8 prototypes that had been tested and evaluated during December 1942. The major difference was replacement of the Twin Wasp powerplant by a more powerful Wright R-1820-56 Cyclone engine with turbocharger, offering a higher speed and rate of climb to an optimum altitude some 50 per cent greater than that of the FM-1. Other changes included the introduction of a taller fin and rudder to maintain good directional stability with the more powerful engine, and efforts were made to reduce airframe weight as much as possible so that this version of the Wildcat, intended for operation from small escort carriers, should have the best possible take-off performance. In fact, to enhance take-off

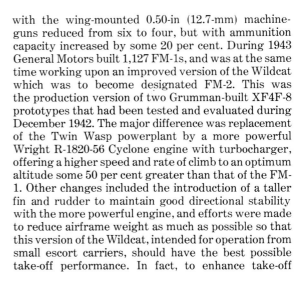

General Motors FM Wildcat

performance the powerplant of later production FM-2s had a water-injection system, permitting a higher short-term power output for this critical phase of carrier operation.

General Motors received an initial contract for the supply of 1,265 FM-2s, but by the time that production was terminated, in August 1945, a total of 4,777 had been delivered: of these, more than 3,000 were manufactured during 1944. FM-1s and -2s were also supplied to the UK under Lend-Lease, with 312 FM-1s received in 1943: these were designated briefly Martlet Vs and redesignated Wildcat V in January 1944. The 370 FM-2s received during 1944 were known as Wildcat VIs.

Both of these latter versions saw widespread service with the FAA, operated mainly from light escort carriers to provide fighter protection for convoys, and working also in collaboration with rocket-carrying Fairey Swordfish in attacks on German submarines. During 1944-5 more than 15 FAA squadrons were equipped with Wildcats, including Nos 813, 816, 819, 821, 824, 825, 832, 835, 842, 845, 849, 851, 853 and 1832.

History and Notes
Type: single-seat carrier-based fighter-bomber
Powerplant (FM-2): one 1,350-hp (1 007-kW) Wright R-1820-56 Cyclone 9 radial piston engine
Performance: maximum speed 332 mph (534 km/h) at 28,800 ft (8 780 m); cruising speed 164 mph (264 km/h); service ceiling 34,700 ft (10 575 m); range 900 miles (1 448 km)
Weights: empty 5,448 lb (2 471 kg); maximum take-off 8,271 lb (3 752 kg)
Dimensions: span 38 ft 0 in (11.58 m); length 28 ft 10¾ in (8.81 m); height 9 ft 11 in (3.02 m); wing area 260 sq ft (24.15 m²)
Armament: four forward-firing 0.50-in (12.7-mm) machine-guns, plus two 250-lb (113-kg) bombs or six 5-in (127-mm) rockets on underwing racks
Operators: RN, USN

General Motors P-75 Eagle

History and Notes
For many years high-flying bomber aircraft had not too much to fear from intercepting fighters, for with no vast disparity in speed between one and the other, the fighter had little chance of climbing from the ground to intercept and attack the enemy, having regard to the amount of notice that was likely to be available. If the enemy aircraft was a fighter or fighter-bomber, with performance more or less equal to that of the interceptor, it could regard itself as more or less immune from attack. Before the introduction of radar (to provide significant warning of an air attack, giving range, altitude and direction to make interception practicable), the only effective but costly solution was to fly standing patrols within the airspace in which an attack was anticipated.

Faced with this problem, especially in the Pacific theatre where the USAAF had neither sophisticated equipment nor high-performance fighters to counter Japanese attacks, it was a matter of extreme urgency to procure a fighter with an unprecedented rate of climb. In 1942 General Motors submitted a proposal for this requirement which suggested the use of the most powerful inline engine then available, installed in a polygenetic airframe, contrived from 'off the shelf' major assemblies, so that the time lag between contract and first flight would be the absolute minimum.

Attracted by the proposal, the USAAF ordered two prototypes in October 1942 under the designation XP-75, and the first of these made its initial flight on 17 November 1943. Its airframe united in adapted form the outer wing panels of the Curtiss P-40, tail unit of the Douglas A-24, and landing gear main units of the Vought F4U; the powerplant installation was similar to that of the Bell P-39, the engine being located in the fuselage aft of the pilot, with a shaft drive coupling the engine and nose-mounted reduction gear. The powerplant of the prototype was a 2,600-hp (1 939-kW) Allison V-3420-19 inline engine, driving two contra-rotating propellers via co-axial shafts. Before the first flight of the XP-75, however, the USAAF had decided as a result of its experience in the first few months of combat operations in the Pacific, that it needed a long-range interceptor far more than one in which all else had been subordinated to rate of climb. Consequently, six additional XP-75s had been contracted on 6 July 1943, these to be suitably modified for evaluation in a long-range role; simultaneously, 2,500 P-75A Eagle production aircraft were ordered.

By early 1944 all of the prototypes had flown and were involved in the development programme, which had a considerable number of apparently intractable problems. Thus it was not until the autumn of that year that the difficulties were more or less resolved, and the first P-75A production aircraft was ready for test, prior to being handed over to the USAAF. By that

A slim nose, forward-sited cockpit and mid-fuselage exhausts indicate the P-75's unusual configuration.

General Motors P-75 Eagle

time, however, the USAAF already had available good long-range fighters, such as the North American Mustang and Republic Thunderbolt, and the P-75 contract was cancelled except for a small batch of five production aircraft for evaluation. Manufacturer's trials were never completed, so that no accurate performance figures are available.

Specification

Type: single-seat long-range escort fighter
Powerplant (P-75A): one 2,885-hp (2 151-kW) Allison V-3420-23 inline piston engine

Performance (approximate): maximum speed 420 mph (676 km/h) at 20,000 ft (6 095 m); cruising speed 310 mph (499 km/h); service ceiling 36,000 ft (10 970 m); range 2,000 miles (3 219 km)
Weights: empty 11,495 lb (5 214 kg); maximum take-off 18,210 lb (8 260 kg)
Dimensions: span 49 ft 4 in (15.04 m); length 40 ft 5 in (12.32 m); height 15 ft 6 in (4.72 m) wing area 347 sq ft (32.24 m²)
Armament: six wing-mounted and four fuselage-mounted 0.50-in (12.7-mm) machine-guns, plus two 600-lb (272-kg) bombs carried externally
Operator: USAAF (for evaluation only)

Grumman F4F Wildcat

History and Notes

The US Navy's requirement of 1936 for a new carrier-based fighter resulted in a design competition in which the Brewster Aeronautical Corporation received an order for a prototype of its Model 39 under the designation XF2A-1. This was to become the US Navy's first monoplane fighter in squadron service, but so tentative was the US Navy in its decision to order this aircraft that it ordered also a prototype of Grumman's competing biplane design under the designation XF4F-1. This was regarded as an each-way bet in the event that the monoplane design was not a winner. However, a more careful study of the performance potential of Brewster's design, plus the fact that Grumman's earlier F3F biplane was beginning to demonstrate performance almost comparable to that estimated for the developed XF4F-1, brought second thoughts, leading to cancellation of the biplane prototype and the initiation of an alternative Grumman monoplane design. Following evaluation of this new proposal, the US Navy ordered a single prototype on 28 July 1936 under the designation XF4F-2.

Flown for the first time on 2 September 1937, the XF4F-2 was powered by a 1,050-hp (783-kW) Pratt & Whitney R-1830-66 Twin Wasp engine, and was able to demonstrate a maximum speed of 290 mph (467 km/h). Of all-metal construction, with its cantilever monoplane wing set in a mid-position on the fuselage, and provided with retractable tailwheel type landing gear, it proved to be marginally faster than the Brewster prototype when flown during competitive evaluation in the early months of 1938. Speed, however, was its major credit: in several other respects it was decidedly inferior, with the result that Brewster's XF2A-1 was ordered into production on 11 June 1938.

Clearly the US Navy had, as we would currently describe it, a gut-feeling about the XF4F-2 prototype, for instead of consigning it to the scrap heap it was returned to Grumman in October 1938, together with a new contract for its further development. The company adopted major changes before this prototype flew again in March 1939 under the designation XF4F-3. These included the installation of a more powerful

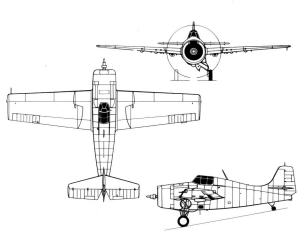

Grumman F4F-4 Wildcat

version of the Twin Wasp, the XR-1830-76 which had a two-stage supercharger enabling it to maintain a power output of 1,000 hp (746 kW) up to an altitude of 19,000 ft (5 790 m). In addition wing span and area had been increased, the tail surfaces redesigned, and the machine-gun installation modified. When tested in this new form the XF4F-3 was found to have considerably improved performance. A second prototype was completed and introduced into the test programme before it ended. This aircraft differed in having a redesigned tail unit in which the tailplane was moved higher up the fin, and the profile of the fin and rudder was changed again. In this final form the XF4F-3 was found not only to have good handling characteristics and manoeuvrability, but was able to demonstrate a maximum speed of 335 mph (539 km/h) at 21,300 ft (6 490 m). Faced with such performance, the US Navy had no hesitation in ordering 78 F4F-3 production aircraft on 8 August 1939.

With war seemingly imminent in Europe, Grumman lost no time in offering their new G-36A design for export, receiving orders for 81 and 30 aircraft for the French and Greek governments respectively. The first of those intended for the French navy, powered by a

Grumman F4F Wildcat

Grumman F4F-3 Wildcat of VF-7, US Navy, aboard USS *Wasp* in December 1940.

Grumman Martlet Mk II of No. 888 Squadron, Fleet Air Arm, aboard HMS *Formidable* in November 1942 and wearing US markings for Operation 'Torch'.

Grumman F4F-4 Wildcat of VF-41, US Navy, aboard USS *Ranger* in early 1942.

Grumman F4F-4 Wildcat of VGF-29, US Navy, aboard USS *Santee* for Operation 'Torch' in November 1942.

Grumman F4F Wildcat

1,000-hp (746-kW) Wright R-1820 Cyclone radial engine, flew on 27 July 1940 but by then, of course, France had already fallen. instead, the British Purchasing Commission agreed to take these aircraft, simultaneously increasing the order to 90, and the first of these began to reach the UK in July 1940 (after the first five off the line had been supplied to Canada), becoming designated Martlet I. They first equipped No. 804 Squadron of the Fleet Air Arm, and two of the aircraft flown by this squadron were the first American-built fighters in British service to destroy a German aircraft during World War II. This was but the begining of a distinguished service record in the FAA, the Martlets gaining immense respect from their pilots for excellent performance, reliability, and potent firepower. Subsequent Grumman-built versions to serve with the FAA included Twin Wasp-powered folding-wing Martlet IIs; 10 F4F-4As and the Greek contract G-36As as Martlet IIIs; and Lend-Lease F4F-4Bs with Wright GR-1820 Cyclone engines as Martlet IVs. In January 1944 they were all redesignated as Wildcats, but retained their distinguishing mark numbers.

The first F4F-3 for the US Navy was flown on 20 August 1940, and at the beginning of December the type began to equip Navy Squadrons VF-7 and VF-41, followed in early 1941 by Navy Squadrons VF-42 and VF-71, and Marine Squadrons VMF-121, VMF-211 and VMF-221. Some 95 F4F-3As were ordered by the US Navy, these being powered by the R-1830-90 engine with single-stage supercharger, and deliveries beginning in 1941. An XF4F-4 prototype was flown in May 1941, this incorporating refinements which resulted from Martlet combat experience in Britain, including six-gun armament, armour, self-sealing tanks, and wing-folding. Delivery of production F4F-4 Wildcats,

as the type had then been named, began in November 1941, and by the time that the Japanese launched their attack on Pearl Harbor a number of US Navy and US Marine Corps squadrons had been equipped. Of these, however, it was Marine Squadron VMF-211 which is remembered with especial honour for its courageous defence of Wake Island against all comers—with just four F4Fs; the remaining eight of the squadron's aircraft had been destroyed in the first Japanese strike. As increasing numbers of Wildcats entered service they equipped additional US Marine and US Navy squadrons including VF-5 (USS *Saratoga*); VF-6 and VF-10 (USS *Enterprise*); VF-72 (USS *Hornet*); and VMF-112, VMF-212, VMF-223 and VMF-224. Involved with conspicuous success in the battles of the Coral Sea and Midway, and the operations in Guadalcanal, they were at the centre of all significant action in the Pacific until superseded by more advanced aircraft in 1943, and also saw action with the US Navy in North Africa during late 1942.

The final production variant built by Grumman was the long-range reconnaissance F4F-7 with increased fuel tankage, no armament and camera installations in the lower fuselage. Only 20 of these were built, but Grumman were to manufacture an additional 100 F4F-3s, and two XF4F-8 prototypes served as the basis of a new variant built by General Motors as FM-2. This company also continued to build a variant of the F4F-4 as FM-1.

A flight of Grumman F4F-4 Wildcat naval fighters wearing the insignia (red border to the national markings) used only between July and September 1943. Modelled closely on the F4F-3, the F4F-4 differed mainly in having manually folding wings.

Grumman F4F Wildcat

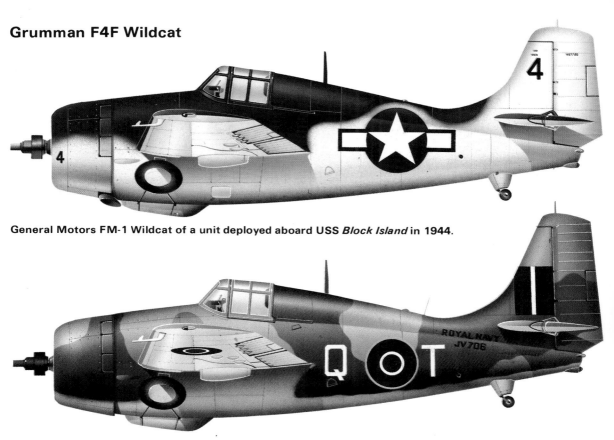

General Motors FM-1 Wildcat of a unit deployed aboard USS *Block Island* in 1944.

General Motors Wildcat Mk VI of No. 835 Squadron, Fleet Air Arm, aboard HMS *Nairana* in August 1944.

Specification
Type: single-seat carrier-based fighter-bomber
Powerplant (F4F-4): one 1,200-hp (895-kW) Pratt & Whitney R-1830-86 Twin Wasp radial piston engine
Performance: maximum speed 318 mph (512 km/h) at 19,400 ft (5 915 m); cruising speed 155 mph (249 km/h); service ceiling 39,400 ft (12 010 m); range 770 miles (1 239 km)

Weights: empty 5,758 lb (2 612 kg); maximum take-off 7,952 lb (3 607 kg)
Dimensions: span 38 ft 0 in (11.58 m); length 28 ft 9 in (8.76 m); height 9 ft 2½ in (2.81 m); wing area 260 sq ft (24.15 m²)
Armament: six 0.50-in (12.7-mm) machine-guns and two 100-lb (45-kg) bombs
Operators: RCAF, RN, USMC, USN

Grumman F6F Hellcat

History and Notes
Developed from a project started by Grumman to evolve a successor to the F4F Wildcat, the Hellcat's design was to benefit from the early operational experience of US Navy pilots in the Pacific theatre, as well as from a feedback of information from the European Allies who had then been involved in war against the Axis for some 18 months. This project was not, of course, a 'cold start' so far as Grumman was concerned, but more accurately an advanced development of the F4F. The family resemblance was unmistakable, but there was one major change to confound spotters in the 'it's one of ours' category: the mid-wing configuration of the Wildcat gave place to a new low-wing layout and this, in turn, made it possible for the main landing gear units to retract into the wing

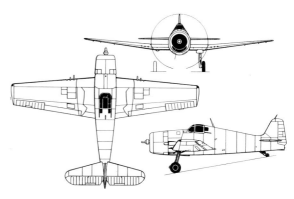

Grumman F6F-3 Hellcat

Grumman F6F Hellcat

centre-section instead of the fuselage. This was not merely cosmetic, for it meant that the main landing gear units could be mounted further outboard from the fuselage, providing a much more stable wide-track undercarriage. The other improvements which resulted from combat feedback included the provision of armour for the pilot, and increased ammunition capacity.

An evaluation by the US Navy of Grumman's design proposal resulted in an order, dated 30 June 1941, covering four prototypes, each with a different engine installation to permit competitive evaluation of the flight envelope. These consisted of the XF6F-1 with a two-stage turbocharged 1,700-hp (1 268-kW) Wright R-2600-10 Cyclone 14; the XF6F-2 with a turbocharged R-2600-16; the XF6F-3 with a two-stage turbocharged 2,000-hp (1 491-kW) Pratt & Whitney R-2800-10 Double Wasp; and the XF6F-4 with the R-2800-27 and two-speed turbocharger.

On 26 June 1942, just under a year from order date, the XF6F-1 flew for the first time. By then there was great urgency to reinforce the Wildcats then in service, an urgency highlighted by the mauling which the US Navy's Douglas TBD Devastators had received during the Battle of Midway. It resulted in the decision to install the most powerful engine then available, the Pratt & Whitney R-2800-10, into the first airframe, and this made its second 'first flight' as the XF6F-3 on 30 July 1942. But even before the first prototype flight, the Grumman design had been ordered into production as the F6F-3 Hellcat, and from that moment the design similarity of the F4Fs and F6Fs was to pay immense dividends in terms of production, for company personnel were able to switch smoothly from one type to the other. The first production F6F-3 flew for the first

time on 4 October 1942; 10 had been completed by the end of the year; US Navy Squadron VF-9 on board USS *Essex* began to equip on 16 January 1943; and on 31 August 1943 Squadron VF-5 on board USS *Yorktown* was the first to join combat with the Japanese. It was a record that many manufacturers would have wished to equal, let alone exceed.

Of all-metal construction with flush-riveted skins, the Hellcat's wings housed the main landing gear units flush in the undersurface of the centre-section, were provided with wide-span trailing-edge split flaps, and had outer panels which folded for carrier stowage. Standard armament comprised six 0.50-in (12.7-mm) machine-guns mounted in the leading edge of the wings. Fuselage and tail unit were conventional in structure and differed little, except in size, from those of the F4F. All three units of the landing gear were retracted hydraulically, the main units turning through 90° to lie fore and aft when housed in the centre-section. A retractable arrester hook was standard. The pilot was accommodated in a cockpit high, and almost directly, above the wing trailing edge. It was an inescapable fact that however large the pilot, he appeared to be far too small to fly such a big aeroplane: for some idea of scale the three-blade propeller was 13 ft 1 in (3.99 m) in diameter.

Hellcat production was superb, with well over 2,500 delivered during 1943, making it possible to re-equip F4F squadrons rapidly with this more potent fighter

Hectic scene aboard USS *Yorktown* before the launch of Grumman F6F-3 Hellcats in June 1944. It was this carrierborne fighter that turned the scales against the Mitsubishi A6M in the Pacific campaign.

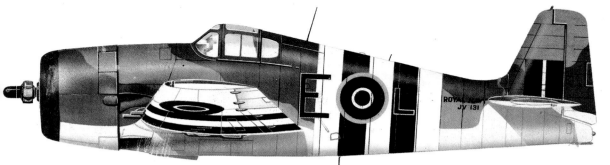

Grumman Hellcat Mk I of No. 800 Squadron, Fleet Air Arm, aboard HMS *Emperor* in the Mediterranean during summer 1944.

Grumman F6F-5 Hellcat of a US Navy Reserve squadron, New York, in the late 1940s.

Grumman F6F-5 Hellcat of Escadrille 1F, Aéronavale (Free French naval air arm), aboard *Arromanches* in the Mediterranean during November 1953.

Grumman F6F-5 Hellcat of the Escuela de Especializacion Aeronaval (Uruguayan naval air arm) in the 1950s.

Grumman F6F Hellcat

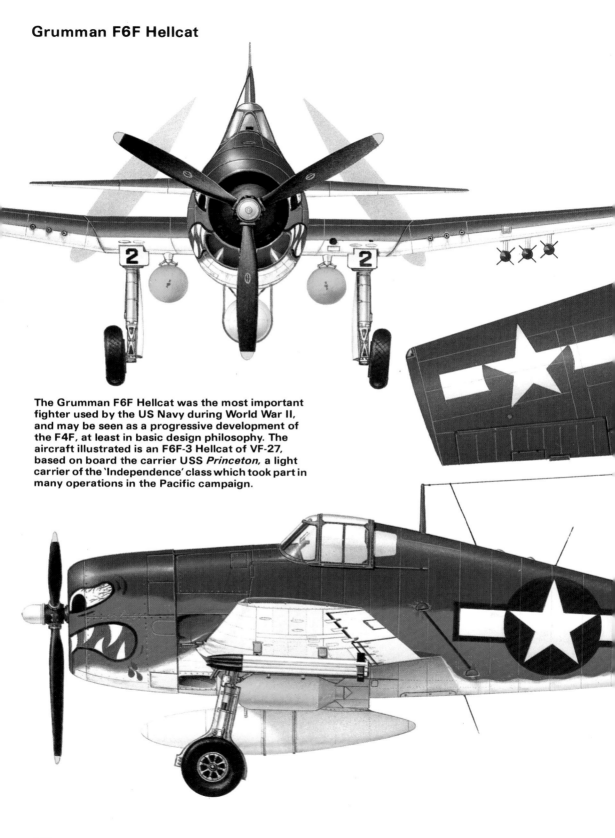

The Grumman F6F Hellcat was the most important
fighter used by the US Navy during World War II,
and may be seen as a progressive development of
the F4F, at least in basic design philosophy. The
aircraft illustrated is an F6F-3 Hellcat of VF-27,
based on board the carrier USS *Princeton,* a light
carrier of the 'Independence' class which took part in
many operations in the Pacific campaign.

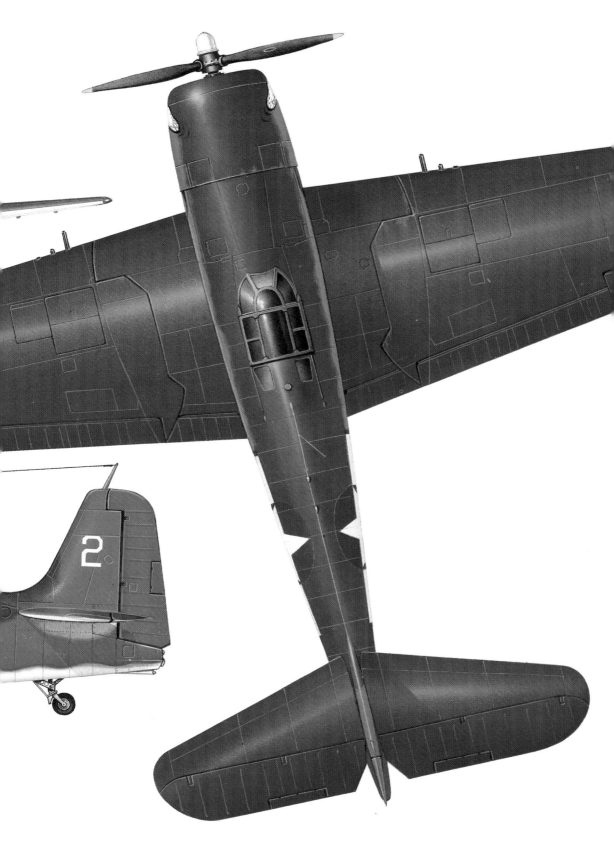

Grumman F6F Hellcat

Grumman F6F-5 Hellcat of the Escuela de Especializacion Aeronaval (Uruguayan naval air arm) in the 1950s.

Grumman Hellcat Mk II of No. 800 Squadron, Fleet Air Arm, aboard HMS *Emperor* in Malayan waters during September 1945.

for, although it had been deployed with considerable success, the F4F was inferior in several aspects of performance by comparison with contemporary Japanese fighters and, especially, the Mitsubishi A6M Zero. The Hellcat was to remain in first-line service with the US Navy for the remainder of World War II. Even when the more advanced Vought F4U Corsair joined the fleet in mid-1944 the Hellcat was not displaced. Instead, the two fighters worked side by side as a team that was extremely low in Japanese popularity ratings: the Hellcat finally acclaimed with 4,947 enemy aircraft destroyed in air-to-air combat, the Corsair with a ratio of kills to losses of 11:1.

F6F-3 Hellcats also began to arrive in the UK during 1943, in the form of 252 originally designated Gannet I (later Hellcat I) and equipping first the Fleet Air Arm's No. 800 Squadron. It was this unit, aboard the light escort-carrier HMS *Emperor*, which first saw action off Norway in December 1943.

By the time F6F-3 production ended in mid-1944, a total of 4,423 Hellcats had been built. Their numbers included 18 F6F-3E night fighters with APS-4 radar mounted in a pod beneath the starboard wing, and 205 generally similar F6F-3N night fighters with APS-6 radar.

Despite their production achievement, Grumman had been 'burning the midnight oil' to develop an improved version, which was duly ordered into production as the F6F-5. This had aerodynamic improvements which included a redesigned engine

cowling, new ailerons and strengthened tail surfaces. The same powerplant as that used in the F6F-3s was retained, but this had the designation R-2800-10W, and the final letter of the designation pinpointed the subtle difference: 'W' indicated a water injection system, and this provided an additional 10 per cent of power for limited periods during take-off and combat. And it also meant that take-off could be made at a higher gross weight, which gave scope to increase both armour and armament without any danger of penalising performance. In addition to the standard six 0.50-in (12.7-mm) guns, therefore, two 1,000-lb (454-kg) bombs or six 5-in (127-mm) rockets could be carried beneath the wing centre-section. Late production models were able to have the inner 0.50-in (12.7-mm) machine-gun in each wing replaced by a 20-mm cannon.

First flown on 4 April 1944, F6F-5s began to enter service with the US Navy very shortly after this date, and 930 were supplied to the UK under Lend-Lease, the type being designated Hellcat II in FAA service. Of this number, some 70 were equivalent to the US Navy's F6F-5Ns, equipped to serve in a night fighter role, and were identified easily by the addition of a small radome on the starboard wing, and the standard camouflage paint scheme replaced by all-over midnight blue.

By far the majority of the FAA's Hellcats equipped the squadrons which served with the British Pacific Fleet in the Far East. These included No. 800 Squadron (aboard HMS *Emperor*), No. 808 Squadron

Grumman F6F Hellcat

Grumman F6F-6 Hellcat of VF-12, US Navy, aboard USS *Randolph* in early 1945.

(HMS *Khedive*), No. 888 Squadron (HMS *Empress*), Nos. 1839 and 1844 Squadrons (HMS *Indomitable*), and No. 1840 Squadron (HMS *Indefatigable*). By VJ-Day no fewer than 12 squadrons were equipped with Hellcat Is or IIs. They saw extensive action over Sumatra, playing an important close escort role in attacks on Sumatra's oil refineries in early 1945, and in operations against Pangkalan Brandan and Palembang. The 70-odd night fighter variants supplied to Britain, and which were designated Hellcat NF. II, were used to equip two squadrons. These were Nos. 891 and 892, formed at Eglinton, Northern Island, in June and April 1945 respectively, but neither of them was to see operational service before the end of World War II.

When production ended in November 1945, a total of 12,275 Hellcats had been built, this number including one XF6F-4 prototype with a Pratt & Whitney R-2800-27 engine and two XF6F-6 prototypes with 2,100-hp (1 567-kW) R-2800-18W engines. Of the original prototypes, those intended as XF6F-2 and XF6F-4 were completed subsequently and delivered as production F6F-5s. F6F-5s and F6F-5Ns remained in service with the US Navy for several years after VJ-Day. Of these,

some were modified for use as F6F-5K target drones and equipped with cameras for service in a reconnaissance role under the designation F6F-5P.

Specification

Type: single-seat carrier-based fighter/fighter-bomber
Powerplant (F6F-5): one 2,000-hp (1 491-kW) Pratt & Whitney R-2800-10W Double Wasp radial piston engine
Performance: maximum speed 380 mph (612 km/h) at 23,400 ft (7 130 m); cruising speed 168 mph (270 km/h); service ceiling 37,300 ft (11 370 m); range with a 150-US gallon (568-litre) drop tank 1,530 miles (2 462 km)
Weights: empty 9,153 lb (4 152 kg); maximum take-off 15,413 lb (6 991 kg)
Dimensions: span 42 ft 10 in (13.06 m); length 33 ft 7 in (10.24 m); height 13 ft 6 in (4.11 m); wing area 334 sq ft (31.03 m²)
Armament: six 0.50-in (12.7-mm) machine-guns (some late models had two machine-guns replaced by 20-mm cannon), plus two 1,000-lb (454-kg) bombs or six 5-in (127-mm) rocket projectiles
Operators: RN, USMC, USN

Grumman F7F Tigercat

History and Notes

The US Navy's contract of 30 June 1938 for a prototype twin-engined carrier-based fighter (XF5F-1), a development of Grumman's G-34 Skyrocket, had not led to the construction of production aircraft. However, in the process of its evolution, trials and modification, the company had gained a far wider appreciation of the problems involved in the creation of such a machine. In early 1941 work was initiated on the design of a new twin-engined fighter for operation from carriers in the future, the larger 45,000-ton (45 720-tonne) 'Midway' class. Identified by the company as the G-51 Tigercat, there was little resemblance to its predecessor, for the US Navy by then wanted to procure a high-performance fighter with unprecedented firepower.

Grumman's proposal resulted in the award of a contract for two XF7F-1 prototypes on 30 June 1941. The first of the prototypes made its initial flight in December 1943. Of all-metal construction, the Tigercat was of cantilever shoulder-wing monoplane configura-

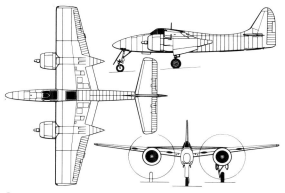

Grumman F7F-2N Tigercat

tion, the outer panels of the wings folding for carrier stowage. Fuselage and tailplane were conventional, but the retractable landing gear was of the tricycle type. A retractable deck arrester hook was mounted in

Grumman F7F Tigercat

the aft fuselage. The powerplant comprised two Pratt & Whitney R-2800-22W Double Wasp engines, installed in large underwing nacelles, and each driving a three-blade constant-speed propeller. To satisfy the requirement for heavy armament, four 20-mm cannon were located within the wing roots, and four 0.50-in (12.7-mm) machine-guns in the fuselage nose, a torpedo could be carried beneath the fuselage, and there was provision for two 1,000-lb (454-kg) bombs below the inner wing sections.

Before the first flight of the prototype, Grumman had received a contract for 500 production aircraft under the designation F7F-1 for supply to the US Marine Corps which, by then, was already engaged in landing operations on Japanese-held islands in the Pacific. Operated from land bases, these aircraft would provide the US Marines with their own close-support but, in fact, the Tigercat materialised too late to see operational service with the USMC before the end of World War II.

The first production F7F-1 was generally similar to the prototypes, as were the 33 aircraft which followed, and delivery of these began in April 1944. The 35th aircraft on the production line was modified for use in a night fighter role, under the designation XF7F-2N, and 30 production examples followed as F7F-2Ns during 1944, with another batch of 34 being built by March 1945. These differed from the F7F-1 by deletion of the aft fuselage fuel tank, to provide space for the radar operator's cockpit, and removal of the nose armament. There followed production of a new single-seat version, the F7F-3 Tigercat, of which 189 were built. This differed from the F7F-1 in the installation of R-2800-34 engines, to provide increased power at altitude, in slightly increased vertical tail surface areas to cater for this, and in a seven per cent increase in fuel capacity. These aircraft terminated production of the

original contract, with the balance cancelled after VJ-Day.

Postwar production included 60 F7F-3N and 13 F7F-4N night fighters, both with a lengthened nose housing advanced radar, the latter 13 aircraft being the only examples with strengthening, arrester hook, and specialised equipment for carrier-based operation. A small number of F7F-3s were modified after delivery for use in electronic (F7F-3E) and photo-reconnaissance (F7F-3P) roles. Some squadrons remained in service with the US Marines in the immediate postwar years, but were soon displaced by higher-performance turbine-powered aircraft.

Specification
Type: twin-engined carrier-based fighter-bomber
Powerplant (F7F-3): two 2,100-hp (1 566-kW) Pratt & Whitney R-2800-34W Double Wasp radial piston engines
Performance: maximum speed 435 mph (700 km/h) at 22,200 ft (6 765 m); cruising speed 222 mph (357 km/h); service ceiling 40,700 ft (12 405 m); normal range 1,200 miles (1 931 km)
Weights: empty 16,270 lb (7 380 kg); maximum take-off 25,720 lb (11 666 kg)
Dimensions: span 51 ft 6 in (15.70 m); length 45 ft 4½ in (13.83 m), length (F7F-3N) 46 ft 10 in (14.27 m); height 16 ft 7 in (5.05 m); wing area 455 sq ft (42.27 m²)
Armament: four 20-mm cannon in wing roots and four 0.50-in (12.7-mm) machine-guns in nose, plus one torpedo beneath the fuselage and up to 1,000 lb (454 kg) of bombs under each wing
Operators: USMC, USN

The Grumman F7F-1 Tigercat was a potent fighter-bomber for the US Marine Corps, but was just to late to see operational service in World War II.

Grumman F8F Bearcat

History and Notes

Last of the line of piston-engined carrier-based fighters which Grumman had initiated with the FF of 1931, the F8F Bearcat was designed to be capable of operation from aircraft carriers of all sizes. It was required to serve primarily as an interceptor fighter, a role which demanded excellent manoeuvrability, with good low-level performance and a high rate of climb. To achieve these capabilities for the two XF8F-1 prototypes ordered on 27 November 1943, Grumman adopted the big R-2800 Double Wasp that had been used to power the F6F and F7F, but ensuring that the smallest and lightest possible airframe was designed to accommodate the specified armament, armour and fuel.

First flown on 21 August 1944, the XF8F-1 was not only smaller than the US Navy's superb Hellcat, but was some 20 per cent lighter, resulting in a rate of climb about 30 per cent greater than that of its predecessor. Grumman had achieved the specification requirements, but was also to crown this by starting delivery of production aircraft in February 1945, only six months after the first flight of the prototype.

A cantilever low-wing monoplane of all-metal construction, the F8F-1 had wings which folded at about two-thirds span (aileron/flap junction point) for carrier stowage, retractable tailwheel type landing gear, armour, self-sealing fuel tanks, and armament which comprised four 0.50-in (12.7-mm) machine-guns. By comparison with the prototypes, a very small dorsal fin had been added. The powerplant of these production aircraft was the Pratt & Whitney R-2800-34W.

Shortly after initiation of the prototype's test programme, on 6 October 1944 the US Navy placed a contract for 2,023 production F8F-1s, and the first of these began to equip US Navy Squadron VF-19 on 21 May 1945. This squadron, and other early recipients of Bearcats, were still in the process of familiarisation

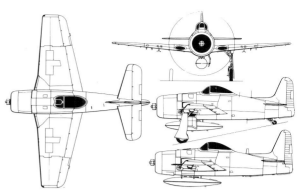

Grumman F8F-1 Bearcat (lower side view: F8F-2)

with their new fighters when VJ-Day put an end to World War II. It also cut 1,258 aircraft from Grumman's contract and brought complete cancellation of an additional 1,876 F3M-1 Bearcats contracted from General Motors on 5 February 1945. When production ended in May 1949, Grumman had built 1,266 Bearcats: 765 F8F-1s; 100 F8F-1Bs, which differed by having the four 0.50-in (12.7-mm) machine-guns replaced by 20-mm cannon; 36 F8F-1Ns equipped as night fighters; 293 F8F-2s with redesigned engine cowling, taller fin and rudder, plus some changes in detail design, and adoption of the 20-mm cannon as standard armament; 12 night fighter F8F-2Ns; and 60 photo-reconnaissance F8F-2Ps, this last version carrying only two 20-mm cannon. In late postwar service, some aircraft were modified to serve in a drone control capacity under the

Having formidable performance, agility and gun armament, the Grumman F8F-1 Bearcat entered US Navy service just too late for World War II.

Grumman F8F Bearcat

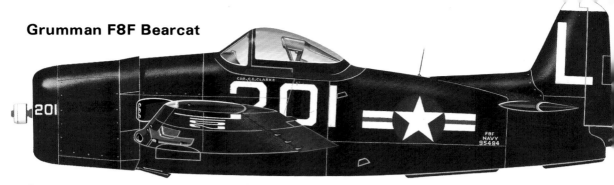

Grumman F8F-1 Bearcat of Lieutenant Commander C.E. Clarke of VF-72, US Navy, based aboard USS Leyte in 1949-5.

designations F8F-1D or F8F-2D.

By the time production ended, Bearcats were serving with some US 24 Navy squadrons, but all had been withdrawn by late 1952. Some of these, with a modified fuel system, were supplied to the French Armée de l'Air for service in Indo-China under the designation F8F-1D. One hundred similar F8F-1Ds and 29 F8F-1Bs were supplied to the Thai air force.

Specification

Type: single-seat carrier-based interceptor fighter
Powerplant: one 2,100-hp (1 566-kW) Pratt & Whitney R-2800-34W Double Wasp radial piston engine

Performance: maximum speed 421 mph (678 km/h) at 19,700 ft (6 005 m); cruising speed 163 mph (262 km/h); service ceiling 38,700 ft (11 795 m); range 1,105 miles (1 778 km)
Weights: empty 7,070 lb (3 207 kg); maximum take-off 12,947 lb (5 873 kg)
Dimensions: span 35 ft 10 in (10.92 m); length 28 ft 3 in (8.61 m); height 13 ft 10 in (4.22 m); wing area 244 sq ft (22.67 m²)
Armament (F8F-1B): four 20-mm cannon, plus underwing hardpoints for two 1,000-lb (454-kg) bombs, or four 5-in (127-mm) rocket projectiles, or two 150-US gallon (568-litre) fuel drop tanks
Operator: USN

Grumman G-21A Goose

History and Notes

In late 1936 Grumman began the design of a small six/seven-seat commercial amphibian flying boat, identified as the G-21. First flown in June 1937, it was of high-wing cantilever monoplane configuration. Of all-metal basic construction, part light alloy-/part fabric-covered, the wing included split trailing-edge flaps. The all-metal two-step hull incorporated retractable tailwheel-type landing gear, all units retracting into recesses formed in the hull. Stabilising floats were strut-mounted and wire-braced beneath the outer wing panels. The powerplant comprised two wing-mounted Pratt & Whitney R-985 Wasp Junior engines with variable-pitch propellers. The pilot's compartment, forward of the wing, seated two side-by-side; the aft cabin had standard seating for four passengers, a lavatory and room for baggage.

These small flying boats were of immediate interest to the US armed services, the US Army losing little time by ordering 31 under the designation OA-9, and these began to enter service in an observation/communications capacity during 1938. Three additional G-21As were purchased from private owners in 1942, and these became designated OA-13A. Two US Navy JRF-5s, transferred in 1945, became designated OA-13B.

US Navy procurement also began in 1938, but a more cautious approach was made by a contract for one prototype XJ3F-1 for evaluation. Successful testing

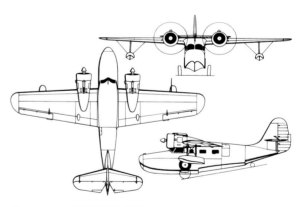

Grumman JRF-5/OA-13B Goose

resulted in an order for 20 production examples under the changed designation JRF-1 (utility transport), 10 of which entered service in late 1939. Subsequent deliveries and procurements resulted in five JRF-1As for target towing and photographic survey; seven production JRF-2s and three JRF-3s for the US Coast Guard, the latter with autopilot and anti-icing equipment for deployment in northern areas; and 10 received from Grumman production, plus two impressed G-21As, as JRF-4s, all equipped with two underwing bomb racks immediately outboard of the engine nacelles to make them suited for deployment in an anti-submarine role.

Grumman G-21A Goose

Major production variant was the JRF-5, of which 190 were built, and these were equipped for primary use in a photographic survey role. Six of these, however, were supplied to the US Coast Guard as JRF-5G, being used for search and rescue. Final production version was the JRF-6, generally similar to the -5 except for updating of electrical and radio equipment, and JRF-6Bs equipped as navigational trainers.

Of the JRF-5/6 versions, some were supplied to the Portuguese naval air service; six went to the RAF as Goose Is (JRF-5) and 29 similar aircraft were delivered to the Royal Canadian Air Force. At least 50 JRF-6Bs were assigned to the UK under Lend-Lease, but records do not confirm that all these were received. They were designated Goose IA, and served with the RAF for navigational training and air-sea rescue, and with the Air Transport Auxiliary for crew ferrying duties.

Specification
Type: multi-role amphibian flying boat
Powerplant (JRF-4): two 450-hp (336-kW) Pratt & Whitney R-985-AN6 Wasp Junior radial piston engines
Performance: maximum speed 201 mph (323 km/h) at 5,000 ft (1 525 m); cruising speed 191 mph (307 km/h) at 5,000 ft (1 525 m); service ceiling 21,000 ft (6 400 m); range 640 miles (1 030 km)
Weights: empty 5,425 lb (2 461 kg); maximum take-off 7,955 lb (3 608 kg)
Dimensions: span 49 ft 0 in (14.94 m); length 38 ft 4 in (11.68 m); height 15 ft 0 in (4.57 m); wing area 375 sq ft (34.84 m²)
Armament: two 250-lb (113-kg) depth bombs on underwing racks
Operators: Portugal, RAF, RCAF, USAAF, USCG

Grumman G-44 Widgeon

History and Notes
The success of the Grumman Goose eight-seat commercial amphibian, and the obvious market for a smaller, cheaper version, led directly to the development of the five-seat G-44 Widgeon, powered by two 200-hp (149-kW) Ranger L-440C-5 engines. The prototype Widgeon was test-flown by Roy Grumman and Bud Gillies at Bethpage on 28 June 1940, and 10 examples had been sold to civil buyers before the first production aircraft was delivered on 21 February 1941. The initial production batch of 44 Widgeons was intended for the civil market, although 11 which had been built to fulfil a Portuguese contract were acquired by the USAAF before delivery in 1942 with the designation OA-14, and deployed in the Caribbean.

The second production run, of 25 aircraft, was earmarked for the US Coast Guard, and designated J4F-1, these aircraft were delivered between 7 July 1941 and 29 June 1942. In August 1942, a Widgeon of Coast Guard Squadron 212, based at Houma, Louisiana, sank the U-boat *U-166* off the Passes of the Mississippi, scoring the first Coast Guard kill of an enemy submarine. J4F-2 production for the US Navy totalled 131, and these aircraft were delivered between 13 July 1942 and 26 February 1945. Operated by a crew of two and with up to three passengers in the utility transport role, the J4F-2 was also used for coastal patrol and anti-submarine duties. Fifteen J4F-2s were supplied under Lend-Lease to the Royal Navy and used for communications, principally in the West Indies where (at Piarco, Trinidad for example) the Royal Navy maintained observer training schools. Royal Navy J4F-2s were known originally as Goslings.

The G-44A was introduced in 1944, making its first flight on 8 August. It featured a revised hull with a deeper keel and 76 were built, the last being delivered on 13 January 1949. Some were later re-engined with Continental W-670s or Lycoming T-435As. During

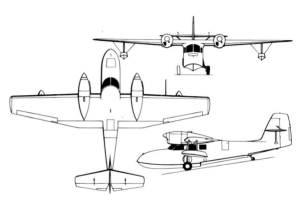

Grumman J4F/OA-14 Widgeon

1948-9 G-44As were built under licence by Société de Construction Aéro-Navale at La Rochelle, France as the SCAN 30; these were powered by 200-hp (149-kW) de Havilland Gipsy Queen 2 engines.

Specification
Type: five-seat light transport, or coastal or anti-submarine patrol aircraft
Powerplant: two 200-hp (149-kW) Ranger L-440C-5 inline piston engines
Performance: maximum speed 153 mph (246 km/h); cruising speed 138 mph (222 km/h); service ceiling 14,600 ft (4 450 m); maximum range 920 miles (1 481 km)
Weights: empty 3,189 lb (1 447 kg); maximum take-off 4,500 lb (2 041 kg)
Dimensions: span 40 ft 0 in (12.19 m); length 31 ft 1 in (9.47 m); height 11 ft 5 in (3.48 m); wing area 245 sq ft (22.76 m²)
Armament (J3F): one 325-lb (147-kg) depth charge
Operators: RN, USAAF, USCG, USN

Grumman JF Duck

History and Notes

Grumman's FF-1 and F2F carrier-based fighters for the US Navy, the company's first production aircraft, had introduced some new ideas, including retractable tailwheel type landing gear, making them the first of their kind to enter US Navy service when the initial examples of the FF-1 began to join Squadron VF-5B on 21 June 1933. At the time the FF-1 was also the fastest fighter operational with the US Navy, but because of the limited procurement that was possible in the 'between wars' years, was acquired in only small numbers. At an even earlier date the US Navy had acquired the first of a series of Loening OL observation amphibians, and although these also were only procured in comparatively small numbers, they served efficiently and reliably for a period of several years.

With the FF-1 nearing production, Grumman began the development of a new utility amphibian which would combine the better features of the FF-1 and Loening OL, and in late 1932 submitted its proposal for review by the US Navy. This resulted in the award of a contract for the supply of an XJF-1 prototype, and this flew for the first time on 4 May 1933. Flight testing found no serious problems, and an uncomplicated evaluation by the US Navy resulted in an initial production order for 27 JF-1s, the first of these being delivered in late 1934.

Intended to fulfil a general utility role (indicated by the letter J of the designation), the type was used first to replace ageing Loening OL-9 observation and general-purpose aircraft in US Navy service, and it was not until 1936 that they began to reach VJ squadrons. Their performance by comparison with the similarly configured Loenings was quite staggering, with maximum speed, rate of climb and service ceiling

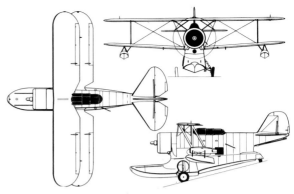

Grumman J2F-6 Duck

all increased by more than 40 per cent, 50 per cent and 65 per cent respectively, but there were significant aerodynamic improvements in the design. The single-bay biplane wings were of equal span, the basic structure of light alloy and fabric-covered; Frise type ailerons on both upper and lower wings provided good roll control. The fuselage was a conventional stressed-skin/light alloy structure, with the large monocoque central float built integrally, and typically Grumman main wheel units were attached to, and retracted into, recesses in this float structure. Stabiliser floats were strut-mounted beneath each lower wing. A crew of two or three could be carried in the tandem cockpits, pilot forward and observer aft as standard, but a radio

A head-on view emphasizes the blocky appearance of the Grumman J2F-5 Duck, with its main float built onto the underside of the fuselage. This example was operated by Utility Squadron VJ-2 during 1942.

Grumman JF Duck

operator could also be accommodated in the observer's cockpit. The powerplant of the prototype and the first batch of JF-1 production aircraft consisted of a 700-hp (522-kW) Pratt & Whitney R-1830 Twin Wasp engine.

The contract which followed was for 14 JF-2s for the US Coast Guard, these having equipment changes and 750-hp (559-kW) Wright R-1820 Cyclone engines. Four of these were transferred subsequently to the US Navy, which service also acquired five new aircraft with similar powerplant under the designation JF-3. There were few major changes in later production examples: the 20 J2F-1s of 1937 and following 21 J2F-2s, 20 J2F-3s and 32 J2F-4s differed in only minor detail. Nine J2F-2As for Marine Squadron VMS-3 were armed with machine-guns and carried underwing bomb racks.

The last version to be built by Grumman was ordered in 1940, and comprised 144 J2F-5s, which were the first to carry the name Duck officially. Generally similar to previous utility models, they were powered by the 850-hp (634-kW) Wright R-1820-50 engine. The final production version, the J2F-2, was built by Columbia Aircraft Corporation of Long Island, New York, from which company the US Navy ordered 330

after the USA had become involved in World War II. These were also generally similar to the Grumman-built Ducks except for the installation of a more powerful R-1820-54 engine.

Most of the JF/J2F Ducks remained in service throughout the war, operated both from carriers and land bases in a variety of roles, including patrol, photo-survey, rescue and target towing.

Specification
Type: two/three-seat utility amphibian
Powerplant (J2F-6): one 900-hp (671-kW) Wright R-1820-54 Cyclone 9 radial piston engine
Performance: maximum speed 190 mph (306 km/h); cruising speed 155 mph (249 km/h); service ceiling 25,000 ft (7 620 m); range 750 miles (1 207 km)
Weights: empty 4,400 lb (1 996 kg); maximum take-off 7,700 lb (3 493 kg)
Dimensions: span 39 ft 0 in (11.89 m); length 34 ft 0 in (10.36 m); height 13 ft 11 in (4.24 m); wing area 409 sq ft (38.00 m²)
Armament: normally none, but provision for two 325-lb (147-kg) depth bombs
Operators: USCG, USMC, USN

Grumman TBF Avenger

Grumman TBF Avenger of the US Navy.

History and Notes
The Battle of Midway, which began on 4 June 1942, was among the early and very vital actions of the Pacific war. In this savage encounter the Japanese fleet lost four aircraft carriers, the *Akagi*, *Hiryu*, *Kaga* and *Soryu*, these representing the spearhead of Admiral Yamamoto's naval strike force. Without them the Japanese fleet was deprived of its initiative and became a defensive rather than the innovative offence force which had brought chaos to Pearl Harbor less than seven months earlier. As in the Battle of the Coral Sea, fought rather less conclusively some four weeks earlier, the attack and defence of both the Japanese and United States naval forces had been almost entirely the prerogative of carrier-based aircraft. US Navy carriers involved in the Battle of Midway comprised the USS *Hornet*, *Enterprise* and *Yorktown*,

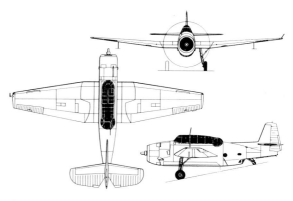

Grumman TBF-1C Avenger

Grumman TBF Avenger

the last being so severely damaged in the action that she became an easy target for a Japanese submarine two days later.

Forming a proportion of the naval aircraft which were involved in this action were the first of the US Navy's monoplane torpedo-bombers, the Douglas TBD Devastator, and the first operational examples of the newly developed torpedo-bomber intended to replace it, the Grumman TBF Avenger. For both of these aircraft types the operation was a disaster: two squadrons of Devastators were almost completely destroyed, and of the six Avengers in action only one survived. The Devastators were no match for the Japanese aircraft against which they were ranged and were thereafter withdrawn from operational service. The Avengers, which had been intended to reinforce Squadron VT-8's TBD-1s aboard USS *Hornet*, arrived at Pearl Harbor after the carrier had sailed. Instead, they flew to Midway Island and went into action from there.

Clearly, the Devastator was obsolete; the Avenger's crews lacked adequate experience, for they had been equipped with these aircraft at Norfolk Naval Air Station only four weeks earlier. To highlight this latter point, it is necessary to point out that the Avenger continued in operational use without the need for any significant modifications as a result of their deployment at Midway, remaining in US Navy service for some 15 years.

The TBF had originated in early 1940 when the US Navy initiated a contest to procure a more modern torpedo-bomber to replace the Douglas TBD Devastator, ordering two XTBF-1 prototypes from Grumman on 8 April 1940, and two competing XTBU-1 prototypes from Vought a couple of weeks later. This latter aircraft was to enter production, built by Consolidated as the TBY Sea Wolf, but only about 180 of them were built. The XTBF-1 represented something of a challenge for the Grumman design team headed by Bill Schwendler, for although the company had produced a number of successful carrier-based fighters, this was the first attempt to evolve a torpedo-bomber.

First flown on 1 August 1941, the prototype was seen to be a hefty mid-wing monoplane of all-metal construction, except for fabric-covered control surfaces. Leading-edge slots forward of the ailerons and split trailing-edge flaps were provided to ensure good low-speed handling characteristics, despite a comparatively high wing loading. On the Avenger the wing loading was to be as much as 37 lb/sq ft (180 kg/m²) by comparison with 24 lb/sq ft (118 kg/m²) for the TBD Devastator. For carrier stowage the outer wing panels could be folded, or unfolded, with the locking pins actuated in correct sequence by hydraulic power, controlled from the pilot's cockpit. The fuselage and tail unit were of conventional construction, the retractable landing gear of the tailwheel type. Attachment points for catapult launch, and an electrically actuated arrester hook were standard. Later versions had RATO (rocket assisted-take off) provision. The powerplant consisted of a 1,700-hp (1 268-kW) Wright R-2600-8 Cyclone 14 radial engine, driving a three-blade constant-speed propeller. Accommodation was for a crew of three (pilot, bomb-aimer and radio-operator/gunner) with a long transparent canopy covering all positions. All-important, of course, was the armament: this comprised a 0.30-in (7.62-mm) machine-gun firing forward through the propeller disc and controlled by the pilot; a ventral machine-gun of the same calibre under control of the bomb-aimer; an 0.50-in (12.7-mm) gun in a power-operated dorsal turret controlled by the radio-operator; and a large fuselage weapon bay with hydraulically actuated doors to accommodate a 22-in (559-mm) torpedo or up to 2,000 lb (907 kg) of bombs.

Flight testing of the prototype by Grumman was followed by US Navy evaluation which ended satisfactorily in December 1941. But 12 months prior to that the US Navy had placed its first production order for 286 TBF-1s, and the first of these began to enter service on 30 January 1942. Despite the inauspicious start to the Avenger's career at Midway, the US Navy procured this aircraft in large numbers, and between first delivery at the end of January 1942 and December 1943, Grumman were to build a total of 2,293. These included TBF-1s, basically the same as the prototypes, and TBF-1Cs, which differed by having two additional 0.50-in (12.7-mm) machine-guns mounted in the wings, and provision for the carriage of drop tanks. Grumman also built XTBF-2 and XTBF-3 prototypes with XR-2600-10 and R-2600-20 engines respectively.

Of the above the Royal Navy received 402 aircraft under Lend-Lease, mostly procured as TBF-1Bs for this purpose, with No. 832 Squadron of the Fleet Air Arm being the first to be equipped on 1 January 1943, the aircraft being designated Tarpon I in British service until January 1944, when they became re-designated Avenger I. No. 832 Squadron took delivery of its aircraft at Norfolk Naval Air Station where crews completed familiarisation with the type before embarking aboard the USS *Saratoga* in April 1943. Two months later they were deployed operationally in support of US Marine Corps landing in the middle of the Solomon Islands chain, and this is regarded as the first occasion that FAA aircraft were flown into action from a US Navy carrier. Subsequent to the equipping of No. 832 Squadron, seven more FAA squadrons took delivery of their aircraft at US Navy air stations in the USA, completing training before embarking on escort carriers for the long journey to Britain. The TBF-1 version was also supplied to the Royal New Zealand Air Force, which received a total of 63 aircraft.

With the demand for Avengers considerably exceeding Grumman's productive capacity, the Eastern Division of General Motors, which had established production lines for the Grumman F4F Wildcat, was selected as a second source of supply. Avengers with the designations TBM-1 (equivalent to TBF-1) and TBM-1C (TBF-1C) began to flow from their lines in September 1942, and the company had produced a

Grumman TBF Avenger

total of 7,546 of these and subsequent versions when its production lines closed down in June 1945. Of these early versions from General Motors, the Royal Navy received 334 TBM-1s, these duly being designated Avenger IIs.

General Motors had produced an XTBM-3 prototype with an R-2600-20 engine, generally similar to the XTBF-3 built by Grumman. It differed, however, by having strengthened wings to allow the carriage of rocket projectiles, drop tanks or radar pod, and many were supplied without the heavy power-operated dorsal turret. Designated TBM-3, delivery of this version began in April 1944, and of these the Royal Navy acquired 222 which were identified as Avenger III. These latter aircraft were particularly useful to the FAA which, despite the torpedo-bomber categorisation, had rarely deployed its Avengers in such a role. Instead they had been deployed on anti-submarine patrol, bombing missions, as rocket-firing strike aircraft and, occasionally, for mine-laying operations.

FAA Avengers thus saw a variety of action and proved a valuable and reliable addition to the Royal Navy's carrier-based forces. Avengers were involved in the Arctic convoys to supply the Russian ally, during which Avengers of No. 846 Squadron shared in the sinking of two German submarines (*U-288* and *U-355*); took part in D-Day preparation and operations, primarily in an anti-shipping strike role; but their major contribution came in the Far East, operating as a component of the East Indies and Pacific Fleets. Almost certainly their most important actions were the two attacks made by 48 aircraft of Nos. 820, 849,

854 and 857 Squadrons on the Japanese oil refineries at Palembang, Sumatra, on 24 and 29 January 1945, reducing the output of the two plants to the merest trickle at a moment when every drop of fuel was critical for both the Japanese army and navy. Many people are unaware of the co-operation and support which the FAA gave to the Americans in these closing stages of the Pacific War: Avengers from the carriers HMS *Formidable*, *Illustrious*, *Indefatigable*, *Indomitable* and *Victorious* gave intensive support in bombing operations against such targets as Formosa and the Japanese home islands. No. 820 Squadron actually launched an attack on Tokyo.

Apart from the main production stream of TBF-1/-1C and TBM-1C/-3 aircraft, there were also small numbers of special versions which resulted from modification programmes introduced during 1944-5. These included the TBF-1D with special radar equipment, the TBF-1E with electronics equipment, the TBF-1L which carried a searchlight in the bomb bay, and the TBF-1CP equipped with cameras for a reconnaissance role. More or less equivalent versions were modified from General Motors production aircraft under the designations TBM-3D, TBM-3E, TBM-3L and TBM-3P respectively. There was, in addition, a TBM-3H version equipped with search radar. General Motors also built the prototype of what had been

One of the US Navy's most important aircraft in World War II, the Grumman Avenger is seen here in its TBF-1 form with the rudder stripes used only between January and May 1942.

Grumman TBF Avenger

anticipated as the next production version and this, designated XTBM-4, differed primarily by having a strengthened fuselage. No production aircraft were built, however, following contract cancellations after VJ-Day.

But the termination of production and the end of World War II did not bring to a halt the career of the Avenger, which still had valuable service to offer during the difficult postwar years. The major operational version in US Navy service was the TBM-3E, by then carrying even more advanced radar equipment for the search and location of submarines, and it is in this area of activity that the Avengers were used extensively until the introduction into service of newly developed and highly specialised aircraft. These were developed to cater for the growing threat from deep-diving nuclear-powered submarines which, in the nuclear-weapons age, were then becoming regarded, somewhat prematurely, as the ultimate weapon. Thus TBM-3Ws and TBM-3W-2s carried APS-20 radar in a large underbelly radome and, in US Navy service, were paired eventually with TBM-3S and TBM-3S-2 strike aircraft to create submarine hunter-killer teams. Other postwar conversions included TBM-3Ns for night and all-weather operation, TBM-3Qs for electronic countermeasures against enemy radar emitters, TBM-3R seven-seat Avengers for COD (Carrier On-board Delivery) of priority personnel or urgently needed supplies, and target-towing TBM-3Us.

In Royal Navy service the wartime Avengers remained operational in only small numbers after the end of World War II, serving last with No. 828 Squadron until 3 June 1946, but even that did not end the Royal Navy's operational use of Avengers. As in the USA, there was a growing awareness of the threat from new-generation submarines, leading to the acquisition from 1953 of TBM-3Es which entered service first with Nos. 815 and 824 Squadrons under the designation Avenger AS.4. These aircraft, supplied under the Mutual Defense Aid Program (MDAP) were followed by later examples, modified more specifically to Royal Navy requirements, which entered service as Avenger AS.4s or AS.5s, these remaining in first-line service until replaced by Fairey Gannets in 1955.

Also under the MDAP, Avenger variants were supplied after the war for service with the French Aéronavale, Japanese Maritime Self-Defense Force, Royal Canadian Navy, and the Royal Netherlands Navy.

Specification
Type: three-seat carrier-based torpedo-bomber
Powerplant (TBM-3): one 1,750-hp (1 305-kW) Wright R-2600-20 Cyclone 14 radial piston engine
Performance: maximum speed 267 mph (430 km/h) at 15,000 ft (4 570 m); cruising speed 147 mph (237 km/h); service ceiling 23,400 ft (7 130 m); range 1,130 miles (1 819 km)
Weights: empty 10,700 lb (4 853 kg); maximum take-off 18,250 lb (8 278 kg)
Dimensions: span 54 ft 2 in (16.51 m); length 40 ft 0 in (12.19 m); height 16 ft 5 in (5.00 m); wing area 490 sq ft (45.52 m²)
Armament: two forward-firing 0.50-in (12.7-mm) machine-guns, one 0.50-in (12.7-mm) gun in dorsal turret and one 0.30-in (7.62-mm) gun in ventral position, plus up to 2,000 lb (907 kg) of weapons in bomb bay, and provision for rocket projectiles, drop tanks or radar pod to be carried externally
Operators: RN, RNZAF, USN

Hall PH

History and Notes
With a distinguished pedigree, the small number of flying boats built by Hall during the 1930s were to represent the end of an era when withdrawn from active service during World War II, for they were the last of the US Navy's biplane flying boats. Their origin ran back in a direct line to the Curtiss *America* of 1914, the hull design of which owed much to Britain's John C. Porte who, subsequently, as a squadron commander in the Royal Naval Air Service, was responsible for further development of the Curtiss H type into the famous Felixstowe F series of flying boats of World War I.

Their evolution continued in the USA during the postwar years and in 1927 the Hall Aluminum Company (later Hall Aluminum Aircraft Corporation) was established to develop for the US Navy the prototype of a new flying boat based on the hull design of the Felixstowe F series. The XPH-1 prototype, ordered on 29 December 1927, appeared in classic flying boat form, with an elegant stepped hull, strutted and

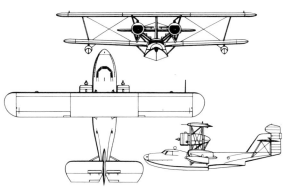

Hall PH-3

braced biplane wings with the engines mounted in nacelles between the centre-sections of the wings, wingtip stabilising floats and a well-strutted and wire-braced tail unit.

First flown in December 1929, the prototype was

Hall PH

powered by two 535-hp (399-kW) Wright GR-1750 Cyclone 9 radial engines, and armament comprised four 0.30-in (7.62-mm) Lewis guns on ring mounts, two each in open bow and dorsal positions. Accommodation for the pilot and co-pilot of the prototype was side-by-side in an open cockpit, but the nine PH-1s ordered on 10 June 1930 had a rudimentary cockpit enclosure and 620-hp (462-kW) Wright R-1820-86 Cyclone engines. Subsequent orders covered five PH-2s for the US Coast Guard with the gun positions deleted, the cockpit more effectively enclosed, and with more powerful Wright R-1820-F51 engines. The final order was for seven generally similar PH-3s, with aerodynamic improvements which included a redesigned cockpit enclosure; delivered during 1939, these also were for service with the US Coast Guard. Soon after the USA became involved in World War II, the PH-3s were re-armed and used by the Coast Guard for anti-submarine patrol, search and rescue duties during the early war years, until replaced by more effective monoplane flying boats.

Specification

Type: six-seat patrol, and search and rescue flying boat
Powerplant (PH-3): two 750-hp (559-kW) Wright R-1820-F51 Cyclone 9 radial piston engines
Performance: maximum speed 159 mph (256 km/h); at 3,200 ft (975 m); cruising speed 138 mph (222 km/h); service ceiling 21,350 ft (6 505 m); range 1,937 miles (3 117 km)
Weights: empty 9,614 lb (4 361 kg); maximum take-off 17,679 lb (8 019 kg)
Dimensions: span 72 ft 10 in (22.20 m); length 51 ft 0 in (15.54 m); height 19 ft 10 in (6.05 m); wing area 1,170 sq ft (108.69 m²)
Armament: up to 1,000 lb (454 kg) of depth bombs
Operators: USCG, USN

Howard DGA-15

History and Notes

Ben Howard, who had designed and built his first aeroplane, identified as the DGA-1 (Damned Good Airplane 1) in 1923, established the Howard Aircraft Corporation on 1 January 1937. A series of successful designs linked those years, and his DGA-6 *Mister Mulligan* of 1934, a four-seat cabin monoplane, won all three of the major American air races during the following year. Direct developments of *Mister Mulligan* led to the DGA-15 which, in 1941, was ordered by the US Navy for service in a general utility role under the designation GH-1.

Of mixed construction, the GH-1 was a braced high-wing monoplane: the wing structure was of wood, with plywood and fabric skins, the fuselage and tail unit were frameworks of welded steel tube with light alloy and/or fabric covering. Landing gear was of the fixed tailwheel type, with low-pressure main wheels enclosed in streamlined fairings. The cabin seated four (in two pairs), and full blind-flying instrumentation and radio equipment was standard. The powerplant comprised a Pratt & Whitney R-985 Wasp Junior engine, driving a two-blade constant-speed propeller.

The GH-1s, which began to enter service with the US Navy in late 1941, were used primarily as utility transports, but with availability of larger numbers following the receipt of 131 GH-2s and 115 GH-3s, a number were used in an air ambulance role and named Nightingale. Final production for the US Navy covered 205 NH-1 instrument trainers.

Following the USA's entry into World War II, the USAAF commandeered 20 civil DGA-15s for deployment in a light transport/communications role under the designation UC-70. These included 11 UC-70 (five seat/R-985-33 engine); two UC-70A (four-seat/Jacobs R-915-1 engine); four UC-70B (five-seat/R-915-10) one UC-70C (four-seat/Wright R-760-1 engine); and two UC-70D (four-seat/Jacobs R-830-1 engine). These reliable and rugged aircraft were to remain in both US Army and US Navy service for several years.

Specification

Type: four/five-seat cabin monoplane
Powerplant (GH-1): one 450-hp (336-kW) Pratt & Whitney R-985 Wasp Junior radial piston engine
Performance: maximum speed 201 mph (323 km/h) at 6,000 ft (1 830 m); cruising speed 191 mph (307 km/h) at 12,000 ft (3 660 m); service ceiling 21,500 ft (6 555 m); range 1,260 miles (2 028 km)
Weights: empty 2,700 lb (1 225 kg); maximum take-off 4,350 lb (1 973 kg)
Dimensions: span 38 ft 0 in (11.58 m); length 25 ft 8 in (7.82 m); height 8 ft 5 in (2.57 m); wing area 210 sq ft (19.51 m²)
Armament: none
Operators: USAAF, USN

The US Navy received some 115 examples of the Howard GH-3 for use in the light utility role.

Interstate L-6 Grasshopper

History and Notes

In 1940 the Interstate Aircraft and Engineering Corporation of El Segundo, California, entered the expanding market for two-seat high wing monoplanes exemplified by the Piper Cub and produced the Model S-1 Cadet which was offered with a choice of four engines: a Continental of 65 hp (48 kW), or Franklin engines of 65, 85 or 90 hp (48, 63 or 67 kW). These initial variants all bore the designation S-1A with an appropriate suffix, respectively -65C, -65F, -85F or -90F.

Interstate, which produced bomb shackles, hydraulic systems and other items for military aircraft, saw that a military order for the S-1 could help to launch the design into mass production and produced the S-1B with the transparent canopy extended aft to assist in the military liaison and air observation post roles. A 102-hp (76-kW) Franklin O-200-5 four-cylinder air-cooled engine was fitted, and the prototype selected for US Army evaluation was designated XO-63 (it was the last aircraft to receive an Observation designation) although this was later changed to XL-6.

As a result of successful evaluation, the US Army ordered a production batch of 250, to be designated L-6 Grasshopper (the name was common to the aircraft used in this role, the others being the Aeronca L-3 and L-16, Piper L-4 and Taylorcraft L-2). Although produced for the US Army in considerably smaller numbers than its contemporaries, the L-6 was also built for civil market; a number of these plus a few ex-military models are still flying in the USA and Canada.

Specification

Type: two-seat observation and liaison aircraft
Powerplant: one 102-hp (76-kW) Franklin O-200-5 inline piston engine
Performance: maximum speed 114 mph (183 km/h); cruising speed 105 mph (169 km/h); service ceiling 16,500 ft (5 030 m); range 540 miles (869 km)
Weights: empty 1,103 lb (500 kg); maximum take-off 1,650 lb (748 kg)
Dimensions: span 35 ft 6 in (10.82 m); length 23 ft 5½ in (7.15 m); height 7 ft 0 in (2.13 m); wing area 173.8 sq ft (16.15 m²)
Armament: none
Operator: USAAC/USAAF

The Interstate L-6 was the least prolific of the civil lightplanes procured by the USAAC for liaison.

Laister-Kauffman TG-4A

History and Notes

In both the UK and the USA, planning staffs had been alerted by the successful deployment of airborne troops in early operations in Europe, and had appreciated the need to acquire suitable gliders both for training and operational use.

In 1941 the US Army began the procurement of several training gliders, emphasising to the suppliers the extreme urgency of the requirement, and selecting companies with experience in the construction of aircraft of this type. As a result of this, Laister-Kauffman were contracted to supply three trainers, based on their Yankee Doodle civil design and allocated the designation XTG-4. They were of mixed construction, with the centre-section of the mid-set cantilever monoplane wing of welded steel tube, and built integrally with the fuselage which was similarly constructed. The outer wing panels and tail unit were all-wood, and covering was a mixture of fabric or fabric over a light wooden skin. Spoilers were mounted on the upper surface of the wing, inboard of the ailerons, and were deployed by operation of the wheel brake of the monowheel mounted beneath the fuselage centre-line. The spoilers could be operated in the air, to steepen the angle of glide, and on the landing run to act as lift dumpers. Accommodation was provided for instructor and pupil in tandem cockpits, one fore and one aft of the centre-section spar, with a long transparent canopy over both.

Successful performance of these prototypes on test resulted in a contract being awarded for 150 production TG-4As, and these were all delivered during 1942. They differed from the prototype only by having radio to allow for communication between the glider and the

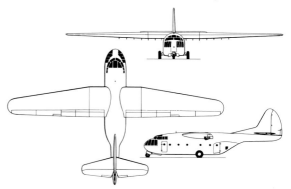

Some 150 production examples of the Laister-Kauffman TG-4 were used as trainers by the USAAF, with two prototypes and one ex civil glider.

Laister-Kauffman TG-4A

towing aircraft. A single example of a civil Laister-Kauffman LK-10 was commandeered by the US Army and designated TG-4B. Used quite extensively for training in the early stages of the war, all had however been declared surplus some time before VJ-Day.

Specification
Type: two-seat training glider
Powerplant: none

Performance: no data available
Weights: empty 475 lb (215 kg); maximum take-off 875 lb (397 kg)
Dimensions: span 50 ft 0 in (15.24 m); length 21 ft 3 in (6.48 m); height 4 ft 0 in (1.22 m); wing area 166 sq ft (15.42 m²)
Armament: none
Operator: USAAF

Lockheed Model 10-A Electra

History and Notes
Lockheed's reputation was founded on its outstanding series of fast single-engined Vega, Air Express, Sirius, Altair and Orion monoplanes, but the impact of twin-engined designs, such as Boeing's Model 247 and the Douglas DC-1, forced the Burbank, California-based Lockheed Aircraft Corporation to follow suit. Work on such a design began at the end of 1932, and the prototype Model 10 Electra was rolled out on 23 February 1934, making its first flight later that day. Flight testing occupied almost three months, at the end of which the Bureau of Air Commerce awarded the Model 10 Approved Type Certificate No. 551 on 11 August 1934.

Powered by two Pratt & Whitney Wasp Junior SB engines, and accommodating 10 passengers and a crew of two, the all-metal L 10-A Electra aroused the interest of Northwest Airlines and Pan American Airways before the prototype had been built. Both ordered the Electra, and Northwest received the first two of its order, for training and route-proving, before certification.

Pan American had held in store a number of surplus Wasp SC engines and these were fitted to the airline's eight L 10-Cs. Eastern Airlines' selection of the 450-hp (336-kW) Wright R-975E-3 Whirlwind radial produced the L 10-B, while Electras with the 550-hp (410-kW) Wasp R-1340-S3H1 engine were designated L 10-E. Total production of 149 aircraft comprised 107 10-As, 19 10-Bs, 8 10-Cs and 15 10-Es.

In 1936 two Electras were supplied to the US Navy and US Coast Guard, designated XR2O-1 and XR3O-1 respectively. The former was a Model 10-A for the use of the Secretary of the Navy and the latter a Model 10-B used by the Secretary of the Treasury. Four Lockheed 10-As were also purchased by the US Army Air Corps; three were delivered in 1936 under the designation Y1C-36 and the fourth, in 1937, as a Y1C-37 for the National Guard Bureau. In 1942 26 civil Electras were impressed for communications duties, comprising 15 10-As which became designated UC-36A, four 10-Es (UC-36B) and seven 10-Bs (UC-36C).

The Royal Air Force similarly received eight impressed aircraft and the Royal Canadian Air Force 15 machines, while other military users included the Argentine army and navy, the Brazilian air force and the Honduran air force.

Of particular importance was the single XC-35 which, powered by two 550-hp (410-kW) Pratt & Whitney XR-1340-43 engines, was first flown on 7 May 1937. Built to USAAC order for experiments with cabin pressurisation and engine supercharging, this had a redesigned fuselage of circular cross-section and with standard cabin windows eliminated. For the sponsorship of this development, and its use in what proved to be a valuable research programme, the US Army was later awarded the Collier Trophy.

Specification
Type: twin-engined light transport aircraft
Powerplant (Y1C-36): two 450-hp (336-kW) Pratt & Whitney R-985-13 Wasp Junior radial piston engines
Performance: maximum speed 201 mph (323 km/h) at 5,000 ft (1 525 m); cruising speed 194 mph (312 km/h) at 5,000 ft (1 525 m); service ceiling 24,000 ft (7 315 m); range 666 miles (1 072 km)
Weight: empty 7,100 lb (3 221 kg); maximum take-off 10,500 lb (4 763 kg)
Dimensions: span 55 ft 0 in (16.76 m); length 38 ft 7 in (11.76 m); height 10 ft 1 in (3.07 m); wing area 458.3 sq ft (42.58 m²)
Armament: none
Operators: Argentina, Brasil, Honduras, RAF, RCAF, USAAC/USAAF, USCG, USN

The Lockheed XR2O-1 was a sole example of the civil Model 10-A procured by the Navy in 1936 as the personal transport of the Secretary of the Navy.

Lockheed Model 12-A

History and Notes
The improved financial stability that the Model 10 Electra had conferred upon Lockheed enabled the company to embark upon a new project in 1935. This evolved as another twin-engined design, outwardly similar to the Electra but of slightly smaller dimensions. Identified as the Model 12, it was one of the first aircraft intended for business use, seating six passengers in a well-furnished cabin.

From the start the company was committed to a first flight target date of 27 June 1936, to enable the type to compete in the 1936 Department of Commerce design competition. That target was achieved, test pilot Marshall Headle taking off at 12.12 on 27 June, and the aircraft subsequently won the competition. The Electra's Pratt & Whitney Wasp Juniors were retained, endowing the Model 12-A with a maximum speed of 240 mph (386 km/h), an improvement of 30 mph (48 km/h) compared with the performance of the earlier model. The 12-A was certificated on 14 October 1936, finding a ready market with both corporate and private owners, and with the airlines; 114 had been built when production ceased in 1942.

In 1937-8 the US services acquired Lockheed 12-As for transport duties, initially in seven-seat configuration. The US Navy took a single JO-1 and two JO-2s, three additional examples of the latter serving with the US Marine Corps. Three for the US Army Air Corps were designated C-40 and 10 five-passenger machines were known as C-40As. Impressed civil aircraft taken on charge in 1942 became UC-40Ds. The US Navy and US Air Corps each operated a single 12-A trainer with fixed tricycle type landing gear, designated respectively XJO-3 and C-40B.

An armed version was evolved in 1938 and the prototype flew in the following February, featuring a 0.50-in (12.7-mm) machine-gun in the nose, another in a dorsal turret, and with underfuselage racks for up to eight 250-lb (113-kg) bombs. Sixteen were ordered by the Royal Netherlands Indies Army's Air Division for crew training duties in support of the division's Martin 139 bombers, and to form a Java-based maritime patrol squadron. Thirteen Lockheed 12-A transports had also been ordered, and eight of these were still undelivered when the Dutch East Indies fell into Japanese hands. Together with four similarly undelivered aircraft intended for KNILM (Royal Dutch East Indies Airways), they were used instead by the Netherlands Military Flying School at Jackson, Mississippi.

Specification
Type: six/seven-seat commercial and military light transport, and gunnery trainer
Powerplant: two 450-hp (336-kW) Pratt & Whitney R-985-SB Wasp Junior radial piston engines
Performance: maximum speed 225 mph (362 km/h); cruising speed 212 mph (341 km/h); service ceiling 22,900 ft (6 980 m); range 800 miles (1,287 km)
Weights: empty 5,960 lb (2 703 kg); maximum take-off 9,200 lb (4 173 kg)
Dimensions: span 49 ft 6 in (15.09 m); length 36 ft 4 in (11.07 m); height 9 ft 9 in (2.97 m); wing area 352 sq ft (32.70 m²)
Armament (military trainer): two 0.50-in (12.7-mm) machine-guns (one in nose and one in dorsal turret), plus up to 2,000 lb (907 kg) of bombs carried externally
Operators: Netherlands, RAF, RCAF, USAAC/USAAF, USMC, USN

USAAC orders for the Lockheed Model 12-A totalled three C-40s and 10 C-40As (later UC-40s and UC-40As).

Lockheed Model 18 Lodestar

History and Notes
Last in the line of Lockheed twin-engined commercial transports, the Model 18 Lodestar was developed from the Model 14 and the prototype, flown for the first time on 21 September 1939, was a conversion from a standard Lockheed 14-H2. With a longer fuselage seating 14 passengers and with a crew of three, the Lodestar was built in a number of versions, principally the Models 18-07 with 750-hp (559-kW) Pratt & Whitney S1E-3G Hornet engines, 18-08 with 900-hp (671-kW) Pratt & Whitney SC-3G Twin Wasps, 18-14 with 1,050-hp (783-kW) S4C-4G Twin Wasps, 18-40 with 900-hp (671-kW) GR-1820-G102A Wright Cyclones, 18-50 with 1,000-hp (746-kW) GR-1820-G202A Cyclones, and 18-56 with similarly powered GR-1820-G205A Cyclones.

US military interest in the Lodestar was first shown in 1940, when the US Navy ordered a single XR5O-1 and two R5O-1 command transports, a similar aircraft being delivered to the US Coast Guard. These were powered by Wright R-1870 engines, as were 12 R5O-4s, 41 R5O-5s and 35 R5O-6s. These last three models were, respectively, four- to seven-seat executive transports, 12/14-seat personnel transports, and 18-seat

Lockheed Model 18 Lodestar

troop carriers used by the US Marine Corps for paratroop operations. Pratt & Whitney R-1830 engines powered the single R5O-2 and three R5O-3s built for the US Navy.

The US Army Air Corps ordered a single Wright R-1820-29-engined aircraft in May 1941, under the designation C-56, this being the military version of the civil Model 18-50. At the same time three Model 18-14s were ordered, with Pratt & Whitney R-1830-53 engines. These were supplemented by later orders for seven and three aircraft respectively, all 13 machines being designated C-57.

An impressed civil aircraft was designated C-57A, seven troop carriers were known as C-57Bs and three of the later model C-60As were re-engined with Pratt & Whitney R-1830-43 radials to become C-57Cs; one of these three became a C-57D with R-1830-92 engines.

The greater number of aircraft taken over from the US internal airlines from December 1941, however, were given designations in the C-56 series, comprising one C-56A, 13 C-56Bs, 12 C-56Cs, seven C-56Ds and two C-56Es.

Ten Model 18-07s and 15 Model 18-56s were acquired under the Defense Aid programme as C-59s and C-60s respectively, later supplemented by another 21 C-60s and 325 C-60As, one of the latter becoming a CB-60B with an experimental hot-air de-icing system. A single Model 18-10, with 1,200-hp (895-kW) R-1830-53 engines and seats for 11 passengers, was purchased in 1942 and designated C-66.

Royal Air Force Lodestars were used principally in the Middle East and Near East, serving with Nos. 162, 173, 216 and 267 Squadrons in the air-ambulance and general-transport roles. Some had been acquired from BOAC, to be designated Lodestar I, while 10 C-59s and 15 C-60s were received and operated as Lodestar IAs and IIs.

Specification

Type: 17-seat personnel and cargo transport
Powerplant (C-56): two 1,200-hp (895-kW) Wright R-1820-71 radial piston engines
Performance: maximum speed 253 mph (407 km/h); cruising speed 200 mph (322 km/h); service ceiling 23,300 ft (7 100 m); range 1,600 miles (2 575 km)
Weights: empty 11,650 lb (5 284 kg); maximum take-off 17,500 lb (7 938 kg)
Dimensions: span 65 ft 6 in (19.96 m); length 49 ft 10 in (15.19 m); height 11 ft 1 in (3.38 m); wing area 550 sq ft (51.10 m²)
Armament: none
Operators: Netherlands, RAAF, RAF, SAAF, USAAC/USAAF, USMC, USN

The elegant, high-performance Lockheed Model 18 soon commended itself to the US forces as a fast transport, but it was only after the start of hostilities that large USAAF orders were placed for variants designated C-56, C-57, C-59 and C-60.

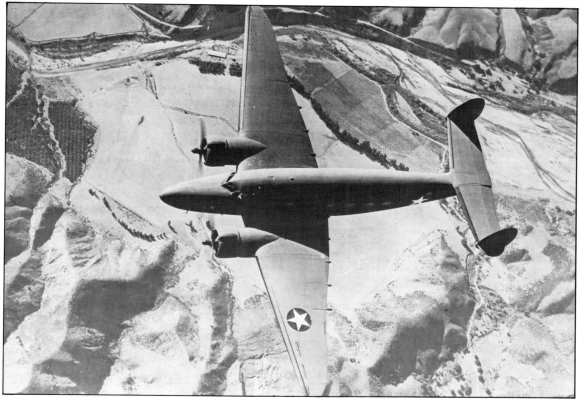

Lockheed A-28/A-29 Hudson

History and Notes

The first American-built aircraft to be used operationally by the RAF during World War II, the Lockheed Hudson was a design which stemmed from the urgent British requirement for a maritime patrol/navigational trainer aircraft. The initial procurement of 200 of these aircraft by the British Purchasing Commission in the summer of 1938 caused a great deal of critical and caustic comment, for those were the days when 'Buy British' was regarded as general, rather than unusual. Nonetheless, the Hudsons were to prove important reinforcements to the nation's somewhat limited air power when World War II began on 1 September 1939.

Faced with the problem of producing these aircraft as quickly as possible for British requirements, Lockheed offered a militarised version of the Lockheed 14 Super Electra civil airliner, duly identified by Lockheed as their Model 414. While its general configuration differed little from that of the civil airliner, there were a number of specific changes to provide suitable armament, plus the installation of more powerful engines, and within a few days of the approach from the purchasing commission it was possible for Lockheed to demonstrate a hastily-prepared mock up of the proposed fuselage. With approval accorded, production went ahead to such good effect that the first of these aircraft, to which the RAF had given the name Hudson, was able to record its initial flight on 10 December 1938.

Of all-metal construction, the wings were set in a mid position and featured a Lockheed-modified version of the Fowler type trailing-edge flap. The fuselage was conventional, but included a bomb bay to provide internal stowage for up to 1,400 lb (635 kg) of bombs, a new bomb-aimer's position in the fuselage nose, and provision for two forward-firing machine-guns mounted in the upper fuselage forward of the pilot, plus a power-operated dorsal turret sited well aft, near the tail unit which had twin vertical fins and rudders. Landing gear was of the tailwheel type, with the main units retracting aft to be housed in the engine nacelles. The powerplant, wing-mounted in long nacelles, comprised two 1,100-hp (820-kW) Wright GR-1820-G102A Cyclone engines, each driving a three-blade variable-pitch propeller. Standard accommodation was for a crew of four, comprising pilot, navigator, bomb-aimer and radio operator/air gunner.

The first Hudsons to reach the UK were delivered by sea to Liverpool, arriving there on 15 February 1939, and Lockheed were to establish a base at Speke Airport, near Liverpool, for the assembly of Hudsons which soon began to arrive in ever increasing numbers. During assembly the Boulton Paul two-gun power-operated turret was installed. These Hudson Is began to equip Nos. 224 and 233 Squadrons in the summer of 1939, and a third squadron (No. 220) was in the process of converting to the type when war broke out. Almost immediately they were deployed on maritime patrol, and as ever increasing numbers were received their service use was extended to make them a regular sight over the Atlantic, as well as in the Middle and Far East.

The initial order was soon supplemented by others for an additional 150 Hudson Is, 20 Hudson IIs which introduced constant-speed propellers and accommodation for a crew of five, and 414 Hudson IIIs with more powerful 1,200-hp (895-kW) Wright GR-1820-G205A engines, and one ventral and two beam gun positions. This last version was the first to be delivered by air across the Atlantic, the first arriving in Northern Ireland on 11 November 1940; subsequent deliveries were all made by air. Procurement for the RAF and RAAF amounted to something like 1,500 aircraft

Lend-Lease supplies of the Lockheed Hudson for the RAF were contracted under the USAAF designations A-28 and A-29. Some of the latter were repossessed, and with dorsal turrets removed served as trainers.

Lockheed A-29 (repossessed Hudson Mk III) of the USAAF, based at Portland (Oregon) for West Coast anti-submarine patrol in April 1942.

Lockheed Hudson Mk III of No. 279 Squadron, RAF, based at Sturgate (UK) for air-sea rescue operations (with droppable ventral lifeboat) during 1942.

Lockheed Hudson Mk V of No. 48 Squadron, RAF Coastal Command, based at Stornoway in the Outer Hebrides islands (UK) in late 1941.

before introduction of the Lend-Lease programme, and subsequent requirements were procured by the USAAF for supply to Britain.

The first USAAF order was for 52 A-28s, all supplied to Britain, as were the 450 A-28As which followed, and these were designated Hudson IV and VI respectively by the RAF. The A-28s had 1,050-hp (783-kW) Pratt & Whitney R-1830-45 Twin Wasp engines, but the A-28As not only had 1,200-hp (895-kW) R-1830-67 Twin Wasps, but also had convertible interiors for alternative use as troop transports. The Hudson V was the last variant supplied to the RAF before the inception of Lend-Lease, and this (309 total) had 1,200-hp (895-kW) Twin Wasps. One hundred similar aircraft went to the Royal Australian Air Force, which designated

them Hudson I, and these had the 1,050-hp (783-kW) Twin Wasp engines. Other versions procured by the USAAF included 416 A-29s with 1,200-hp (895-kW) Wright R-1820-87 engines and 384 A-29As with the same engines, the latter having the convertible interiors of the A-28A. All were allocated to the RAF but only 382 were received by that service and designated Hudson IIIA. The balance were impressed for service by the USAAF for bomber crew training and anti-submarine patrol, and 20 went to the US Navy which designated them PBO-1. Some 24 of the US Army's A-29s were converted in 1942 for use in a photo-reconnaissance role, and were accordingly redesignated A-29B. The designation C-63 was allocated by the USAAF for a projected cargo variant, but this

Lockheed A-28/A-29 Hudson

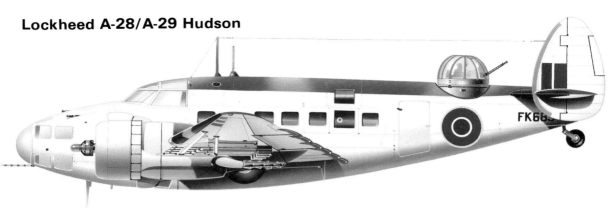

Lockheed Hudson Mk VI of Coastal Command, RAF, in the markings adopted in August 1941.

was cancelled before any were built. The final variant procured by the USAAF was required as a trainer for air gunners, or as a target tug. These were generally similar to A-29As and were equipped with a Martin dorsal turret. Lockheed built 217 of them as AT-18s, followed by 83 AT-18As, as navigation trainers, with different internal equipment and the turret deleted.

Despite its derivation from a peaceful civil airliner, the Hudson was to achieve some surprising 'firsts'. A Hudson of No. 224 Squadron, for example, was the first RAF aircraft operating from Great Britain to destroy an enemy aircraft on 8 October 1939; a Hudson of No. 220 Squadron located and directed naval forces to the German prison ship *Altmark* in February 1940; one from No. 269 Squadron accepted the surrender of German submarine *U-570* in the Atlantic on 27 August 1941; No. 280 Squadron was the first to be equipped with airborne lifeboats and deployed the first in the North Sea in early May 1943; and in the same month a Hudson of No. 608 Squadron became the first aircraft in RAF service to sink an enemy submarine by rocket fire. One of the USAAF's A-29s was the first US Army aircraft to sink a German U-boat in World War II, and the PBO-1s of the US Navy's VP-82 Squadron based at Ventia, Newfoundland, sank the first two U-boats

credited to that service in World War II on 1 and 15 March 1942. Well over 2,500 were built, and many remained in service in a secondary role until the end of the war.

Specification
Type: twin-engined maritime patrol-bomber
Powerplant (A-29): two 1,200-hp (895-kW) Wright R-1820-87 Cyclone 9 radial piston engines
Performance: maximum speed 253 mph (407 km/h) at 15,000 ft (4 570 m); cruising speed 205 mph (330 km/h); service ceiling 26,500 ft (8 075 m); range 1,550 miles (2 494 km)
Weights: empty 12,825 lb (5 817 kg); maximum take-off 20,500 lb (9 299 kg)
Dimensions: span 65 ft 6 in (19.96 m); length 44 ft 4 in (13.51 m); height 11 ft 11 in (3.63 m); wing area 551 sq ft (51.19 m²)
Armament: two 0.30-in (7.62-mm) machine-guns in fuselage nose, one 0.30-in (7.62-mm) gun in ventral position and two 0.30-in (7.62-mm) guns in optional dorsal turret, plus up to 1,600 lb (726 kg) bombs
Operators: Netherlands, RAF, RAAF, RCAF, RNZAF, SAAF, USAAF, USN

Lockheed B-34/PV Ventura/Harpoon

History and Notes
The early success of its Hudsons in operational service with the RAF induced Lockheed to initiate the design of a more advanced version. This was based on the company's somewhat larger Model 18 Lodestar civil airliner, but the resulting aircraft was very similar in appearance to the Hudson. The design was duly investigated and approved by the British Purchasing Commission, and 675 were ordered from Lockheed. Produced by the company's Vega division, the first of these flew on 31 July 1941. Its design had benefitted from the company's manufacturing and British operational experience with the Hudson, so there were no major problems discovered during the flight test programme, and the first examples began to enter service with the RAF's No. 21 Squadron in October 1942.

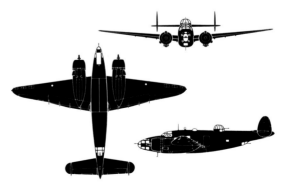

Lockheed B-34A/PV-1 Ventura

Lockheed B-34 PV Ventura/Harpoon

Designated Ventura I, the new aircraft differed from the Hudson in having a wider and deeper fuselage with a length increase of 16 per cent, more powerful Pratt & Whitney engines each rated at 2,000 hp (1 491 kW), the introduction of a proper ventral gun position with two 0.303-in (7.7-mm) machine-guns and, because of the more voluminous fuselage and greater engine power, a bomb bay able to accommodate a maximum bomb load of 2,500 lb (1 134 kg). Two fuel drop tanks, two 500-lb (227-kg) bombs or two depth charges could be carried beneath the wings, outboard of the engine nacelles, and the bomb bay was long enough to contain a standard short 22-in (559-mm) torpedo. After 188 Ventura Is had been delivered, changes in armament and the installation of R-2800-31 engines brought redesignation as Ventura II.

Venturas were first used operationally by RAF Bomber Command on 3 November 1942, and during this and subsequent daylight raids the new bombers were not considered to be well suited to such a role, although an additional 200 had been ordered under Lend-Lease as Ventura IIA by this time. Consequently, those then in service were transferred for operation with Coastal Command under the designation Ventura GR.I and outstanding orders were cancelled. Thus only about 300 of the total on order were delivered.

When the Ventura was included in the Lend-Lease programme it was procured initially for the RAF under the designation B-34, but with the cancellation of British orders the outstanding aircraft began to enter USAAF service initially as R-37s, based on the manufacturer's model number, and later as B-34A Lexingtons, and these were used mainly for maritime patrol. In addition a small number were procured as navigational trainers, and these were allocated the designation B-34B. In 1942 the US Army ordered 550 examples of a new version with 1,700-hp (1 268-kW) Wright R-2600-31 Cyclone 14 engines, for use in a patrol/reconnaissance role under the designation O-56, but only 18 of these aircraft had been received, becoming instead B-37, before the balance of 532 on order were cancelled.

The two cancellations would suggest that there was no further use for Lockheed's Model 37, but this was not the case, for towards the end of 1942 the US Navy requisitioned 27 of the Lend-Lease B-34s then still in the process of supply to the UK. These entered service with the US Navy under the designation PV-3, for training and familiarisation, pending the receipt of the first quantities of a larger batch of aircraft then in production for the USAAF as B-34s, and which the US Navy had designated PV-1. Following the RAF and USAAF cancellations of B-34s and O-56s respectively, all subsequent production was transferred to the US Navy, and it was this service which continued also to procure Lend-Lease requirements.

The first of the US Navy's PV-1s were delivered in December 1942, these entering service first with US Navy Squadron VP-82 and replacing the PBO-1s (Hudsons) which had been requisitioned from Lend-Lease production for the UK in the autumn of 1941. Approximately 1,600 PV-1s were procured by the US Navy, of which 388 were supplied to the UK under Lend-Lease. These were designated Ventura GR.Vs, and the majority served with the Commonwealth air forces in Australia, Canada, New Zealand and South Africa. Some of the US Navy's PV-1s were modified subsequently to serve in a reconnaissance role, under the designation PV-1P.

Designed as an improved version of the Hudson, the Lockheed PV-1 Ventura was based on the Model 18.

Lockheed B-34/PV Ventura/Harpoon

On 30 June 1943 the US Navy ordered a new version under the changed designation PV-2 and with the name Harpoon. While retaining the same general appearance as the earlier Model 37, it differed in several respects. General configuration and powerplant was unchanged, but the wing span was increased by 9 ft 5 in (2.87 m), this giving a wing area of 686 sq ft (63.73 m²). Other changes included increased fuel capacity, greater fin and rudder area, and much improved armament. In the basic PV-2 this consisted of five 0.50-in (12.7-mm) forward-firing machine-guns in the fuselage nose and two flexibly-mounted 0.50-in (12.7-mm) guns in both dorsal turret and ventral position, plus up to four 1,000-lb (454-kg) bombs carried in the bomb bay, and two similar bombs carried externally. The final production version, of which 33 were built, had the designation PV-2D and in these aircraft the nose armament was increased to a total of eight 0.50-in (12.7-mm) machine-guns.

Orders for the PV-2 totalled 500, and initial delivery of these to US Navy squadrons began in March 1944. One of the aims of the increased wing span of this version was to provide considerably increased fuel capacity by use of the wing structure to form integral tanks, but great difficulty was experienced in making these fuel-tight. The first 30 aircraft were withdrawn from service and the integral tanks in the outer wings sealed off: the aircraft were then used in a training role with the designation PV-2C. The problem of the leaking outer tanks was beyond solution at that time, and all of the 470 production PV-2s had leak-proof fuel cells installed within the integral tanks.

The PV-2 served primarily in the Pacific theatre as a patrol-bomber, until VJ-Day brought its withdrawal from front-line service. However, operated by US Navy Reserve units, the type remained in use for several years after the war.

Specification
Type: four/five-seat medium bomber
Powerplant (B-34A): two 2,000-hp (1 491-kW) Pratt & Whitney R-2800-31 Double Wasp radial piston engines
Performance: maximum speed 315 mph (507 km/h) at 15,500 ft (4 725 m); cruising speed 230 mph (370 km/h); service ceiling 24,000 ft (7 315 m); range 950 miles (1 529 km)
Weights: empty 17,275 lb (7 836 kg); maximum take-off 27,250 lb (12 360 kg)
Dimensions: span 65 ft 6 in (19.96 m); length 51 ft 5 in (15.67 m); height 11 ft 11 in (3.63 m); wing area 551 sq ft (51.19 m²)
Armament: two forward-firing 0.50-in (12.7-mm) machine-guns in fuselage nose and six 0.30-in (7.62-mm) guns on flexible mounts in nose, dorsal and ventral position, plus up to 2,500 lb (1 134 kg) of bombs in bomb bay
Operators: RAAF, RAF, RCAF, RNZAF, SAAF, USAAF, USN

Lockheed C-69 Constellation

History and Notes
Lockheed's C-69 Constellation military transport was destined to become perhaps the most elegant and certainly one of the most successful airliners ever produced by the US industry. Indeed, the aircraft was designed originally to meet a Transcontinental and Western Airlines (TWA) specification for a pressurised airliner capable of flying the US transcontinental routes without a stop, carrying a 6,000-lb (2 722-kg) payload over a range of 3,500 miles (5 633 km) and cruising at up to 300 mph (483 km/h) at an altitude of 20,000 ft (6 095 m).

The Constellation was launched with an initial TWA order for nine aircraft and construction of the prototype was initiated in 1940. Completed in December, the first Model 49 was quickly readied for its maiden flight which took place on 9 January 1943, flown by Lockheed chief test pilot Miro Burcham and his Boeing opposite number Eddie Allen. Powered by four 2,200-hp (1 641-kW) Wright R-3350-35 Cyclone 18 two-row radials, the prototype lifted from the runway at Burbank and, after an hour in the air, landed at Muroc Field (now known as Edwards Air Force Base). The test programme quickly established that design performance had been achieved, with a maximum speed of 347 mph (558 km/h), a cruising speed of 275 mph (443 km/h) achieved on 52.5 per cent power at 1 US gallon per mile (2.4 litres per km) fuel consumption.

TWA's order had been increased to 40, and was followed by one from Pan American Airways for a similar number of a transoceanic version. However, both airlines waived their rights to the first batches of aircraft from the production line in favour of the USAAF, which needed fast troop transports. The

Commandeered on the production line, the first Lockheed L-49 flew as the prototype C-69 on 9 January 1943. Only 22 production examples followed before 1946.

Lockheed C-69 Constellation

prototype, despite its civil registration, had been painted with military green upper surfaces and grey undersides, and was handed over to the USAAF on 28 July 1943.

The first production C-69, although bearing its military serial, was painted in TWA livery and, at 03.56 on 17 April 1944, co-captains Howard Hughes and Jack Frye of TWA took off from Burbank on delivery to the USAAF, flying to Washington National in a new transcontinental record time of a few seconds under 6 hours 58 minutes. On 23 January 1945 TWA's Intercontinental Division was contracted to see the C-69 into regular service with USAAF Air Transport Command. During this period the first Constellation transatlantic flight took place, nonstop from New York to Paris, in a record time of 14 hours 12 minutes.

After VJ-Day the USAAF's transport requirements were drastically reduced, however, and the C-69 programme was cancelled. Twenty C-69-1-LO or -5-LO 63-seat troop transports and one C-69C 43-seat personnel transport were completed to USAAF order, although some did not enter service. With the exception of one which crashed at Topeka, Kansas on 18 September 1945 and one destroyed in static tests at

Wright Field, all C-69s were sold to civil operators and many aircraft which had been at an advanced state of construction when cancellation took place were completed as civil Model 049s. This version was awarded the US Civil Aeronautics Board Approved Type Certificate No. 763 on 11 December 1945 after only 27 hours of test flying. Total Constellation production, of all models and developments, was eventually 856.

Specification

Type: cargo and passenger transport
Powerplant: four 2,200-hp (1 641-kW) Wright R-3350-35 Cyclone 18 radial piston engines
Performance: maximum speed 330 mph (531 km/h); cruising speed 300 mph (483 km/h); service ceiling 25,000 ft (7 620 m); range 2,400 miles (3 862 km)
Weights: empty 50,500 lb (22 906 kg); maximum take-off 72,000 lb (32 659 kg)
Dimensions: span 123 ft 0 in (37.49 m); length 95 ft 2 in (29.01 m); height 23 ft 8 in (7.21 m); wing area 1,650 sq ft (153.29 m²)
Armament: none
Operator: USAAF

Lockheed P-38 Lightning

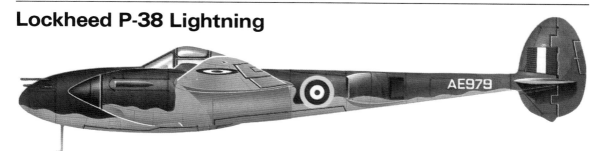

Lockheed L-322 Lightning Mk I evaluated at the Aircraft and Armament Experimental Establishment, Boscombe Down (UK) during spring 1942.

History and Notes

In early 1937 the US Army Air Corps announced a design competition which was intended to lead to the procurement of a long-range interceptor fighter, one which would have the capability of attacking enemy aircraft at a considerable distance from its base and of outperforming them at high altitude. The specification called for a maximum speed of 360 mph (579 km/h) and the ability to climb to a height of 20,000 ft (6 095 m) within six minutes. To highlight this requirement in a more easily assimilated way, a comparison with the Hawker Hurricane I and the Messerschmitt Bf 109E, which entered service in 1937 and 1938 respectively, shows the problem which confronted contenders in this competition. The Hurricane had a top speed of 324 mph (521 km/h) at optimum altitude and needed 9.8 minutes to climb to 20,000 ft (6 095 m); comparable figures for the Bf 109E were 354 mph (570 km/h) and 8 minutes. Both of these single-engined aircraft were powered by advanced inline engines, and at that time

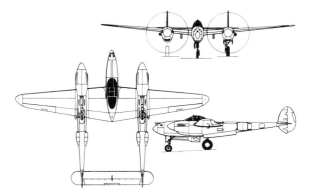

Lockheed P-38F/P-38G Lightning

no designer would have expected to better such performance using any alternative powerplant.

At the beginning of 1937, when Lockheed received

Lockheed P-38 Lightning

the USAAC's Request for Proposals, the company had no previous experience in the design and construction of military aircraft. This, at first, would seem to be a disadvantage; in this particular instance it was probably the reverse, for it enabled H. L. Hibbard and his design team to look at the project objectively, without any preconceived ideas. The result was a solution that at the time was little short of revolutionary. The design team probably began by conducting a very simple exercise like that above, confirming that no single engine then in existence could provide the speed and rate of climb required by the specification, but that two engines almost certainly would, at the same time providing a better payload margin for armament and fuel.

Having decided on twin-engined configuration (then a rather way-out proposal for a fighter), the designers were faced with the problem of how best to mount the two engines. Among the possibilities which they considered were conventional wing-mounted engines, in both tractor and pusher arrangements; engines within the fuselage driving two tractor propellers via co-axial shafts, or wing-mounted propellers via complex transmission; and a twin-boom structure between which could be mounted a central nacelle with push and pull engines. This last proposal was the most attractive, for if the tail unit was designed to link the aft end of the booms the resulting structure could be comparatively light in weight and yet immensely strong; and why not put the engines in the forward end of the booms and the pilot in a central nacelle?

This was the basic layout of Lockheed's Model 22, details of which were submitted to the USAAC in the spring of 1937. On 23 June 1937, the company was awarded a contract for the construction of a prototype under the designation XP-38, and was also required to

construct a mock-up to finalise equipment and cockpit layout to the requirements of the Army Air Corps. Following inspection and final approval of the mock-up Lockheed was able to begin detail design, but it was not until just over a year later, in July 1938, that construction of the prototype was initiated. Rolled out just after Christmas that year, with engine ground tests being made in early January, the XP-38 was flown for the first time on 27 January 1939. The powerplant of the XP-38 comprised two 960-hp (716-kW) Allison V-1710-11/15 inline engines, these being 'handed' so that the propellers rotated in opposite directions to neutralise the effect of engine torque: in this installation, when viewed from behind the port propeller turned clockwise.

Initial testing of the XP-38 progressed well, so well in fact that on 11 February 1939 the aircraft was despatched on a transcontinental flight from March Field, California to Mitchell Field, New York, a journey accomplished with two en-route refuelling stops in 7 hours 2 minutes. However, this breathtaking performance was to end in disaster, for as the result of an undershoot on the final approach to Mitchell Field the XP-38 was damaged so severely that it was a write-off. Fortunately, earlier testing and the trans-America flight had given some indication of the potential of the design, so that the USAAC had no qualms in ordering a service test batch of 13 YP-38s on 27 April 1939, and followed by the first production order for 66 P-38s on 10 August.

As a result of the early tests it was possible to

The twin booms of the Lightning, with their massive radiator baths and turbocharger installations, are well shown in this photograph of a P-38F, which had underwing payload increased to 3,200 lb (1452 kg).

Lockheed P-38 Lightning

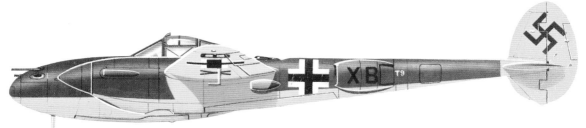

Lockheed P-38E Lightning used by Sonderkommando Rosarius for demonstration flights to Luftwaffe units during 1943-4.

Lockheed P-38F of the 347th Fighter Group, USAAF, based in New Caledonia but detached to the 13th Air Force on Guadalcanal during February 1943.

Lockheed P-38F Lightning of the Esquadrilha OK, Portuguese Air Force, in 1942.

Lockheed F-5 Lightning of the Royal Australian Air Force in 1943.

Lockheed P-38J Lightning of the 401st Fighter Squadron, 370th Fighter Group, USAAF, based at Florennes (Belgium) in November 1944.

Lockheed P-38 Lightning

incorporate certain modifications to the pre-production aircraft (Lockheed Model 122), this including the introduction of 1,150-hp (858-kW) V-1710-27/29 engines, installed with propeller rotation opposite to that of the XP-38. In addition, the nose-mounted armament was changed from the 23-mm cannon and four 0.50-in (12.7-mm) machine-guns of the prototype to one 37-mm cannon, two 0.50-in (12.7-mm) machine guns and two 0.30-in (7.62-mm) machine-guns. There was, however, a long wait before the first YP-38 flew on 16 September 1940, the aircraft being handed over to the USAAC for evaluation in early March 1941; all 13 YP-38s had been delivered by early June.

There was little doubt that Lockheed had produced a remarkable aeroplane, for early testing showed a maximum speed of 405 mph (652 km/h) at 20,000 ft (6 095 m), an altitude to which it could climb within the required 6 minutes. One major problem was discovered during service trials, namely buffeting of the tail unit, and early attempts to eliminate this by the introduction of upswept booms, to lift the tail unit above the disturbed airflow aft of the engines and wings, were completely unsuccessful. The solution lay in an adjustment of tailplane incidence and changes to elevator mass balancing, these modifications being incorporated on the last 36 aircraft of the first production order, which were designated P-38D.

The first 30 production aircraft were delivered as P-38s (Lockheed Model 222), with delivery starting in mid-1941. These were generally similar to the YP-38 but had armament much the same as that of the prototype (except that the 23-mm cannon was replaced by one of 37-mm calibre) plus the provision of armour protection for the pilot. Their basic structure was typical of the whole family of P-38s which were to follow, with production continuing throughout World War II until terminated finally by the contract cancellations after VJ-Day. Of all-metal construction, the wing was mid-set and consisted of a centre-section on which was mounted the central nacelle structure to accommodate the pilot, and nosewheel unit when retracted. Mountings for the engine nacelles/tail booms were integral with the wing centre-section structure, and outboard of these were the wing outer panels. Fuel tanks were housed in the wing centre section, and trailing-edge flaps of the Lockheed-Fowler type were installed. The tail booms, which could be considered to start from the fireproof bulkheads aft of each engine, provided housing for the main landing gear units when retracted, the General Electric engine turbochargers, cooling radiators and, at their extremities, the twin fins and rudders; they were rigidly braced together at the aft end by a continuous tailplane structure, with the single elevator inset in the tailplane trailing edge between the rudders. The powerplant comprised two Allison V-1710 inline engines, each driving a constant-speed and fully-feathering propeller.

The P-38s were followed into service by the 36 P-38Ds, beginning in August 1941, and these could be regarded as the first combat-worthy examples of the P-38. In addition to the tail unit modifications mentioned above, they were provided with a low-pressure oxygen system, a retractable landing light, and self-sealing fuel tanks. The gap in designations between P-38 and P-38D is accounted for by a single XP-38A (Lockheed Model 622) prototype with a pressurised cockpit nacelle, while the P-38B and P-38C identifications were allocated to projects which failed to materialise.

The Lockheed P-38L Lightning was the ultimate model of this distinctive fighter, combining good weapon load with more than adequate performance.

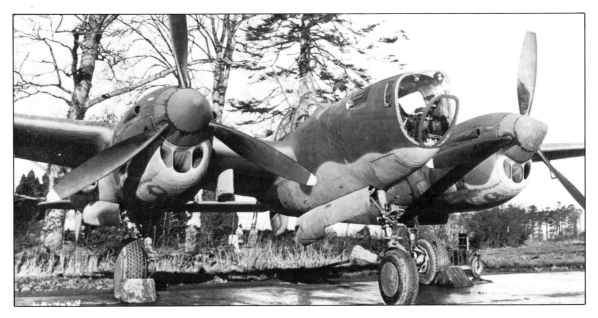

Lockheed P-38 Lightning

Lockheed P-38J Lightning of the 432nd Fighter Squadron, 475th Fighter Group, USAAF, based in New Guinea in winter 1943.

Lockheed P-38J Lightning of the 338th Fighter Squadron, 55th Fighter Group, USAAF, based at Nuthampstead (UK) in spring 1944.

Lockheed P-38J Lightning of the 79th Fighter Squadron, 20th Fighter Group, USAAF, based at Kingscliffe (UK) in spring 1944.

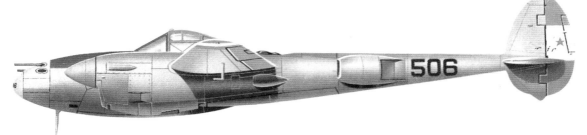

Lockheed P-38L Lightning of the Fuerza Aerea Hondurena (Honduran air force), based at Tocontin Air Base (Honduras) in 1948.

Lockheed F-5E Lightning of the Chinese Nationalist air force in summer 1945.

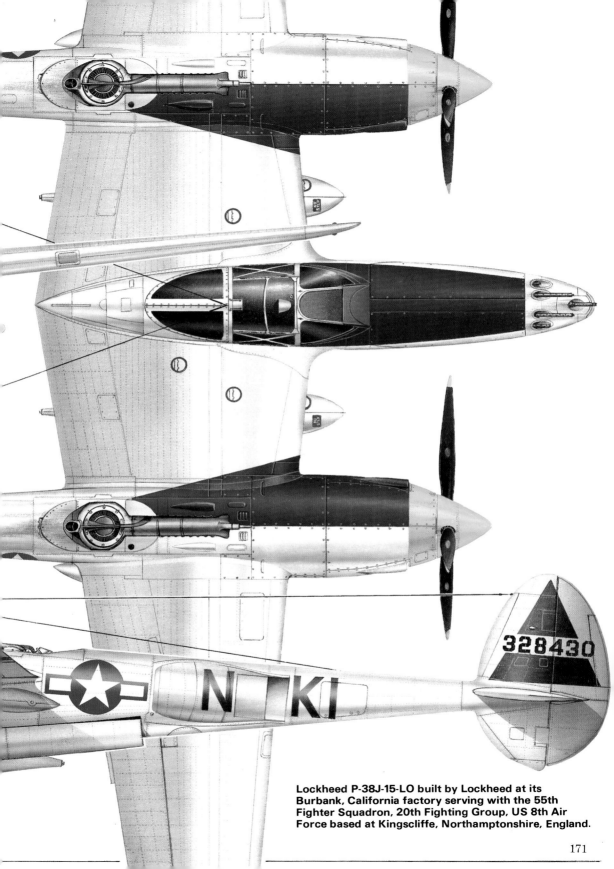

328430

Lockheed P-38J-15-LO built by Lockheed at its
Burbank, California factory serving with the 55th
Fighter Squadron, 20th Fighting Group, US 8th Air
Force based at Kingscliffe, Northamptonshire, England.

Lockheed P-38 Lightning

The P-38Ds were also the first to bear the name Lightning, which was the designation allocated to this aircraft when ordered for the RAF by the British Purchasing Commission of 1940. Some 667 were ordered, these being Lockheed Model 322s, and either because of an oversight on the part of the Commission (and this seems the more likely explanation), or because of an export ban on the engine/turbocharger combination, the first three examples, supplied to Britain as Model 322-61s, were considered to have inadequate performance when tested and the entire order was cancelled. These Lightning Is, as designated by the RAF, had two 1,150-hp (858-kW) Allison V-1710-C15 (R) engines without turbochargers and, as indicated by the R suffix, both were of right-hand rotation. Testing by the USAAF, following their acceptance of the 140 outstanding on the first British order, confirmed the RAF's findings and they were used only for various training and experimental purposes under the designation P-322. The balance of 524, representing the second British order, which were to have had the standard P-38 engine installation (Lockheed Model 322-60 and allocated the British designation Lightning II), were absorbed into USAAF contracts and were produced as either P-38F or P-38G Lightnings.

The P-38E was the last version to enter production for the USAAF before the attack at Pearl Harbor, and the type differed from the P-38D in having changed electric and hydraulic systems, the 37-mm nose cannon replaced by one of 20-mm calibre, and the provision of additional ammunition capacity for the nose guns. Some 210 of this version were contracted, but before they were completed 99 were converted for a photo-reconnaissance role under the designation F-4, with the nose armament replaced by a cluster of four cameras. In a similar manner, 20 of the ensuing P-38Fs were converted as photo-reconnaissance F-4As, but the 507 P-38Fs, which began to enter service in February 1942, had a number of changes which were introduced progressively to different production batches. These included the installation of 1,225-hp (913-kW) V-1710-49/53 engines, changed oxygen equipment, and the provision of so-called 'manoeuvring flaps'. In fact, these were the standard trailing-edge flap installation, but a readily-selected setting of 8° extension made possible much tighter turns. Provision was made also for the carriage of external weapons, and racks beneath the inner wings could accommodate two 75- or 150-US gallon (284- or 568-litre) drop tanks, 1,000-lb (454-kg) bombs, 22-in (559-mm) torpedoes, or smoke producing installations.

By this time, of course, USAAF Lightnings had become actively engaged in the war, and a P-38 of the 50th Fighter Squadron, a unit of the 342nd Composite Group based in Iceland, recorded the first operational success for a Lightning a few days after arriving on the island on 14 August 1942, destroying a German Focke-Wulf Fw 200 Condor over the Atlantic. In November 1942 P-38Fs saw their first large-scale use during the North African campaign. Their wholesale destruction of German cargo and transport aircraft over the Mediterranean quickly earned the nickname *der Gabelschwanz Teufel* ('the fork-tailed devil') from the Luftwaffe, but in this theatre also came the first appreciation that in its intended fighter role there were shortcomings. Not only was the wide span, twin-engined P-38 less manoeuvrable than the Messerschmitt fighters against which it was in combat, but it did not take long for Axis pilots to discover that high performance was limited to high altitude. If the P-38s could be forced to fight at altitudes between 10,000 and 15,000 ft (3 050 and 4 570 m), then the margin of performance was in favour of the enemy's Bf 109s.

Prior to the above encounters, however, the next production version had begun to enter service, during June 1942. This was the P-38G (1,082 built), which introduced the 1,325-hp (988-kW) V-1710-51/55 engine. Improved turbochargers, oxygen system and radio were incorporated at various stages throughout the production run. Of the P-38G total, 181 were converted for a reconnaissance role as F-5As, and because the V-1710-51/55 powerplant tended to overheat, an additional 200 F-5Bs were provided with intercoolers. With larger numbers of P-38F/-38Gs becoming available, the type was soon in service with USAAF squadrons in all theatres. In August 1942 the 1st Fighter Group's 71st and 94th Fighter Squadrons arrived in England after ferrying their aircraft across the Atlantic, and were joined almost immediately by the 37th, 49th and 50th Squadrons of the 14th Fighter Group. The 94th Fighter Squadron was, incidentally, the famous 'Hat in the Ring' squadron which had fought alongside the Allies in Europe during World War I.

The last 200 P-38Gs off the production line introduced underwing racks with a combined stores capacity of 3,200 lb (1 451 kg), and this became standard on the P-38Hs which followed. Production of this version totalled 601, of which 128 were converted for photo-reconnaissance duties as F-5Cs. All had 1,425-hp (1 063-kW) V-1710-89/91 engines, and late production examples had improved turbochargers. There were, however, extensive changes in the P-38Js (Lockheed Model 422) which followed, for the wider use of P-38s in zones where high temperatures were commonplace had made it essential to eliminate the engine overheating problem. This resulted in the introduction of 'chin' radiators at the base of the engine nacelles, enabling this version to develop its full take-off power to a height of 26,500 ft (8 075 m). Other progressive improvements included increased fuel capacity; the introduction of small electrically-actuated dive flaps, beneath the undersurface of the outer wing panels, to overcome a nose-down pitching movement at high speed; and in the first application of power-boosted controls to a fighter aircraft, the provision of hydraulically powered aileron boosters which required the pilot to provide only 17 per cent of the force needed for aileron operation. P-38J production reached 2,970,

Lockheed P-38 Lightning

and this version was in extensive use by early 1944.

Three P-38 fighter groups were operational in the Pacific, where Lightnings were accredited with the destruction of more Japanese aircraft than any other fighter in USSAF service. They are well recorded in the air force's history for a string of memorable actions, including the interception and destruction, some 550 miles (885 km) from their base, at Guadalcanal, of the Mitsubishi G4M carrying Japan's Admiral Isoroku Yamamoto, a skilful action carried out by aircraft from the 70th, 112th and 339th Fighter Squadrons. And, of course, the USAAF's 'ace of aces' of World War II, Major Richard I. Bong, scored all of his 40 confirmed victories while flying P-38s in the Pacific theatre. In Europe P-38s served mainly with the 9th Air Force, used extensively on long-range fighter escort duties in support of 8th Air Force daylight bombing missions against German targets.

There were no production P-38Ks, the designation XP-38K being allocated to a single P-38J powered by Allison V-1710-75/77 engines with larger diameter propellers, and this was followed by the most exensively built version, the P-38L. Lockheed produced no fewer than 3,810 of the model, while an additional 113 were built by Consolidated-Vultee: the P-38L differed from the P-38J in having 1,475-hp (1 100-kW) V-1710-111/113 engines which had a combat rating of 1,600 hp (1 193 kW) at 26,500 ft (8 075 m). Some batches of this version had a mounting, dubbed 'Christmas tree', beneath each outer wing panel for the carriage of five 5-in (127-mm) rocket projectiles.

Well over 700 reconnaissance aircraft with the designations F-5E/-5F/-5G were converted from P-38J/-38Ls, and 75 P-38M night fighters were also derived from P-38Ls. These latter, used operationally during the closing stages of war in the Pacific, carried a radar operator (seated aft of the pilot), his equipment,

and retained the full weapon load of the P-38L. Variants of the P-38 included some P-38Js modified in Europe to serve as 'lead-bombers' or 'Pathfinders'. In these aircraft the standard nose was replaced by one with a bomb-aimer's position, which had a transparent nose: others for a similar role carried BTO (Bomb Through Overcast) radar in the nose, making it possible to attack a target that was obscured by cloud. There were also a few TP-38L two-seat trainers derived from P-38Ls.

With the end of the war and the inevitable cancellations after VJ-Day most of the USAAF's Lightnings rapidly disappeared from the scene, but a few F-38J/-38Ls remained in service until 1949.

Specification
Type: single-seat long-range escort fighter
Powerplant (P-38J): two 1,425-hp (1 063-kW) Allison V-1710-89/91 inline piston engines
Performance: maximum speed 414 mph (666 km/h) at 25,000 ft (7 620 m); cruising speed 290 mph (467 km/h); service ceiling 44,000 ft (13 410 m); range with internal fuel 475 miles (764 km)
Weights: empty 12,780 lb (5 797 kg); maximum take-off 21,600 lb (9 798 kg)
Dimensions: span 52 ft 0 in (15.85 m); length 37 ft 10 in (11.53 m); height 9 ft 10 in (3.00 m); wing area 327.5 sq ft (30.42 m²)
Armament: one 20-mm cannon and four 0.50-in (12.7-mm) machine-guns, plus up to 3,200 lb (1 451 kg) of bombs
Operators: China, FFAF, RAF (evaluation only), USAAC/USAAF

The Lockheed P-38M was a night-fighter derivative of the P-38L, with radar under the nose and its operator in a special cockpit behind the pilot.

Lockheed P-80 Shooting Star

History and Notes

Although the Lockheed P-80 was not the first jet aircraft to enter service with the US Army Air Force, it was its first operational jet fighter. Design work began in June 1943, when a team headed by Clarence L. Johnson was tasked with the development of an aircraft to be powered by the 3,000-lb (1 361-kg) static thrust de Havilland H-1 turbojet. Lockheed's submission offered prototype completion within 180 days, and the USAAF awarded a contract for three prototypes and 13 service trials aircraft.

Built at Burbank, the first prototype, designated XP-80, was flown at the Muroc Dry Lake test centre on 8 January 1944, just 143 days after the start of work. This was the only aircraft to have the H-1 engine, which was originally to have been manufactured under licence by Allis-Chalmers as the J36. The General Electric J33, rated as the J36. The General Electric J33, rated at 3,750-lb (1 701-kg) thrust, was instead adopted and installed initially in the first of two prototype XP-80As, flown on 10 June by Tony Le Vier. The engines in both XP-80As were, in fact, developed General Electric I-40s, but the pre-series YP-80As were powered by production J33-GE-9 or -11 turbojets.

Increases in wing span and fuselage length, a stronger landing gear and a taller fin were features of the XP-80As, intended to allow for the increased weight and power of the new engine. Inevitably, gross weight also rose, from the 8,916 lb (4 044 kg) of the XP-80 to 13,700 lb (6 214 kg), and a weight reduction programme was initiated to achieve 11,500 lb (5 216 kg) for the YP-80As, which carried six 0.50-in (12.7-mm) machine-guns in the nose, instead of five as in the prototype. The first of 13 YP-80s was delivered in October 1944 and, shortly before the end of the war in Europe, two of them were assigned to Italy for operational trials.

Planned production was for some 5,000 aircraft, including 1,000 P-80Ns which were to have been built by North American at Kansas City. After VJ-Day, however, more than 3,000 were cancelled, including all the P-80Ns. Deliveries of 525 P-80As began in December 1945, these being very similar to the YP-80As but with wing-tip fuel tanks and provision for underwing bombs or fuel tanks, raising the gross weight to 14,500 lb (6 577 kg).

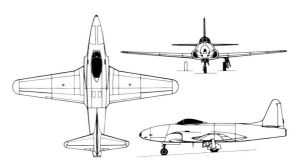

Lockheed P-80C Shooting Star

The J33-A-17 engine fitted in the later P-80As was replaced by the 5,200-lb (2 359-kg) thrust J33-A-21 in the P-80B, which also had a thinner wing and provision for JATO equipment, and 240 were produced in 1946-7. The final production interceptor/fighter bomber version was the P-80C, initially with a J33-A-23 engine and later with the 5,400-lb (2 449-kg) J33-A-35, and 798 were manufactured in 1948-9. With cameras replacing the nose armament, one of the YP-80As had been completed as the XF-14, a reconnaissance version, of which 152 were produced as FP-80As.

Specification

Type: single-seat fighter, fighter-bomber and reconnaissance aircraft
Powerplant (P-80A): one 4,000-lb (1 814-kg) static thrust General Electric J33-A-11 turbojet
Performance: maximum speed 558 mph (898 km/h); cruising speed 410 mph (660 km/h); service ceiling 45,000 ft (13 715 m); range 540 miles (869 km)
Weights: empty 7,920 lb (3 592 kg); maximum take-off 14,500 lb (6 577 kg)
Dimensions: span 39 ft 11 in (12.17 m); length 34 ft 6 in (10.52 m); height 11 ft 4 in (3.45 m); wing area 238 sq ft (22.11 m²)
Armament: six 0.50-in (12.7-mm) machine-guns (plus two 1,000-lb/454-kg bombs or 10 5-in/127-mm rockets on F-80C)
Operators: USAAF, USN

Martin 167 Maryland

History and Notes

Used in action as a three-seat reconnaissance bomber, the Martin Maryland was designed originally around a US Army Air Corps specification for an attack bomber, but with its failure to win production orders for the type at home the Glenn L. Martin Company looked to Europe for a market.

The prototype Model 167W had flown in February 1939, and faced with the prospect of war France placed an order for 115 aircraft even before the first flight. All were built within six months of the contract being

signed, but delivery was delayed until a US government arms embargo was lifted in October 1939, by which time a further 100 had been ordered.

At the time of the French armistice in June 1940, 140 had been delivered as Model 167Fs, or 167A-3 as they were designated by the French air force, and the remaining 75 aircraft from the second batch were diverted to Britain and given the RAF name Maryland.

Assembled at Burtonwood, after arrival by sea at Liverpool, these aircraft were followed by another 75 which Britain had ordered; all were designated Mk I,

Martin 167 Maryland

and had single-stage supercharged engines. Further British orders followed for an improved model, the Mk II with two-stage superchargers, and a total of 225 Marylands was delivered to the RAF.

French Marylands saw action against both Axis and Allied forces: against the former in France until the armistice in 1940, and against the latter with the Vichy forces in West Africa and the Middle East. Two were left in Vichy service when hostilities in Morocco ceased in November 1942.

Meanwhile the RAF Marylands, their original Cyclone engines replaced by Pratt & Whitney Twin Wasps, had entered service in the UK on non-operational tasks such as target towing and long-range reconnaissance. No. 431 Flight formed at Malta in September 1940 and was the first operational RAF unit to receive the type; it was later renumbered as No. 69 Squadron. No. 39 Squadron changed its Bristol Blenheims for Marylands in the Western Desert in May 1941, and No. 223 Squadron, originally an OTU for Marylands and Douglas Bostons, became operational.

A number of Marylands went to South African Air Force squadrons Nos. 12, 20, 21 and 24. The type was in operational service for about two years in Africa and the Middle East before being replaced, largely by Baltimores, and performed sterling service, its comparatively high speed often standing it in good stead.

Specification

Type: three-seat reconnaissance/bomber aircraft
Powerplant (Mk I): two 1,050-hp (783-kW) Pratt & Whitney R-1830-SC3G Twin Wasp radial piston engines
Performance: maximum speed 304 mph (489 km/h) at 13,000 ft (3 960 m); cruising speed 248 mph (399 km/h); service ceiling 29,500 ft (8 990 m); range 1,300 miles (2 092 km)
Weights: empty 11,213 lb (5 086 kg); maximum take-off 16,809 lb (7 624 kg)
Dimensions: span 61 ft 4 in (18.69 m); length 46 ft 8 in (14.22 m); height 14 ft 11¾ in (4.57 m); wing area 538.5 sq ft (50.03 m²)
Armament: four 0.30-in (7.62-mm) wing-mounted machine-guns and two 0.303-in (7.7-mm) guns (one each in ventral and dorsal positions), plus up to 2,000 lb (907 kg) of bombs
Operators: FAF, FNAF, FVAF, RAF, RN, SAAF

Though it was a useful light bomber, the Martin Maryland Mk I found its real niche as a medium-range high-speed reconnaissance machine.

Martin 187 Baltimore

History and Notes

The Martin Baltimore four-seat light bomber was developed from the Maryland to specific British requirements. Following the passing of the Lend-Lease Act in the USA, the original RAF order for 400 Baltimores, placed in May 1940, was increased by another 575 in June 1941 and a further 600 in July for the following year, making a total of 1,575.

The prototype flew on 14 June 1941, and speed of production was such that deliveries to the UK began only four months later. The initial production batch consisted of 50 Mk Is and 100 Mk IIs, the main difference being in the dorsal armament, the early aircraft having a single hand-operated Vickers 'K' gun while the Mk II had twin Vickers guns.

Baltimore Is entered service with Operational Training Units, while first deliveries of Mk IIs were made in 1942 to two Middle Eastern units: No. 223 Squadron which was operating as an OCU, and No. 55 Squadron, a Bristol Blenheim unit. The latter squadron claimed to have been the first to bomb the Italians in World War II, on 11 June 1940 with Blenheim Is, and to

Martin 187 Baltimore

have been the last Allied aircraft to drop bombs at night in the Italian campaign, on 30 April 1945, with a Boston. After conversion to Baltimores No. 55 Squadron flew 353 operational sorties during the first 10 days of the Battle of Alamein, a greater number than any other Desert Air Force light bomber squadron at that time.

While the early Baltimores had a high performance they could be difficult to handle during take-off, but a more serious problem was the poor protection afforded by the mid-upper gun position. Although the rest of the armament was comparatively heavy, with four 0.303-in (7.7-mm) Brownings in the wings firing forward, four more fixed guns in the belly firing downwards and behind, and another single gun in the ventral hatch, the mid-upper position was plagued with problems of manoeuvrability which made the aircraft extremely vulnerable to attacks from above and behind.

The situation was considerably improved with the introduction of the Baltimore Mk III which was fitted with a Boulton Paul hydraulically operated turret containing four Browning 0.303-in (7.7-mm) guns; following this came the Mk IIIA with a lower profile Martin-built electrically operated turret with two 0.50-in (12.7-mm) Brownings. A total of 250 Mk IIIs and 281 Mk IIIAs were built, but some were lost during transatlantic delivery when two cargo ships carrying them were sunk. The Mk IV which followed was basically similar to the Mk IIIA and 294 were produced before there appeared the final variant, the Mk V with more powerful Wright Cyclone engines. Numerically this was the most prolific Baltimore, the whole of the final order for 600 aircraft being to this standard. Production ended in May 1944.

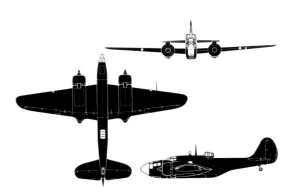

Martin Baltimore Mk II

No. 13 Squadron operated Baltimores between December 1943 and October 1944. During a conversion from Blenheim Vs to Lockheed Venturas the squadron was suddenly switched to Baltimores and flew almost 300 sorties on daylight bombing operations in April 1944 without loss before going on to night bombing.

Other RAF squadrons to use Baltimores included Nos. 52, 69, 500 and 223, the last named serving, as previously mentioned, as the Baltimore OCU for a few months before it became an operational squadron in May 1942. In August 1944 it was redesignated No. 30

A Martin Baltimore Mk III begins to unload over an Italian railway line during February 1944. Maximum bomb load was 2,000 lb (907 kg).

Martin 187 Baltimore

Squadron, South African Air Force, while serving in Italy, reforming very shortly afterwards in England as a bomber support unit, part of No. 100 Group. A few Baltimores were used in the reconnaissance role based at Malta with No. 203 Squadron.

The Baltimore also saw service with No. 454 Squadron, Royal Australian Air Force, from July 1944 to August 1945. Re-equipping with the new aircraft from its former Blenheim Vs, the squadron was initially involved in rather mundane operations such as anti-submarine patrols, attacks on shipping and leaflet raids from bases in Egypt, Libya and Palestine, but from July 1944 when it was transferred to the Mediterranean Allied Air Forces, Italy, things began to liven up and between 2 August 1944 and 23 January 1945 the squadron flew 1,420 sorties in 2,539 hours and dropped 1,013 tons (1 029 tonnes) of bombs. Very precise bombing was necessary because, on some occasions flying at 12,000 feet (3 660 m) the Baltimores had to bomb targets only 800 yards (732 m) ahead of Allied troops.

Several South African Air Force squadrons operated Baltimores, including Nos. 15, 21 and 60, while Greeks flew the type with No. 13 (Hellenic) squadron, returning to southern Greece when that part of the country was reoccupied by the Allied forces in November 1944. Attacks were carried out by this latter squadron against German pockets of resistance in the Greek islands and on shipping.

The Italian Co-Belligerent Air Force received 28 Baltimore IVs and Vs to equip a wing, but had some difficulty in converting to the type and sustained considerable losses during the training period. The wing had an operational life of just over six months, flying its last mission on 4 May 1945. A few Baltimores went to the Turkish air force in the later stages of the war.

Baltimore Vs were supplied to the Free French Air Force in Syria in 1944, where they were used for mapping and survey work until May 1945 when they attacked forces supporting a Syrian insurrection in Damascus. After the unit disbanded in 1946 the handful of surviving Baltimores were flown to France for use at the test centre of the CEV (*Centre d'Essais en Vol*) or Flight Test Centre at Bretigny. Surviving until 1948, these were no doubt the last flying examples of the type.

It can justifiably be said that the Baltimore, after initial problems with its rear armament, served with distinction, and was liked by its pilots for its superior performance and great structural strength. A drawback common to several other types, including the Martin Maryland, Douglas Boston and Handley Page Hampden, was the narrow fuselage which prevented easy movement and made it virtually impossible for crew members to change position in flight if injured.

With a maximum speed of over 300 mph (483 km/h), armament of up to 14 machine-guns and maximum bomb load of 2,000 lb (907 kg), the Baltimore represented a major step up from the admittedly older Blenheim with its 266 mph (4218 (428 km/h), four machine-guns and 1,320 lb (599 kg) bomb load.

Specification
Type: four-seat twin-engined light bomber
Powerplant (Mk IV): two 1,660-hp (1 238-kW) Wright GR-2600-A5B Cyclone 14 radial piston engines
Performance: maximum speed 305 mph (491 km/h) at 11,500 ft (3 505 m); cruising speed 225 mph (362 km/h) at 23,300 ft (7 100 m); service ceiling 23,300 ft (7 100 m); range 1,082 miles (1 741 km) with 1,000 lb (454 kg) of bombs; maximum range 2,800 miles (4 506 km)
Weights: empty 15,460 lb (7 013 kg); maximum take-off 22,600 lb (10 521 kg)
Dimensions: span 61 ft 4 in (18.69 m); length 48 ft 5¾ in (14.78 m); height 17 ft 9 in (5.41 m); wing area 538.5 sq ft (50.03 m²)
Armament: four wing-mounted 0.303-in (7.7-mm) machine-guns, two or four 0.303-in (7.7-mm) guns in dorsal turret, two 0.30-in (7.62-mm) guns in ventral position and provision for four fixed rear-firing 0.30-in (7.62-mm) guns, plus up to 2,000 lb (907 kg) of bombs
Operators: FFAF, Greek Air Force, ICoBAF, RAAF, RAF, SAAF, Turkey

Martin AM-1 Mauler

History and Notes
As 1943 drew to a close the US Navy began to plan a new breed of attack aircraft, drawing on its experience of two years active service carrier operations and reflecting the changing demands upon naval air power. The intention was to combine the formerly separate roles of the scout and torpedo bombers, as exemplified by the Curtiss SB2C Helldiver and the Grumman TBF Avenger.

The emphasis was to be on load-carrying capability and performance, and the resulting specification was for a single-seater, designed around the most powerful

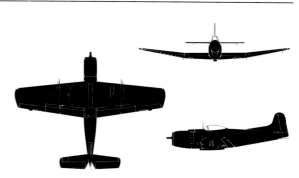

Martin AM-1 Mauler

Martin AM-1 Mauler

engine then available, which was to carry its offensive load on external hardpoints rather than in internal weapons bays, saving both in airframe weight and complexity.

Among the designs submitted to the US Navy was Martin's Model 210, two prototypes of which were ordered on 31 May 1944. Designated XBTM-1, the first aircraft was flown on 26 August 1944, powered by a 3,000-hp (2 237-kW) Pratt & Whitney XR-4360-4 engine. Four 20-mm cannon were mounted in the folding outer panels of the wings, which also incorporated 14 hardpoints that, together with one beneath the fuselage, could carry up to 4,500 lb (2 041 kg) of bombs and rockets.

A production order for 750 BTM-1s was placed on 15 January 1945, but the designation was changed to AM-1 before the first production aircraft flew on 16 December 1946. Type acceptance trials were flown during 1947 and the first deliveries to an active squadron were made to Attack Squadron VA-17A on 1 March 1948.

Only 149 of these 2,975-hp (2 218-kW) Cyclone-powered Maulers were completed before production ceased in October 1949, and units flying the type passed their equipment on to US Navy Reserve squadrons. All first-line attack squadrons were then equipped with the Douglas AD-1 Skyraider which, designed to the same specification, became the US Navy's standard attack bomber. A small number of Maulers operated in the electronic countermeasures role under the designation AM-1Q.

Specification
Type: single-seat carrier-based attack aircraft
Powerplant: one 2,975-hp (2 218-kW) Pratt & Whitney R-3350-4 Cyclone 18 radial piston engine
Performance: maximum speed 367 mph (591 km/h) at 11,600 ft (3 535 m); cruising speed 189 mph (304 km/h); service ceiling 30,500 ft (9 295 m); range 1,800 miles (2 897 km)
Weights: empty 14,500 lb (6 577 kg); maximum take-off 23,386 lb (10 608 kg)
Dimensions: span 50 ft 0 in (15.24 m); length 41 ft 2 in (12.55 m); height 16 ft 10 in (5.13 m); wing area 496 sq ft (46.08 m²)
Armament: four fixed forward-firing 20-mm cannon, plus up to 4,500 lb (2 041 kg) of assorted bombs and rocket projectiles
Operator: US Navy

The Martin XBTM-1 was the prototype (two built) of the powerful AM-1 Mauler.

Martin B-10/B-12

History and Notes
The Glenn L. Martin Company had been responsible for the supply of a fair proportion of the US Army Air Corps' bomber aircraft in the 1920s, but after being the unsuccessful contender in several design competitions decided in 1930 to design and build an advanced bomber project as a private venture. Identified initially by the company as the Martin 123, this monoplane bomber had a wing span of 62 ft 2 in (18.95 m), was powered by two 600-hp (447-kW) Wright SR-1820-E Cyclone engines, and had open cockpits for a crew of three. Delivered to the US Army for testing at Wright Field on 20 March 1932, it was able to demonstrate a maximum speed of 197 mph (317 km/h) at 6,000 ft (1 830 m) when trials got under way in July of that year. For these trials the aircraft was allocated an experimental designation of XB-907.

Following discussions, the Martin 123 was returned to the factory in the autumn for modifications which included the installation of a front gun turret that, despite only carrying one machine-gun, was the first to be installed in a US bomber. In addition to this work, more powerful 675-hp (503-kW) R-1820-19 engines were installed in long-chord NACA cowlings, and wing span was increased to 70 ft 7 in (21.51 m). When tested

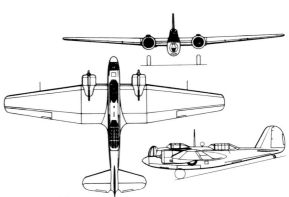

Martin B-10

again by the US Army in late 1932, then designated XB-907A, a speed of 207 mph (333 km/h) was recorded at 6,000 ft (1 830 m), a turn of speed which proved to be faster than that of any contemporary US fighters.

Not surprisingly, the US Army contracted for 48 production aircraft on 17 January 1933 and, at the same time, bought the XB-907A and redesignated it XB-10. Martin's production version (Martin 139) began to enter service in June 1934. It was a mid-wing

Martin B-10/B-12

Martin B-10B of the 28th Bomb Squadron, USAAC, based at Camp Nichols, Luzon (Philippines group) between 1937 and 1941.

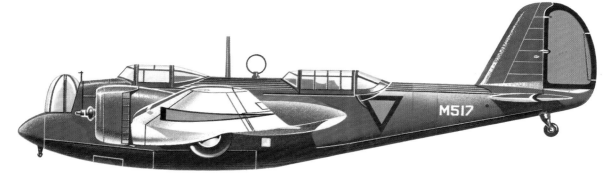

Martin 139WH2 of the Royal Netherlands Army Air Corps.

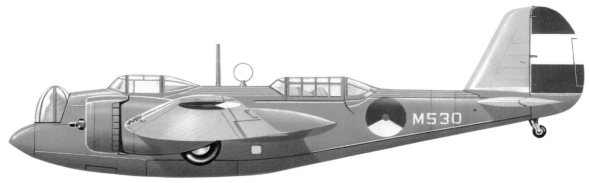

Martin 139WH2 of the Royal Netherlands Indies Army Air Corps in 1941.

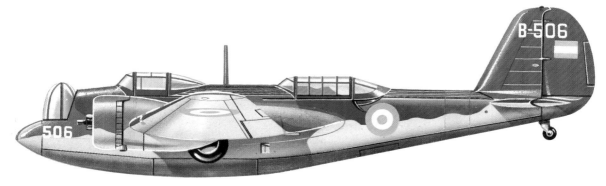

Martin 139WAA of the 1st Bomber Regiment, Argentine Army Aviation Command, in 1938.

monoplane of all-metal construction, except for fabric-covered control surfaces. The tailwheel type landing gear was manually retractable, and the standard powerplant was two Wright Cyclones. The crew of four or five was accommodated in three separate cockpits with the bomb aimer/gunner in the nose, pilot and radio operator just forward of the wing leading edge, and gunners or a gunner and navigator in the aft cockpit. The nose position was covered by a cupola to permit deployment of the gun, and the other two cockpits each had a transparent canopy.

The first 14 production aircraft, with 675-hp (503-kW) R-1820-25 engines, were designated YB-10; the seven YB-12s which followed had 775-hp (578-kW) Pratt & Whitney R-1690-11 Hornet engines. The next 25 production B-12As also had the Hornet engines, and provision was made for an extra fuel tank to be carried in the bomb bay. Late procurement, in 1934 and 1935, resulted in 88 and 15 B-10Bs respectively. Odd variants included a single B-10A with two turbocharged R-1820-31 Cyclones, one XB-14 with two 950-hp (708-kW) Pratt & Whitney YR-1830-9 Twin Wasps and, resulting from late modification programmes, B-10Ms and B-12AMs equipped as target tugs. Several YB-10s and B-12As were modified by the US Army, provided with float installations for operation on coastal defence duties.

These Martin bombers remained in service with the USAAC until replaced by Boeing B-17s and Douglas B-18s in the late 1930s, but Martin also had some success with sales to foreign customers, building 189 for export. Of these, 120 went to the Netherlands East Indies, and these were amongst the first US-built bombers to see operational service at the beginning of World War II. Late examples from this export production had a continuous transparent canopy over the pilot's and rear gunner's cockpits, and were identified by Martin as Martin 166s.

Specification

Type: four/five-seat twin-engined light bomber
Powerplant (B-10B): two 775-hp (578-kW) Wright R-1820-33 Cyclone 9 radial piston engines
Performance: maximum speed 213 mph (343 km/h) at 6,000 ft (1 830 m); cruising speed 193 mph (311 km/h); service ceiling 24,200 ft (7 375 m); range 1,240 miles (1 996 km)
Weights: empty 9,681 lb (4 391 kg); maximum take-off 16,400 lb (7 439 kg)
Dimensions: span 70 ft 6 in (21.49 m); length 44 ft 9 in (13.64 m); height 15 ft 5 in (4.70 m); wing area 678 sq ft (62.99 m²)
Armament: three 0.30-in (7.62-mm) machine-guns (one each in nose and rear turrets and one in ventral position), plus up to 2,260 lb (1 025 kg) of bombs carried internally
Operators: Argentina, China, Netherlands, Philippines, Thailand, Turkey, USAAC

Martin B-26 Marauder

History and Notes

Certainly one of the most elegant bomber aircraft to appear in the early years of World War II, Martin's B-26 Marauder stemmed from a US Army Air Corps high-speed medium bomber specification which had been circulated to US manufacturers in January 1939. This called for a number of characteristics which, together, made the US Army requirement very difficult to meet. To accommodate a crew of five, which meant that it must be fairly large, it was required also to be fast and with good high-altitude performance, to have a range in excess of 2,000 miles (3 219 km), and be able to carry good defensive armament plus a worthwhile load of bombs.

Martin's design, by Peyton M. Magruder, was far in advance of competing submissions, and as the company not only guaranteed that performance would be as good as or better than performance estimates and also promised early production, it was not surprising that this company was chosen to build the USAAC's new bomber. The startling feature of the contract, awarded in September 1939, lay in the fact that it was for a substantial number of production aircraft (201) ordered 'straight off the drawing board', a course then unprecedented in USAAC history. No prototypes or preproduction aircraft were called for, so the first of the Martin 179s, designated B-26 by the US Army, flew

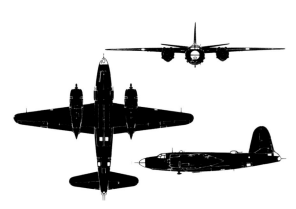

Martin B-26G Marauder

for the first time on 25 November 1940.

As then flown it was a cantilever shoulder-wing monoplane of all-metal construction, except that all control surfaces were fabric-covered, and the conventional but small-area wing had plain trailing-edge flaps. The fuselage was a near perfect aerodynamic cigar-shape form of circular cross-section, marred only by the 'step' of the windscreen, and with a conventional tail unit which had a high-set tailplane. Landing gear was of the retractable tricycle type, the main units

Martin B-26 Marauder

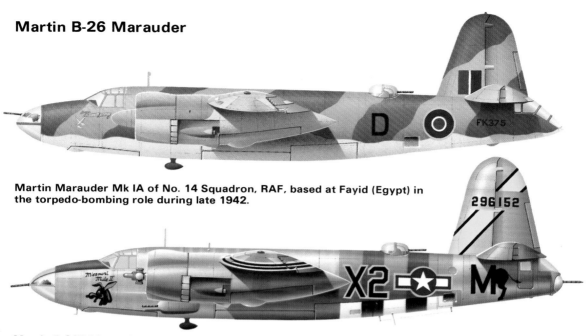

Martin Marauder Mk IA of No. 14 Squadron, RAF, based at Fayid (Egypt) in the torpedo-bombing role during late 1942.

Martin B-26B Marauder of the 598th Bomb Squadron, 397th Bomb Group, USAAF, based at Dreux (France) in September 1944.

retracting forward and upward into the centre of the engine nacelles, and the nosewheel unit aft into the forward fuselage. To provide the necessary performance a new Pratt & Whitney engine had been selected, the 1,850-hp (1 380-kW) R-2800-5 Double Wasp, and the two of these each drove a four-blade constant-speed fully-feathering propeller. An innovation was the use of a 'cuff' at the root end of each propeller blade, this enabling the normally useless area of each blade to provide extra air-flow for improved engine cooling. Initial armament comprised two 0.30-in (7.62-mm) machine-guns, one in the nose position and one in the tailcone, plus two 0.50-in (12.7-mm) guns in an electrically operated dorsal turret, the first powered gun turret to be installed in an American bomber. Maximum bomb load, all carried internally, was as much as 5,800 lb (2 631 kg) for deployment at short range.

Following the first flight, it was not until February 1941 that succeeding production aircraft began to come off the line, and while some of these were diverted for test purposes, there were sufficient available to begin deliveries to the USAAC. This initial equipping of the US Army Air Corps' squadrons was not without problems, for while they had been supplied with an aircraft which attained the desired high-performance specification, this performance had been achieved at the expense of good low-speed handling characteristics, leading to what is usually termed a 'hot' aeroplane. This made conversion training a difficult and slow process, for even at loaded weights well under maximum the aircraft's stalling speed was not far below 100 mph (161 km/h), a very high figure for that period.

In spite of this Marauders, as the B-26 had been

named in preference to the originally chosen Martian, gradually began to equip USAAF squadrons and as experience was gained a number of modifications were considered to be desirable, resulting in the B-26A of which 139 were built. All had engines of the same power as the B-26, but R-2800-5, -9 and -39 units were installed in different batches. The electrical system was changed from 12-volt to 24-volt, two additional fuel tanks were installed in the bomb bay, provision was made for the carriage of a 22-in (559-mm) torpedo and, as a result of combat reports from the war then being fought in Europe, the nose and tail guns of 0.30-in (7.62-mm) calibre were replaced by similar 0.50-in (12.7-mm) installations. The result of these changes, of course, was to increase the gross weight and also, as a consequence, the problems that were soon to come to a head.

Before that, however, the Japanese on 7 December 1941 attacked Pearl Harbor and, on the following day, the USAAF's 22nd Bombardment Group was despatched to the Pacific zone, becoming operational initially from northern Australia in April 1942. This unit's B-26As soon found ready employment in a variety of roles, including unsuccessful torpedo attacks against the Japanese fleet engaged in the Battle of Midway. At about that same time the RAF received three examples of the B-26A for evaluation, these being designated Marauder I. Successful testing resulted in this type being chosen for tactical use in the North African campaigns, and the additional 48 of this version allocated under Lend-Lease were delivered direct to the Middle East and used first to equip No. 14 Squadron.

While these events had been taking place, a special board of investigation had been set up in the USA,

Martin B-26

2107812

KS ✪ J

Martin B-26C *Baby Bumps II* was allocated to the
557th Bomb Squadron, 387th Bomb Group, 9th Air
Force based in the UK just before the Allied
invasion of Normandy, June 1944.

Martin B-26 Marauder

under the chairmanship of Major General Carl Spaatz, to enquire into the abnormally high accident rate associated with the B-26, especially during training, and to decide whether production should be terminated. Fortunately this latter course was not adopted for, with growing experience of how best to handle the Marauder, it was later to have the lowest attrition rate of any American aircraft operated by the US 9th Air Force in Europe. The eventual findings of the investigation board resulted in continuing production, but with some recommendations regarding modifications intended to improve low-speed handling.

During the foregoing enquiry all production had been suspended but soon after it was resumed, in May 1942, Martin began to deliver its first B-26Bs, the major production version of which 1,883 were built. These incorporated initially improvements which combat experience had proved to be necessary, but many other changes were introduced on the line throughout the long manufacturing run. Major items included the installation of 1,920-hp (1 432-kW) R-2800-41 or -43 engines, the introduction of slotted trailing-edge flaps, and a lengthened nosewheel strut to increase wing incidence and so improve take-off characteristics. The most important change, one which had been recommended by the enquiry board, was an increase in wing span/area but this, in fact, achieved nothing because the USAAF immediately upped the gross weight. The comparisons of maximum wing loading are interesting, the B-26's being 53.16 lb/sq ft (259.5 kg/m²), the early B-26B's 56.48 lb/sq ft (275.7 kg/m²), and the late B-26B's 58.05 lb/sq ft (283.4 kg/m²), which all goes to prove that the initial handling problems were largely those of inexperience. Today little is thought of a wing loading of 149 lb/sq ft (728 kg/m²), and that for a civil transport aircraft, not a 'hot' military aeroplane.

The introduction of the larger wing necessitated an increase in vertical tail surface area, achieved by increasing fin and rudder height by 1 ft 8 in (0.51 m). The armament, through a succession of modifications, became almost as potent as that of the USAAF's heavy bombers, with no fewer than 12 0.50-in (12.7-mm) machine-guns. The increasing demand for Marauders resulted in the establishment of a second production line by Martin at Omaha, Nebraska, which built 1,235 aircraft as B-26Cs from late 1942, these duplicating various batches of the B-26Bs built at Baltimore, Maryland. The D and E designations were taken up by two one-off aircraft: the XB-26D was an experiment in thermal wing de-icing; and the XB-26E was a 'weight watchers' version with some 2,000-lb (907-kg) weight reduction and with the dorsal turret moved forward to a position adjacent to the wing leading edge.

The final production versions were the generally similar B-26F (300 built) and B-26G (893), plus 57 TB-26Gs without armament and other purely operational equipment to serve as target tugs or trainers. The major difference between these aircraft and the B-26B/-26Cs which had preceded them lay in a final attempt to improve take-off performance, wing incidence being increased by 3°30', so giving a noticeable nose-in-the-air look to the engines. There were also some armament and fuel system changes. Last of the B-26 designations was taken by a single XB-26H with tandem bicycle type landing gear; each of the main units carried twin wheels and an outrigger, for balancing, was housed in each engine nacelle. This experimental installation was made to evaluate a landing gear of this type which was being developed for the Boeing XB-47.

All of the USAAF's early deployment of the B-26 had been confined to the Pacific theatre, but B-26Bs and B-26Cs began to appear in North Africa during November 1942, equipping 12 squadrons of the 17th, 319th and 320th Bombardment Groups of the 12th Air Force, providing admirable support to the Allied ground forces as they followed the bitter but victorious

A Martin B-26C Marauder of the 9th Air Force passes over landing craft heading for Normandy on 6 June 1944, on a mission to soften up the German defences.

Martin B-26 Marauder

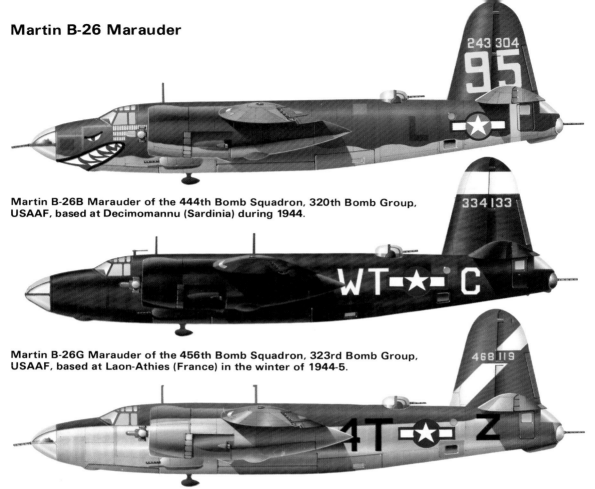

Martin B-26B Marauder of the 444th Bomb Squadron, 320th Bomb Group, USAAF, based at Decimomannu (Sardinia) during 1944.

Martin B-26G Marauder of the 456th Bomb Squadron, 323rd Bomb Group, USAAF, based at Laon-Athies (France) in the winter of 1944-5.

Martin B-26G Marauder of the 585th Bomb Squadron, 394th Bomb Group, USAAF, based at Cambrai-Niergnives (France) in November 1944.

trail to the south of France via Sicily, Italy, Sardinia and Corsica. However, the B-26's first operation with the 8th Air Force in Europe was disastrous, all 11 aircraft sent to make a low-level attack on installations in the Netherlands failing to return to base. Subsequently, in a tactical role, Marauders went from strength to strength in operations with the USAAF's 9th Air Force, also in Europe.

Under Lend-Lease the RAF received a total of 522 Marauders, these comprising the Marauder I mentioned above, plus Marauder IA (B-26B), II (B-26C) and III (B-26F/-26G). Used by the RAF's Nos. 14, 39, 326, 327 and 454 Squadrons and the South African Air Force's Nos. 12, 21, 24, 25 and 30 Squadrons, they were deployed most successfully alongside the B-26s of the US 12th Air Force, after initial failure in a torpedo-carrying role.

In 1943 the USAAF converted 208 B-26Bs and 350 B-26Cs for use as high-speed target tugs, stripping out all armament and operational equipment, and these were redesignated initially as AT-23A and AT-23B respectively, but subsequently TB-26B and TB-26C. Of these the US Navy acquired 225 AT-23Bs which they

designated JM-1, and 47 TB-26Gs, the last Martin production version, as JM-2s.

Specification

Type: seven-seat twin-engined light bomber
Powerplant (B-26G): two 1,920-hp (1 432-kW) Pratt & Whitney R-2800-43 Double Wasp radial piston engines
Performance: maximum speed 283 mph (455 km/h) at 5,000 ft (1 525 m); cruising speed 216 mph (348 km/h); service ceiling 19,800 ft (6 035 m); range 1,100 miles (1 770 km)
Weights: empty 25,300 lb (11 476 kg); maximum take-off 38,200 lb (17 327 kg)
Dimensions: span 71 ft 0 in (21.64 m); length 56 ft 1 in (17.09 m); height 20 ft 4 in (6.20 m); wing area 658 sq ft (61.13 m²)
Armament: 11 0.50-in (12.7-mm) machine-guns (in fixed forward-firing, flexible nose and waist mounts, and in power-operated dorsal and tail turrets), plus up to 4,000 lb (1 814 kg) of bombs
Operators: FAF, FFAF, RAAF, RAF, SAAF, USAAC/USAAF, USMC, USN

Martin JRM Mars

History and Notes

Developed to meet a US Navy specification for a long-range patrol bomber, the Martin Model 170 Mars was at the time of its construction the world's largest flying boat. Ordered on 23 August 1938, the prototype XPB2M-1 made its first flight on 3 July 1943, this event having been delayed by an accident during preparatory tests. In its reconnaissance role, the Mars had provision for power-operated nose and tail turrets, but these were deleted during 1943 and the positions faired over.

With reinforced floors, enlarged hatches and with loading equipment installed, the machine was redesignated XPB2M-1R to denote its transport role. In this form it made its first flight in December 1943, from Patuxent River Naval Air Station to Natal, Brazil, carrying a payload of 13,000 lb (5 897 kg) for the 4,375-mile (7 041-km) journey. Early in the following year it flew a 4,700 mile (7564 km) mission to Hawaii and back, in a time of 27 hours 26 minutes and carrying a load of 20,500 lb (9 299 kg).

In January 1945, 20 purpose-built cargo aircraft were ordered under the US Navy designation JRM-1, although only five were completed. The first JRM-1, named *Hawaiian Mars*, took off on its maiden flight on 21 July 1945 but proved to be short-lived by crashing in Chesapeake Bay on 5 August. The other four JRM-1s were named after Pacific Islands as *Philippine Mars*, *Marianas Mars*, *Marshall Mars* and *Hawaii Mars*.

A sixth production Mars, completed in November 1947, had structural changes to permit operation at a gross weight of 165,000 lb (74 843 kg): this was designated JRM-2 and named *Caroline Mars*. The four surviving JRM-1s were retrospectively converted to JRM-2 standard, becoming redesignated JRM-3. *Marshall Mars* was destroyed in a fire at Honolulu on 5 April 1950, but the remaining aircraft were flown by US Navy Squadron VR-2 from Alameda Naval Air Station, California, until withdrawn in November 1956, the largest flying boats ever to have seen service with the US Navy. In July 1959 they were sold to Canada for use as fire-fighting water bombers, fitted with tanks

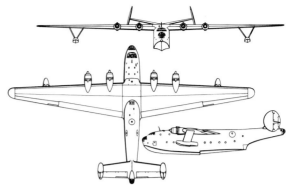

Martin XPB2M-1 Mars

which had a capacity of 7,000 US gallons (26 497 litres). These could be filled in 15 seconds, with the aircraft skimming the surface of a suitable lake.

Specification:

Type: seven-seat long-range passenger/cargo transport flying boat
Powerplant (JRM-2): four 2,200-hp (1 641-kW) Wright R-3350-18 Duplex Cyclone radial piston engines
Performance: maximum speed 221 mph (356 km/h) at 4,500 ft (1 370 m); cruising speed 149 mph (240 km/h); service ceiling 14,600 ft (4 450 m); range 4,945 miles (7 958 km)
Weights: empty 75,573 lb (34 279 kg); maximum take-off 165,000 lb (74 843 kg)
Dimensions: span 200 ft 0 in (60.96 m); length 117 ft 3 in (35.74 m); height 38 ft 5 in (11.71 m); wing area 3,683 sq ft (342.15 m²)
Armament: none
Operator: USN

Designed as a patrol bomber flying-boat, the Martin JRM Mars (JRM-2 shown) was used only as a transport.

Martin PBM Mariner

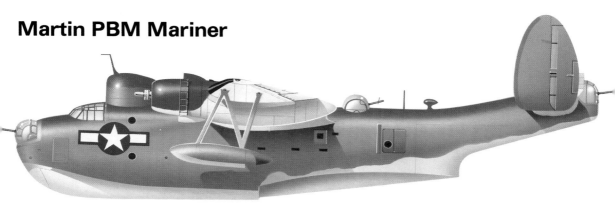

Martin PBM-5 Mariner of a US Navy patrol flying boat squadron in 1944-5.

History and Notes

Developed to replace Martin's earlier open-cockpit P3M flying boats which had served the US Navy as patrol aircraft since 1931, the Martin 162 was designed in 1937 and a single-seat quarter-scale model, known as the 162A, was built to test its flight characteristics. On 30 June 1937 the US Navy ordered a single develop-ment prototype which, as the XPBM-1, was flown on 18 February 1939, powered by two 1,600-hp (1 193-kW) Wright R-2600-6 Cyclone engines. These were mounted in large nacelles which included weapons bays to accommodate a combined total of up to 2,000 lb (907 kg) of bombs or depth charges, and the XPBM-1 had provision for five 0.50-in (12.7-mm) and one 0.30-in (7.62-mm) gun in nose, dorsal and tail turrets, and at waist positions. The underwing floats were retractable and the tailplane was originally without dihedral, although this was later modified to follow the angle of the inboard section of the gull wing.

Twenty production PBM-1s had been ordered on 28 December and these were completed by April 1941, together with a single XPBM-2 with additional fuel capacity and provision for catapult launching. They were initially delivered to US Navy Squadron VP-74, formed from the merged VP-55 and VP-56.

The PBM-3 which followed had 1,700-hp (1 268-kW) R-2600-12 engines in elongated nacelles each able to house four 500-lb (227-kg) bombs or depth charges, and larger non retractable floats were introduced. A total of 379 was ordered on 1 November 1940, but the first 50 aircraft were unarmed PBM-3R transports with strengthened cargo floor, cargo loading doors, and able to seat 20 passengers. These were followed by 274 armed PBM-3Cs and 201 PBM-3Ds, the latter with 1,900-hp (1 417-kW) R-2600-22 engines driving four-bladed propellers. The -3Ds also had increased armour protection for the crew and additional armament which comprised two 0.50-in (12.7-mm) guns in each of the bow, dorsal and tail turrets and a similar weapon in each of the waist positions. Bomb load in the nacelle bays was also increased and could include eight 500-, 1,000- or 1,600-lb (227-, 454- or 726-kg) bombs or eight depth bombs. Alternative offensive loads were four mines or two torpedoes, the latter carried beneath the wings. Fuel tanks were self-sealing, with the exception of auxiliary tanks which could be fitted in the nacelle

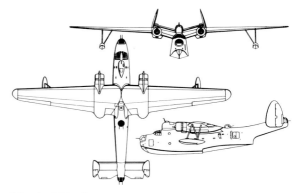

Martin PBM-3D Mariner

bomb bays.

Examples of both -3C and -3D variants were fitted with search radar, mounted dorsally just behind the cockpit, and a specialised long-range anti-submarine version, the PBM-3S, was developed in 1944. Some 156 were built, with R-2600-12 engines and additional fuel capacity to provide increased range. A weight reduction programme resulted in the removal of most of the armour plating and the deletion of the power-operated nose and dorsal turrets, armament being limited to four hand-held 0.50-in (12.7-mm) guns.

The projected PBM-4, which was to have been powered by R-3350-8 engines, was not produced, despite the fact that 180 had been ordered in 1941. However, in 1943 the XPBM-5 was evolved and this version was ordered instead on 3 January 1944. Powered by 2,100-hp (1 566-kW) R-2800-34 engines and with eight 0.50-in (12.7-mm) guns installed, PBM-5s were delivered from August 1944. Some were fitted with APS-15 radar, being redesignated PBM-5E, and others specially equipped for anti-submarine warfare were identified as PBM-5Ss. Production totalled 631.

The last of the Mariner line was the PBM-5A amphibian, which was used principally for air-sea rescue duties by the US Navy and US Coast Guard after World War II, 36 being manufactured before production ceased in April 1949.

Twenty-seven PBM-3Bs, with some British equip-ment, were supplied to the Royal Air Force as

Martin PBM Mariner

Martin PBM-5 Mariner of the Escuela de Especializacion Aeronaval (Uruguayan naval air arm) in the 1950s.

Mariner GR.Is, the first aircraft arriving at Beaumaris, Anglesey in August 1943. A special operational experience and training unit, No. 524 Squadron, was set up on 20 October at Oban, but after only a few weeks it was decided that the Mariner would not be used operationally by the Royal Air Force and the squadron was disbanded on 7 December 1943, its aircraft passing to HQ Transport Command. The remaining Mariners were stored at No. 57 Maintenance Unit, Wig Bay near Stranraer, awaiting return to the United States. The Royal Australian Air Force received 12 Mariners between 1943 and 1946, for operation by No. 41 Squadron.

Postwar deliveries from US Navy stocks included three PBM-5s for the Uruguayan navy, some similar aircraft for the Argentine navy and 17 for the Netherlands navy. The last were supplied in 1955-6 to replace Lockheed PV-2 Harpoons, serving with No. 321 Squadron at Biak in Dutch New Guinea and with No. 8 Squadron (for training and air-sea rescue duties)

at Valkenburg until withdrawn in March 1960 on replacement by Lockheed P2V-7 Neptunes.

Specification
Type: seven/eight-seat patrol flying boat
Powerplant (PBM-3D): two 1,900-hp (1 417-kW) Wright R-2600-22 Cyclone radial piston engines
Performance: maximum speed 211 mph (340 km/h) at 1,500 ft (455 m); service ceiling 19,800 ft (6 035 m); range 2,240 miles (3 605 km)
Weights: empty 33,175 lb (15 048 kg); maximum take-off 58,000 lb (26 308 kg)
Dimensions: span 118 ft 0 in (35.97 m); length 79 ft 10 in (24.33 m); height 27 ft 6 in (8.38 m); wing area 1,408 sq ft (130.80 m²)
Armament: eight 0.50-in (12.7-mm) machine guns (in nose and dorsal turrets and at waist and tail positions), plus up to 8,000 lb (3 629 kg) of bombs or depth charges
Operators: RAAF, RAF, USCG, USN

McDonnell FD/FH Phantom

History and Notes
Although the new McDonnell company had not been involved in large-scale production contracts, and had no previous experience of designing aircraft for shipboard operations, in 1943 the US Navy decided to approach McDonnell to design a new carrier-based fighter. This was to be powered by the newly developing turbojet engine, and while the choice of contractor might seem more than speculative for this untried concept, it was considered that they were likely to approach its design with a far more open mind.

While McDonnell's design team saw no serious problem in finalising a suitable airframe, there was some difficulty in arriving at the most suitable combination of turbojets, which were to be provided by the Westinghouse Electric Corporation. This company had under development a range of engines with diameters of 9.5, 11, 13.5 and 19 in (0.24, 0.28, 0.34 and 0.48 m) and the potential of eight, six, four and two of these engines respectively was investigated to ascertain which combination would prove most effective. Final selection was for two of the maximum

diameter, for not only was this the most simple for installation, control and instrumentation, but also represented a significant weight saving over the alternatives.

A pioneering design, the McDonnell FH-1 Phantom was the US Navy's first all-jet fighter, and as such served for only a short time.

McDonnell FD/FH Phantom

The US Navy contract for three XFD-1 prototypes was placed on 30 August 1943, but it was not until January 1945 that the first airframe was completed, and the first flight was made on 26 January 1945. Of cantilever low-wing monoplane configuration, the XFD-1 had an all-metal wing that folded for carrier stowage, had electrically-actuated split trailing-edge flaps, and had a turbojet engine buried in each wing root; the fuselage and tail unit were conventional light alloy structures; and the landing gear was of the retractable tricycle type, with the main units retracting inward into the undersurface of the wing, the nosewheel unit aft into fuselage. Provision was made for catapult-launching. The pilot's cockpit was well forward of the wing, providing an excellent view.

Successful testing resulted in the US Navy awarding a contract for 100 production FH-1 Phantom aircraft on 7 March 1945, but war-end cancellations resulted in only 60 being built, with deliveries between 23 July 1947 and 27 May 1948. The Phantoms' service was only short, soon displaced by more advanced turbojet-powered aircraft, but they recorded a little batch of firsts: the first US pure-jet to operate on an aircraft-carrier, and the first to serve with the US Marine Corps and US Navy (Squadron VF-17A).

Specification

Type: single-seat carrier-based fighter
Powerplant (FH-1): two 1,600-lb (726-kg) thrust Westinghouse J30-WE-20 turbojets
Performance: maximum speed 479 mph (771 km/h) at sea level; cruising speed 248 mph (399 km/h); service ceiling 41,100 ft (12 525 m); range 980 miles (1 577 km)
Weights: empty 6,683 lb (3 031 kg); maximum take-off 12,035 lb (5 459 kg)
Dimensions: span 40 ft 9 in (12.42 m); length 38 ft 9 in (11.81 m); height 14 ft 2 in (4.32 m); wing area 276 sq ft (25.64 m²)
Armament: four 0.50-in (12.7-mm) machine-guns in fuselage nose
Operators: USMC, USN

McDonnell XP-67

History and Notes

First established in July 1939, the McDonnell Aircraft Corporation started with the objective of securing military contracts, and began immediately the design of a new long-range fighter. it was obviously of importance to the company to evolve an outstanding design and, in order to achieve this, some new features were incorporated. When the design was submitted to the US Army Air Corps for consideration, the powerplant proposal was considered to be unsuitable, but there was sufficient interest in the overall plan for the US Army to request a further submission.

The original powerplant layout had envisaged the installation of a powerful engine (2,500-3,000 hp/ 1 864-2 237 kW) with a two-stage supercharger mounted within the fuselage and driving, via right-angle gearboxes and shafting, two pusher propellers. With the need to adopt a more conventional engine layout, a complete redesign was initiated, leading to a single-seat fighter to be powered by two Continental XIV-1430-1 engines with General Electric turbochargers. To enhance the performance of these engines, the exhaust from each turbocharger was ejected through an annular aperture at the rear of each engine nacelle, providing some thrust augmentation. The pilot was to be accommodated in a pressurised cabin, and armament was to consist of six 0.50-in (12.7-mm) machine-guns and four 20-mm cannon. When resubmitted, this proposal was sufficiently attractive for the US Army to award a contract for two XP-67 prototypes, on 29 July 1941.

Of all-metal construction, the aircraft had a most unusual profile when viewed in planform, for efforts had been made by the design team to ensure that engine nacelles and fuselage were shaped to preserve true aerofoil sections throughout. The wide-track landing gear was of the retractable tricycle type, and the tail unit conventional except that the tailplane and elevators were mounted part way up the fin. During the construction period it was decided to change the armament to comprise an unprecedented six 37-mm cannon, and it was proposed that one prototype should be so armed, the other with a single 75-mm cannon, so that they could be evaluated during service trials to discover which was the more effective.

However, when the first XP-67 made its initial flight on 6 January 1944, it was without armament, cabin pressurisation and oxygen system. Early flight trials were plagued with engine problems and it was not until mid-May 1944 that USAAF pilots were able to fly the XP-67 and to discover that the performance was far from satisfactory. Modifications were made to

The McDonnell XP-67 was an extraordinary attempt to blend a whole airframe into basic aerofoil shapes.

McDonnell XP-67

improve instability problems, but before official trials commenced the first prototype was extensively damaged by fire and the programme was abandoned.

Specification
Type: single-seat long-range fighter
Powerplant: two 1,350-hp (1 007-kW) Continental XIV-1430-17/19 contra-rotating inline piston engines
Performance: maximum speed 405 mph (652 km/h) at 25,000 ft (7 620 m); service ceiling 37,400 ft (11 400 m); maximum range 2,385 miles (3 838 km)
Weights: empty 17,745 lb (8 049 kg); maximum take-off 25,400 lb (11 5231 kg)
Dimensions: span 55 ft 0 in (16.76 m); length 44 ft 9¼ in (13.65 m); height 15 ft 9 in (4.80 m); wing area 414 sq ft (38.46 m²)
Armament (proposed): six 37-mm or one 75-mm cannon
Operator: USAAF (for evaluation only)

Naval Aircraft Factory N3N

History and Notes
The Naval Aircraft Factory (NAF), authorised in 1917 and established in the Philadelphia Navy Yard, Pennsylvania, in 1918, was created to provide the US Navy with its own manufacturing and test facilities. However, in 1918 the most urgent requirement of the US Navy was an adequate supply of aircraft, with the result that the NAF was immediately involved in what, for the size of the facility, was quite large-scale production. This continued until 1922, after which the NAF functioned more or less as had been intended from the beginning, building prototypes only of US Navy-designed aircraft, carrying out modification and overhaul of machines in service, as well as providing testing of anything from components to complete aircraft.

This period was of only short duration, however, for in the mid-1930s the NAF began construction of 10 per cent of the aircraft procured by the US Navy, to keep a check on costs and manufacturing techniques, and with American involvement in World War II the NAF again became involved in the design and construction of aircraft on a large scale.

In 1934 the US Navy had designed a new primary trainer which offered superior performance to the Consolidated NY-2s and NY-3s then in service, and construction of a prototype XN3N-1 resulted in a first flight in August 1935. Following successful tests in both landplane and seaplane configurations, the decision was made to order it into production, and the NAF began immediately the manufacture of the first batch of N3N-1s, of which 179 had been ordered. The N3N was a neat single-bay biplane with staggered wings, conventional fuselage, braced tail unit, and fixed tailwheel type landing gear as standard, and 158 aircraft of this initial order were to be powered by a 220-hp (164-kW) Wright J-5 radial engine which the US Navy had held in store. An additional prototype was ordered as XN3N-2 and one production aircraft was converted to XN3N-3 prototype configuration, both powered by 240-hp (179-kW) US Navy-built versions of the Wright R-760-96 radial engine. This was considered desirable to evaluate the suitability of the R-760 engine as a replacement for the J-5 which was obsolescent, and when found to be suitable the last 20 production N3N-1s were provided with the US Navy-built R-760.

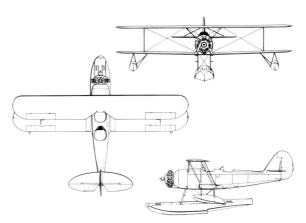

Naval Aircraft Factory N3N-3 Canary

At a later date all remaining N3N-1s had their J-5 engines replaced by 235-hp (175-kW) R-760-2s.

Production of 816 N3N-3s followed from 1938, these having amended tail units and landing gear, and four of these were transferred to the US Coast Guard in 1941. Used extensively in US Navy primary flying training schools throughout World War II, the majority became surplus immediately the war ended. The exception was due to a small number of the seaplane version retained for primary training at the US Naval Academy and these, when retired in 1961, were the last biplanes to be used in US military service.

Specification
Type: two-seat primary trainer
Powerplant (N3N-3): one 235-hp (175-kW) Wright R-760-2 Whirlwind 7 radial piston engine
Performance: maximum speed 126 mph (203 km/h); cruising speed 90 mph (145 km/h); service ceiling 15,200 ft (4 635 m); range 470 miles (756 km)
Weights: empty 2,090 lb (948 kg); maximum take-off 2,792 lb (1 266 kg)
Dimensions: span 34 ft 0 in (10.36 m); length 25 ft 6 in (7.77 m); height 10 ft 10 in (3.30 m); wing area 305 sq ft (28.33 m²)
Armament: none
Operators: USCG, USN

North American B-25 Mitchell

North American B-25A Mitchell of the 34th Bomb Squadron, 17th Bomb Group, USAAF, based at McChord Field (Washington) in November 1941.

History and Notes

North American's response to the US Army Air Corps' Circular Proposal 38-385 for a twin-engined attack bomber was the NA-40, a shoulder-wing design with a tricycle landing gear and capable of carrying a 1,200-lb (544-kg) bomb load. Armament consisted of 0.30-inch (7.62-mm) machine-guns in nose, dorsal and ventral positions. The prototype, built at the Inglewood factory, was first flown by Paul Balfour in January 1939, powered by two 1,100-hp (820-kW) Pratt & Whitney R-1830-S6C3-G engines which were soon replaced by Wright CR-2600-A71 Cyclones each rated at 1,300 hp (969 kW). In this form the aircraft became the NA-40-2 and in March it was delivered to Wright Field for USAAC evaluation, crashing two weeks later as the result of pilot error.

The USAAC was impressed by the promise of the NA-40, however, and North American was asked to continue development of the aircraft for the medium bomber role under the company designation NA-62. September 1939 saw the completion of the basic design of the NA-62 and in that month the type was ordered

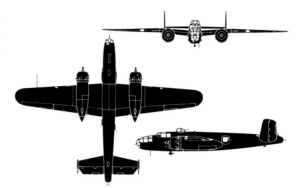

North American B-25J Mitchell

North American B-25Js of the 12th Air Force's 321st Bombardment Group over Italy. Many B-25Js were soon retrofitted with a 'solid' nose carrying eight 0.50-in (12.7-mm) machine-guns for the ground-attack role.

North American B-25 Mitchell

into immediate production under a USAAC contract for 184 aircraft designated B-25. Several improvements were incorporated, including the widening of the fuselage to allow the pilot and co-pilot/navigator to be seated side-by-side in a cockpit faired into the fuselage, rather than in the tandem glasshouse of the NA-40; the relocation of the wing to a mid-position; and an increase operating weights and bomb load. New engines were also specified, these being 1,700-hp (1 268-kW) Wright R-2600-9 Cyclone radials, and a tail gun position was added.

The B-25 was named after the controversial proponent of US air power, William 'Billy' Mitchell, and the first production machine was flown on 19 August 1940. Nine B-25s were completed with the original root-to-tip dihedral before flight tests revealed a degree of directional instability, which was remedied by a reduction in the dihedral angle on the outer wing panels.

The introduction of self-sealing fuel tanks and crew protection armour plating, from aircraft number 25, resulted in redesignation to B-25A. Forty B-25As were built, and this variant was the first to see operational service, with 17th Bombardment Group (Medium) at McChord Field, scoring the type's first kill on 24 December 1941 when a Japanese submarine was sunk off the US west coast.

Some 120 B-25Bs were manufactured, this model having power-operated dorsal and ventral turrets, each with two 0.50-in (12.7-mm) machine-guns. B-25Bs were among the US reinforcements sent to Australia in 1942, serving with the 3rd Bombardment Group's 13th and 19th Squadrons, and were also used for the Tokyo raid, led by Lieutenant Colonel James H. Doolittle, on 18 April 1942. For this attack 16 modified aircraft, with an autopilot, fuel tankage increased by more than 60 per cent to 1,141 US gallons (4 319 litres) and the ventral gun turret and Norden bombsight removed, took off from the carrier USS *Hornet* for an 800-mile (1 287-km) flight to their targets at Tokyo, Kobe, Yokohama and Nagoya, flying on to China where most force-landed.

Two USAAF contracts, for 63 and 300 aircraft, were placed for the B-25C which had an autopilot, R-2600-13 engines and additional bomb-racks under the wings and fuselage which could carry, respectively, eight 250-lb (113-kg) bombs and a 2,000-lb (907-kg) torpedo for anti-shipping strikes; total offensive load was 5,200 lb (2 359 kg).

Other B-25C contracts included a Dutch order for 162, intended for service in the Netherlands East Indies, although these were never delivered there (and probably diverted to the Royal Air Force), and two Defense Aid-financed contracts, each for 150 and intended for delivery to China and the UK. The basically-similar B-25D was built in a US government-owned but North American-operated factory at Kansas City, where the company manufactured two batches of 1,200 and 1,090 aircraft.

Two machines from the B-25C line were modified for experiments into wing de-icing, these being the XB-25E with a hot-air system and the XB-25F which used electrically heated elements.

Developed for attacks on Japanese shipping, the B-25G carried a 75-mm M4 US Army cannon mounted in the nose, the cannon being provided with 21 15-lb (6.8-kg) shells. The armament was supplemented by a pair of 0.50-in guns which were used also to aim the heavier weapon. in addition, the dorsal and fully-retractable ventral turrets each contained two machine-guns. Five B-25Cs were, in fact, completed as B-25Gs, and 400 were subsequently built at Inglewood. This version was initially assigned to the US Far East Air Forces, entering service with the 498th Squadron in February 1944.

The Mitchell with the greatest firepower was the B-25H, of which 1,000 were built at Inglewood. The 75-mm cannon was of the lighter T13E1 model and the four 0.50-in (12.7-mm) guns, also mounted in the nose, were augmented by two similar guns in blisters on each side of the fuselage below the cockpit. The twin-gun dorsal turret was relocated to a position just aft of the cockpit, and armament was completed by a 0.50-in (12.7-mm) gun in each of the waist positions and two in the tail. Additionally, the B-25H could carry a 3,000-lb (1 361-kg) bomb load and a torpedo, as could the B-25J in which the glazed nose with its bomb aiming station was reintroduced, reducing the nose armament to one hand-operated and four fixed 0.50-in (12.7-mm) guns. Some later aircraft had a solid nose with eight 0.50-in (12.7-mm) guns, bringing the total of these weapons to 18. Underwing racks could carry eight 5-in (127-mm) rockets. The USAAF contract was for 4,805 B-25Js, but as the war ended 415 were cancelled and 72 were completed but not delivered; all were manufactured at Kansas City.

For reconnaissance duties the F-10 version was introduced in 1943, 10 being converted from B-25Ds. Armament was removed, additional fuel tanks fitted in the bomb bay, and cameras installed in the rear fuselage and in the nose.

Sixty B-25Ds, B-25Gs, B-25Cs and B-25Js were converted during 1943-4 for use as advanced trainers under the designations AT-25A, AT-25B, AT-25C and AT-25D. They were later redesignated TB-25D, TB-25G, TB-25C and TB-25J; more than 600 of the last model were converted after the war and between 1951 and 1954 117 and 40 Mitchells were respectively converted to TB-25K and TB-25M standard, as flying classrooms for instruction in the use of Hughes E-1 and E-5 fire-control radar. The final training versions were the TB-25L and TB-25N multi-engine conversion trainers, of which Hayes Aircraft Corporation produced 90 and 47 examples respectively.

US Navy Mitchells, of which delivery began in January 1943 with an initial assignment to VMB-413, comprised 50 PBJ-1Cs, 152 PBJ-1Ds , one PBJ-1G, 248 PBJ-1Hs and 255 PBJ-1Js, the letter suffix identifying the equivalent B-25 variant.

The advent of the Mitchell allowed the Royal Air

North American B-25 Mitchell

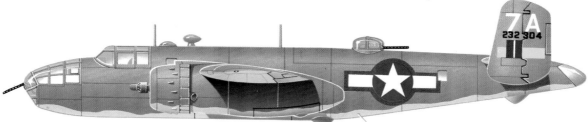

North American B-25C Mitchell of the 487th Bomb Squadron, 340th Bomb Group, 12th Air Force, based in Catania (Sicily) in September 1943.

North American B-25C Mitchell of the 488th Bomb Squadron, 340th Bomb Group, 9th Air Force, based at Sfax (Tunisia) in April 1943.

North American Mitchell Mk II of No. 320 Squadron, RAF Bomber Command, based in the UK during 1943.

North American Mitchell Mk II of No. 226 Squadron, RAF 2nd Tactical Air Force, based in the UK during June 1944.

North American B-25C Mitchell of the 81st Bomb Squadron, 12th Bomb Group, 12th Air Force, based at Gerbini (Sicily) in August 1943.

North American B-25 Mitchell

North American B-25J Mitchell of No. 18 Squadron, Royal Netherlands Indies Army Air Corps, based near Darwin (Australia) in autumn 1944.

North American B-25J Mitchell of No. 2 Squadron, Royal Australian Air Force, based in north west Australia during spring 1945.

Force to replace the Douglas Bostons and Lockheed Venturas flown by No. 2 Group on daylight operations. The first 23 aircraft, delivered in May and June 1942, were B-25B Mitchell Is, three of which were subjected to evaluation and acceptance trials at the Aircraft and Armament Experimental Establishment; of this batch one was retained in Canada and another crashed before delivery. The rest were flown to Nassau in the Bahamas where No. 111 Operational Training Unit had been established on 20 August, based at Windsor and Oakes Fields. Between May 1943 and June 1945, No. 13 OTU also flew Mitchells from Bicester, Finmere and Harwell in Britain.

As deliveries of B-25C Mitchell IIs built up through the second half of 1942, Bahamas-trained crews returned to the United Kingdom to form the first squadrons, originally to have been Nos. 21 and 114. In fact, the first two operational units were Nos. 98 and 180 Squadrons, formed at West Raynham on 12 and 13 September, respectively. The Dutch-manned No. 320 Squadron gave up its Lockheed Hudsons for Mitchells at Methwold in March 1943, and No. 226 replaced its Bostons at Swanton Morley in May. All four squadrons flew Mitchells until after the cessation of hostilities.

After initial problems with the Mitchell's armament had been solved, RAF operations began on 22 January 1943 when six aircraft from No. 98 Squadron and six from No. 180 attacked oil installations at Ghent. The four squadrons of No. 2 Group continued their formation attacks throughout 1943 and 1944, operating increasingly in a tactical role following the Allied invasion of France in June 1944. Nos. 98, 180 and 320 Squadrons moved up to Melsbroek, Brussels in October, while No.

226 took up residence at Vitry-en-Artois. The last No. 2 Group Mitchell operation of the war was flown on 2 May 1945 when 47 aircraft attacked marshalling yards at Itzehoe. RAF Mitchell operations outside Europe included those of Nos. 681 and 684 Squadrons, flying in a photographic reconnaissance role in India from 1943 to 1945.

RAF serial batches covered 886 Mitchells, comprising 23 B-25B Mitchell Is; 432 B-25Cs and 113 B-25Ds, both of which were known as Mitchell IIs; and 316 B-25J Mitchell IIIs. The remaining two were B-25Gs, with the 75-mm gun, and one of them, with armament removed, was probably the last in service in the United Kingdom, flying with the Meteorological Research Flight at Farnborough as late as 1950.

In addition to the Dutch-manned No. 320 Squadron, RAF Mitchell units manned by foreign nationals included No. 305, whose Polish crews converted from Vickers Wellingtons at Swanton Morley in September 1943, and No. 342 (Lorraine) Squadron which exchanged its Bostons for Mitchells at Vitry-en-Artois in March 1945. After disbandment as RAF units both the French and Dutch took their aircraft home.

No. 320 Squadron was reformed at Valkenburg as a Dutch navy patrol/search and rescue unit on 29 March 1949, its initial equipment including Mitchells which, replaced by Lockheed Harpoons when the squadron changed role to maritime patrol, were passed on first to No. 5 Squadron, formed on 7 May 1951, and then to No. 8 Squadron on 10 March 1952.

During the war the Dutch had flown Mitchells at the Royal Netherlands Military Flying School at Jackson, Missouri and with No. 18 (Netherlands East Indies)

North American B-25 Mitchell

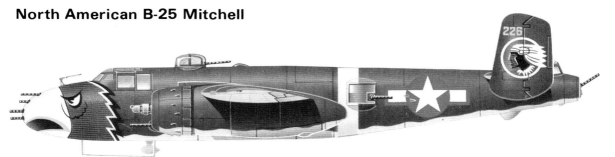

North American B-25J Mitchell of the 498th Bomb Squadron, 345th Bomb Group, 5th Air Force, based at San Marcelius, Luzon island (Philippines group) in April 1945.

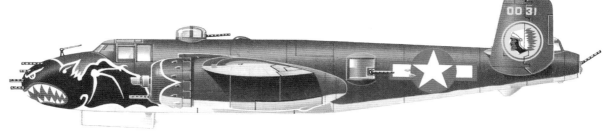

North American B-25J of the 499th Bomb Squadron, 345th Bomb Group, 5th Air Force, based on Ie Shima (Okinawa) in July 1945.

Squadron of the Royal Australian Air Force, formed with Dutch personnel at Canberra on 4 April 1942, and operating throughout the campaigns to recapture the Pacific islands. Control passed to the Netherlands on 15 January 1946 and, based at Bandoeng in Java, the squadron was soon in action again, in the conflict with the Indonesians. After the ceasefire, which resulted in the disbandment of the Netherlands East Indies air force on 21 June 1950, Mitchells were handed over to the new Indonesian government to form the equipment of the bomber flight of No. 1 Squadron. The RAAF acquired 50 Mitchells, including B-25Ds and B-25Js, which were flown by Nos. 2 and 119 Squadrons.

The Mitchells supplied to the Chinese air force remained in service throughout the postwar struggle which led to the communist overthrow of the Chiang Kai-shek government, some captured aircraft being used by the Sino-Communist forces while others escaped to Taiwan. A total of 807 Mitchells was supplied under Lend-Lease to the USSR, although eight were lost in transit.

In Central and South America, Mitchells were supplied to Brazil, Chile, Mexico and Uruguay. Signature of the Rio Pact of Mutual Defense in 1947 resulted in the United States supplying B-25Js to Brazil, Colombia, Cuba, Peru and Venezuela.

Among Commonwealth air forces, the Royal Canadian Air Force received a small number of Mitchell IIs from the Royal Air Force in May 1944 and these, modified to the standard of the USAAF F-10 version with cameras installed in the nose, equipped the Photographic Flight at Rockcliffe, Ottawa. The unit was unofficially designated No. 13 (Photographic)

Squadron, as part of No. 7 (Photographic) Wing, although this title was not formally promulgated until 15 November 1946. The squadron was renumbered as No. 413 (Photographic) Squadron on 1 April 1947 and the Mitchells served alongside Avro Lancaster Xs until withdrawn in October 1948.

Auxiliary squadrons formed after the war includes Nos. 406 and 418 Squadrons, based at Saskatoon and Edmonton respectively. Both were light bomber units, flying Mk II and Mk III Mitchells until they were retired in 1958. VIP-configured Mitchells were used by No. 412 Squadron between 1956 and 1960.

Specification

Type: five-seat medium bomber

Powerplant (B-25J): two 1,700-hp (1 268-kW) Wright R-2600-92 Cyclone radial piston engines

Performance: maximum speed 272 mph (438 km/h) at 13,000 ft (3 960 m); cruising speed 230 mph (370 km/h); service ceiling 24,200 ft (7 375 m); range 1,350 miles (2 173 km)

Weights: empty 19,480 lb (8 836 kg); maximum take-off 35,000 lb (15 876 kg)

Dimensions: span 67 ft 7 in (20.60 m); length 52 ft 11 in (16.13 m); height 16 ft 4 in (4.98 m); wing area 610 sq ft (56.67 m²)

Armament: from 12 to 18 0.50-in (12.7-mm) machine-guns, plus eight 5-in (127-mm) rocket projectiles and up to 3,000 lb (1 361 kg) of bombs

Operators: Brazil, China, FAF, FFAF, Netherlands, RAAF, RAF, RCAF, Soviet Union, USAAC/USAAF, USN

North American P-51 Mustang

History and Notes

In another 20 years time, when the peoples of this world enter the 21st century, aviation enthusiasts may still be in heated argument as to whether or not the North American P-51 Mustang was the greatest single-seat fighter to be evolved by any of the combatant nations during World War II. There is, however, no doubt at all that it can be numbered among the half-dozen which will be remembered for as long as men record and discuss the history of aviation, hopefully into an epoch when the world's problems are settled by words and work, rather than by weapons and woe.

One of the small number of aircraft to be conceived, developed, produced and put into wide-scale use all within the six years of the war, the Mustang had its origins in April 1940, when the British Purchasing Commission negotiated with North American Aviation to design and build an advanced fighter for the RAF. This had to meet British specifications and, because of the serious situation in Europe, with German forces already in Denmark and Norway and likely to move towards western Europe at any moment, it was stipulated that a prototype must be completed within 120 days.

This was not quite so wishful as one might think, for North American had already drawn up the outline of a new fighter based on information from air combat in Europe, and the company's design team, headed by Raymond Rice and Edgar Schmued, began immediately to shape the tentative design to fit the British specification. Designated NA-73X by North American,

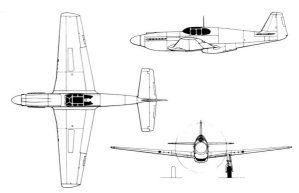

North American Mustang Mk II (side view: NA-73X)

the airframe was completed within 117 days, but the 1,100-hp (820-kW) Allison V-1710-39 engine which was to power it was behind schedule, and it was not until 26 October 1940 that the prototype flew for the first time.

Engineers have for countless years made the comment 'if it looks right, it is right'. This was certainly true of the NA-73X, which at first glance registered as a superb example of aircraft design, and which completed quite rapidly a remarkably trouble-free test programme. On 1 May 1941, just over seven months

North American P-51Ds and one P-51B (background) of the 361st Fighter Group's 375th Fighter Squadron, allocated (but for a short period) to the 8th Air Force.

North American P-51 Mustang

North American Mustang Mk I of an RAF Army Co-operation Command Squadron during mid-1942.

North American Mustang Mk I of No. 613 Squadron, RAF Army Co-operation Command, based at Ringway (UK) during 1942.

North American A-36A of the 27th Fighter-Bomber Group, USAAF, based in Corsica during July 1944.

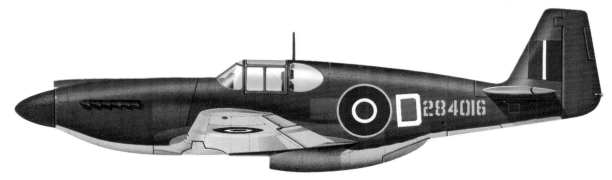

North American A-36A, the sole example of its type transferred to the RAF.

North American P-51 Mustang

after the prototype's maiden flight, the first production example was flown. The second production aircraft was despatched to Britain for evaluation by the RAF, arriving during November 1941, and was soon followed by a steady flow from the initial contract for 320 NA-73s placed by the British Purchasing Commission before the prototype's first flight.

No time was wasted by the RAF in making its evaluation of the new fighter, given the designation Mustang I. It needed no sophisticated equipment to measure the capability of this aircraft; any experienced pilot could determine after only a few minutes in the air that it was fast and highly manoeuvrable at low levels, far superior to any other US fighter then extant. The limitations came at higher altitudes, for the power output of the Allison engine fell off rapidly as it climbed, which meant that the Mustang I in that particular form was unsuited for combat operation in Europe. However, its particular attributes promised well for deployment in a tactical reconnaissance role, and the standard armament of four 0.50-in (12.7-mm) and four 0.30-in (7.62-mm) machine-guns meant also that it had potential for ground-attack.

The RAF's Mustang Is were therefore provided with an obliquely-mounted camera, behind the pilot on the port side, and in this form were used to equip No. 2 Squadron of Army Co-operation Command in April 1942, the first operational sortie being flown on 27 July 1942. Three months later these Mustangs demonstrated their long-range potential, the first RAF single-engine fighter based in Britain to cross the German border, during an attack on the Dortmund-Ems Canal. The Mustang I was soon found to be fulfilling a valuable role, eventually equipping no fewer than 23 squadrons of Army Co-operation Command, leading to a new contract for an additional 300 aircraft of this mark, and with only minor modifications.

North American's development and production of their NA-73 design for the UK had to receive the blessing of the US government before it could become ratified. A condition of this approval was the supply of two examples to the USAAC for evaluation, and two aircraft from the first production batch were delivered and given the designation XP-51. Before that, however, the US Army had already contracted for the procurement of 150 additional aircraft for supply to Britain under Lend-Lease, designating these P-51, and these differed from the earlier version by having self-sealing tanks, and four wing-mounted 20-mm cannon in place of the eight machine-guns. From this batch 93 were supplied to Britain, becoming designated Mustang IA, 55 went to the USAAF as F-6As, equipped with two K-24 cameras for use in a tactical reconnaissance role, and the remaining two also went to the USAAF with different engines, initially as XP-78s, but later brought into the family as XP-51Bs.

USAAF testing of its two XP-51 prototypes had proved highly successful, but at that time the US Army was satisfactorily committed to a large-scale procurement programme involving the Lockheed P-38 and Republic P-47. Their findings confirmed those of the RAF, and it was decided to procure 500 A-36As, these being P-51s provided with dive brakes and underwing racks, to operate as dive-bombers in a close-support role. Armament of this version, first flown in September 1942, comprised six 0.50-in (12.7-mm) machine-guns, and the powerplant consisted of an Allison V-17110-87 engine which was rated at 1,325 hp (988 kW) at 3,000 ft (915 m). These were the first Mustangs to enter operational service with the USAAF, equipping two groups in the Middle East in 1943, and used in support operations during the invasions of Sicily and Italy. At about the same time that the US Army had ordered its A-36As, a second contract had been placed for 310 P-51As with a 1,200-hp (895-kW) V-1710-81 engine, armament of four 0.50-in (12.7-mm) machine-guns, and with underwing racks to accommodate up to 1,000-lb (454-kg) bombs, or two 75- or 150-US gallon (284- or 568-litre) drop tanks. Of the foregoing a single A-36A was supplied to the RAF for evaluation, plus 50 P-51As which became designated Mustang II, and 35 were converted for tactical reconnaissance in USAAF service under the designation F-6B, equipped with two K-24 cameras.

At this point in the Mustang story comes the transition to fulfilment of its design potential, initiated in 1942 soon after the first Mustang Is were received in Britain. In order to provide the all-important performance at high altitude, which was needed for the combat fighter role, it was decided to make experimental installations of Rolls-Royce Merlin 61 and 65 engines in Mustang airframes, and four of the Mk Is were supplied to Rolls-Royce at Hucknall for this purpose. Within six weeks the modifications had been completed and the first tests made, demonstrating such improved performance that the results were communicated immediately to North American. They, following the same lines, installed the 1,430-hp (1 066-kW) US-built Packard Merlin V-1650-3 into two P-51 airframes, these duly becoming the XP-78/XP-51Bs mentioned above. It is worth highlighting here that the Allison V-1710 engine installed in the early P-51s was subject to a rapid loss in power after climbing above a height of about 12,000 ft (3 660 m); the Packard Merlin V-1650-3 installed in the XP-51B prototypes, and which had a two-speed two-stage supercharger with intercooler, was rated at 1,400 hp (1 044 kW) for take-off, and developed 1,450 hp (1 081 kW) at 19,800 ft (6 035 m).

Early testing of the XP-51B in September 1942 confirmed the British findings, a maximum speed of 441 mph (170 km/h) being attained at 29,800 ft (9 085 m). Rate of climb was better than that which the twin-engine P-38 had been built to achieve, 20,000 ft (6 095 m) in six minutes, for one of the first test airframes reached this altitude in 5.9 minutes. The USAAF was suitably impressed, ordering large numbers of Merlin-engined Mustangs. The numbers were so large, indeed, that North American's factory at Inglewood, California, could not cope alone, and a

North American P-51 Mustang

North American F-6A Mustang borrowed from the USAAF and operated by No. 225 Squadron, RAF, based at Souk-el-Khemis (Tunisia) in April 1943.

North American P-51A Mustang flown by Colonel Philip Cochran, commander of the 1st Air Commando Group, USAAF, and based at Hailakandi (India) in March 1944.

North American F-6B Mustang of the 107th Tactical Reconnaissance Squadron, 9th Air Force, based in the UK (later France) during 1944.

North American P-51B Mustang of the 487th Fighter Squadron, 352nd Fighter Group, USAAF, based at Bodney (UK) in May 1944.

North American P-51 Mustang

second production line was thus established in a new plant at Dallas, Texas.

Inglewood began production of the new model in the summer of 1943 as the P-51B, and an identical version from Dallas was designated P-51C. Both of these differed from the earlier P-51/-51As in having a strengthened fuselage, new ailerons, and small variations which were specific to the new powerplant. Armament comprised four 0.50-in (12.7-mm) machine-guns. A total of 1,988 P-51Bs and 1,750 P-51Cs were built before both production lines turned over to construction of the P-51D, more than 2,100 of this total being powered by the V-1650-7 engine which produced 1,450 hp (1 081 kW) for take-off and had a combat rating of 1,695 hp (1 264 kW) at 10,300 ft (3 140 m).

The RAF began to receive its first Lend-Lease allocations from this P-51B/-51C production at the beginning of 1944, the first equipping No.19 Squadron at Ford, Sussex, in February 1944. All designated Mustang III, they comprised 274 aircraft equivalent to P-51B and 636 to P-51C, and were used extensively by no fewer than 21 RAF squadrons, many of which were deployed with the 2nd Tactical Air Force. All of these as delivered had the original sideways opening cockpit canopy which had been standard on all production aircraft until that time, but for air combat the rear view was totally inadequate, and the Mustang IIIs were equipped instead with a modified sliding hood which overcame this shortcoming.

In USAAF service P-51B/-51Cs began to enter service a little earlier than with the RAF, being used operationally by the 8th Air Force in Britain for their first long-range escort mission, against Kiel, on 13 December 1943. By early 1944, and using drop tanks to confer the necessary range, they were regularly accompanying 8th Air Force bombers on daylight missions deep into the German homeland, making the first of many visits to Berlin in March, and becoming operational at about the same time with the 10th Air Force in Burma and the 15th Air Force in Italy. Of the 2,828 P-51B/-51C fighters received by the USAAF, 71 P-51Bs and 20 P-51Cs were modified for the tactical reconnaissance role with the designation F-6C.

By then, of course, North American had already become involved in what was to become the major production version, the P-51D, of which 7,956 were built, 6,502 coming from Inglewood alone. They differed from the P-51B/-51Cs by introducing as standard a bubble canopy to provide the pilot with an excellent all-round view, a modified rear fuselage, and an armament of six 0.50-in (12.7-mm) machine-guns. Of this version 136 were modified to serve as tactical reconnaissance F-6Ds. Later production aircraft introduced as standard a small dorsal fin to compensate for a loss of rear fuselage profile surface resulting from the cockpit modification, and the last 1,100 produced at Inglewood were equipped to launch 5-in (127-mm) rocket projectiles. P-51Ks followed, these differing only by a change in propeller, and of the 1,500 ordered, 163 were completed as tactical reconnaissance F-6Ks. Of the

above versions the RAF was allocated 281 P-51Ds and 594 P-51Ks, all designated Mustang IV.

In 1944 the USAAF had contracted for three XP-51F and two XP-51G versions. These were experiments in lightweight construction, and the opportunity was taken to redesign the airframe and take a new look at the entire project. This resulted in the replacement of the original laminar-flow wing by one of a more advanced low-drag section, its planform changed somewhat, and to reduce drag to a minimum the cockpit canopy was elongated and the bulky oil cooler replaced by a shallower heat exchanger. Unnecessary equipment was deleted and substantial weight reductions were achieved by careful redesign and simplification of the structure, and by the introduction of new lightweight materials, including plastics. The XP-51Fs were powered by 1,695-hp (1 264-kW) Packard Merlin V-1650-7 engines, the XP-51Gs by the 1,410-hp (1 424-kW) Rolls-Royce Merlin 145M. Subsequently two similar XP-51J prototypes were ordered, to be powered by an Allison V-1710-119 engine which was rated at 1,720 hp (1 283 kW) at 20,700 ft (6 310 m), but only one of these was completed.

With the knowledge gained from these lightweight prototypes, North American evolved what was to be the last production version of the Mustang, the P-51H. Powered by the V-1650-9 Packard Merlin, which had a combat rating of 2,218 hp (1 654 kW) with water injection at 10,200 ft (3 110 m), these proved to be the fastest of all the Mustangs, able to attain 487 mph (784 km/h) at 25,000 ft (7 620 m). Generally similar to the XP-51F, production aircraft subsequent to number 13 all had vertical tail surface area increased by the introduction of a taller fin and rudder, and all were of an increased length, had a longer dorsal fin and a shorter bubble canopy. More importantly, they had a 40 per cent weight saving by comparison with the P-51D, making possible enhanced performance.

P-51H production totalled 555 before VJ-Day brought cancellation of the balance of the 2,000 ordered. Also cancelled were 1,700 similar V-1650-11 powered P-51Ls, and 1,628 P-51Ms, which was the Dallas-built version of the P-51H, and of which only a single example was completed. Of these latter versions the RAF received one XP-51F, one XP-51G and one P-51H for evaluation.

On the grand total of 14,819 Mustangs built, production ended in America, but one other source of supply had originated in early 1944 when 314 P-51Ks had been allocated to the RAAF under Lend-Lease. Before any of these were delivered Commonwealth Aircraft Corporation in Australia had begun to tool up to build P-51Ds under licence, beginning by the assembly of 80 aircraft from imported components. Subsequently, a further 120 were built in Australia, but none of the total of 200 was completed in time to be used operationally before VJ-Day. Commonwealth Aircraft production comprised 80 Mustang Mk 20 (P-51Ds), 26 Mk 21 (V-1650-7 engines, 14 later converted to Mk 22), 67 Mk 23 (Merlin 66 or 70 engines) and 13

North American P-51 Mustang

North American P-51B Mustang of the 318th Fighter Squadron, 325th Fighter Group, USAAF, based in Italy during late 1944.

North American P-51B Mustang of the 374th Fighter Squadron, 361st Fighter Group, USAAF, based at Bottisham (UK) in June 1944.

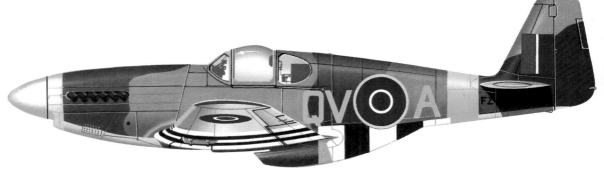

North American Mustang Mk III of No. 19 Squadron, RAF 2nd Tactical Air Force based at Ford (UK) in summer 1944.

North American Mustang Mk IIIB of No. 316 (Polish) Squadron RAF 2nd Tactical Air Force, based at Coltishall (UK) in summer 1944.

North American P-51 Mustang

The North American P-51 Mustang was the answer
to every fighter pilot's dream. Its speedy lines,
smooth finish and classic lines marked it as the

culmination of its breed. It looked simply elegant,
and was elegantly simple, but in the frame of a
light-footed ballet dancer it carried the punch of a
heavyweight boxer. Packed with fuel, and carrying
even more in external tanks, it had the range to
escort bombers to the deepest targets in the Third
Reich. And based on Iwo Jima, Mustangs escorted
the very long-range B-29s in their devastating raids
on the Japanese home islands. The type could
tangle without qualms with anything the Germans
or Japanese could put into the air, and win most of
these contests. The aircraft illustrated is a P-51D-
NA built by North American and allocated to the
375th Fighter Squadron, 361st Fighter Group, 8th
Air Force, based at several airfields in the UK. The
squadron received Mustangs while located at
Bottisham in Cambridgeshire, beginning in May
1944; the unit later moved to Little Walden in
Essex.

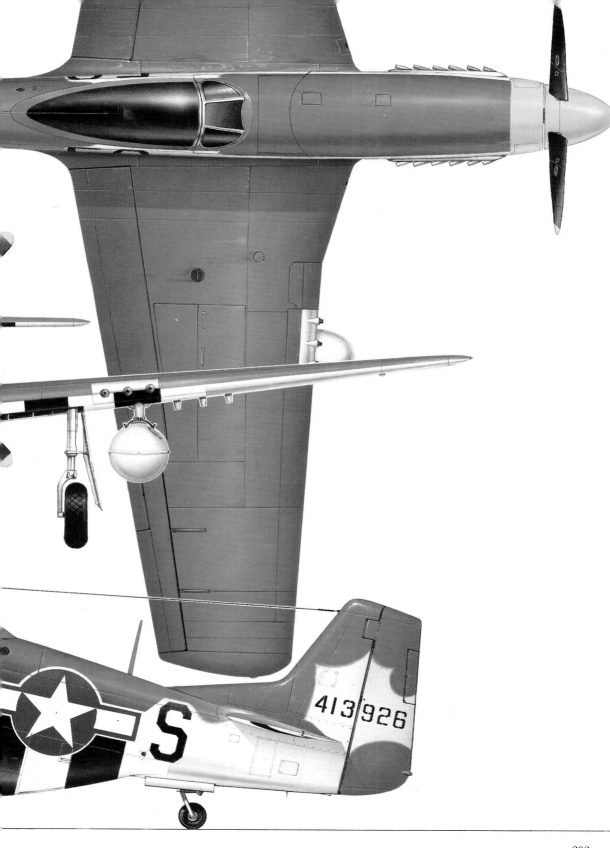

North American P-51 Mustang

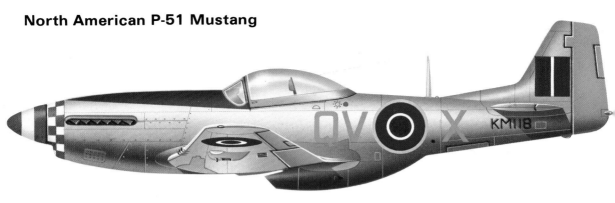

North American Mustang Mk IV of No. 19 Squadron, RAF, based at Acklington (UK) in June 1945.

Mk 22 for tactical reconnaissance.

Under Lend-Lease 50 P-51Ds were supplied to China, 40 to Netherlands forces in the Pacific theatre, and some USAAF P-51s were supplied to the AVG in China. In the immediate postwar years P-51s remained in service, particularly with Strategic Air Command, until 1949, and others served for several more years with US Air Reserve and Air National Guard units, being among the first USAF fighters to see action in the Korean War. In the RAF some remained in service with fighter command until 1946, and war surplus P-51s from both the USA and the UK continued to have some years of postwar employment with over 50 air forces.

Specification

Type: single-seat interceptor or long-range escort fighter
Powerplant (P-51D): one 1,695-hp (1 264-kW) Packard Merlin V-1650-7 inline piston engine
Performance: maximum speed 437 mph (703 km/h) at 25,000 ft (7 620 m); service ceiling 41,900 ft (12 770 m); maximum range 2,080 miles (3 347 km)
Weights: empty 7,125 lb (3 232 kg); maximum take-off 12,100 lb (5 488 kg)
Dimensions: span 37 ft 0¼ in (11.28 m); length 32 ft 3 in (9.83 m); height 8 ft 8 in (2.64 m); wing area 233 sq ft (21.65 m²)
Armament: six 0.50-in (12.7-mm) machine-guns, plus up to two 1,000-lb (454-kg) bombs or six 5-in (127-mm) rocket projectiles
Operators: China, FFAF, Netherlands, RAAF, RAF, RCAF, RNZAF, SAAF, USAAC/USAAF

North American P-64

History and Notes

Both the North American NA-16 design, which had led initially to the production of more than 500 two-seat BT-9/-14 two-seat trainers, and the related NA-26 (AT-6 Texan/Harvard) which was built in thousands for use both by the USA and its allies, are well known to most aviation enthusiasts. Few know of the single-seat members of this family, developed as lightweight fighters for export.

First came the NA-50A, of which seven were built for the Peruvian air force, and all of which had been delivered by May 1939. This was followed in mid-1940 by an order for six generally similar aircraft for the Royal Thai air force, and these had the company identification NA-68. Of low-wing cantilever monoplane configuration, these little fighters had retractable tailwheel type landing gear, were powered by a Wright R-1820 Cyclone 9 radial engine in a neat cowling, and accommodated the pilot in an enclosed cabin with transparent canopy. Armament comprised two machine-guns and two cannon, and four 100-lb

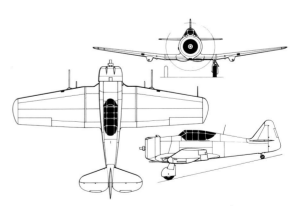

North American NA-68

(45-kg) bombs could be carried on underfuselage racks.

Completed and shipped to Thailand, these six aircraft were on board ship at Hawaii when Japan occupied that nation in December 1941. The aircraft were immediately seized by the US authorities and returned

North American P-64

to the US, where they were taken into USAAC service in late 1940. Designated P-64, their 20-mm cannon armament was removed, and they entered service in an advanced trainer role.

Specification
Type: single-seat fighter-bomber
Powerplant: one 870-hp (649-kW) Wright R-1820-77 Cyclone 9 radial piston engine
Performance: maximum speed 270 mph (435 km/h) at 8,700 ft (2 650 m); service ceiling 27,500 ft (8 380 m);
range 635 miles (1 022 km)
Weights: empty 4,660 lb (2 114 kg); maximum take-off 6,800 lb (3 084 kg)
Dimensions: span 37 ft 3 in (11.35 m); length 27 ft 0 in (8.23 m); height 9 ft 4 in (2.84 m); wing area 227.5 sq ft (21.13 m²)
Armament: two 0.30-in (7.62-mm) machine-guns and two 20-mm cannon, plus up to 400 lb (181 kg) of bombs carried externally
Operators: Peru, USAAC/USAAF

North American P-82 Twin Mustang

History and Notes

The concept of joining two aircraft together to provide more power or space was not new when development of the North American Twin Mustang began in 1944. It had already been used with some success by the five-engined Heinkel He 111Z and by the General Aircraft Twin Hotspur glider.

The Mustang development came as a result of the USAAF requirement for a very long-range escort fighter for Pacific operation. The purpose of having two pilots was as a relief against fatigue on the long overwater missions: to help combat such fatigue, adjustable seats were fitted and there was also provision for uncoupling the rudder pedals. A complete control system was available in each cockpit, although full flight instruments were only fitted in the port fuselage.

The first two prototype XP-82s flown in 1945 had two Packard Merlin V-1650 engines with counter-rotating propellers, while the third XP-82 had two Allison V-1710s with common rotation. It was a variant of the latter engine that was chosen for production aircraft.

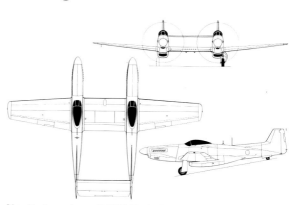

North American P-82E Twin Mustang

The North American P-82B Twin Mustang was the first production model of this escort fighter, whose range could be extended considerably by the carriage of four large drop tanks under the wings. Primary gun armament was six 0.50-in (12.7-mm) weapons.

North American P-82 Twin Mustang

A large USAAF order for 500 P-82Bs was placed, but only 20 had been built when the end of World War II resulted in large-scale cancellations of contracts.

The tenth and eleventh production aircraft were converted to night fighters as the P-82C (SCR-720 radar) and P-82D (APS-4 radar) respectively, the radar being carried in a large nacelle beneath the centre section, while the radar operator was carried in a modified starboard cockpit.

The P-82 was reinstated in 1946 USAAF order books with a batch of 250, comprising 100 P-82E escort fighters, 100 P-82F night fighters (APS-4 radar) and 50 P-52Gs (SCR-720 radar). A change of USAF designations in 1948 led to the models B to G becoming F-82s, and by December of that year the type had completely replaced the Northrop P-61 Black Widow.

Following deployment to Japan, F-82s of the US 5th Air Force were among the first USAF aircraft to operate over Korea, and a pilot from the 68th Fighter (All-Weather) Squadron based at Itazuke shot down the first enemy aircraft in the Korean War.

The last version of the Twin Mustang to see service was the F-82H, a winterised variant of the F-82F and G, which was assigned to Alaska.

Specification
Type: long-range escort and night fighter
Powerplant (F-82G): two 1,600-hp (1 193-kW) Allison V-1710-143/145 inline piston engines
Performance: maximum speed 461 mph (742 km/h) at 21,000 ft (6 400 m); cruising speed 286 mph (460 km/h); service ceiling 38,900 ft (11 855 m); range 2,240 miles (3 605 km)
Weights: empty 15,997 lb (7 256 kg); maximum take-off 25,591 lb (11 608 kg)
Dimensions: span 51 ft 3 in (15.62 m); length 42 ft 5 in (12.93 m); height 13 ft 10 in (4.22 m); wing area 408 sq ft (37.90 m²)
Armament: six wing-mounted 0.50-in (12.7-mm) machine-guns, plus up to four 1,000-lb (454-kg) bombs on underwing racks
Operator: USAAF

North American Texan/SNJ/Harvard

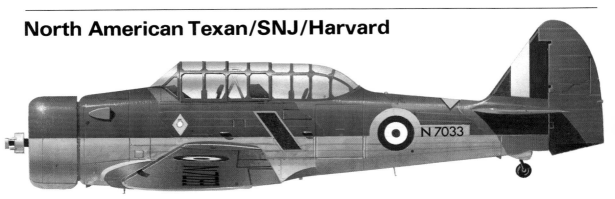

North American Harvard Mk I of an RAF flying training school during 1940-1.

History and Notes
Almost certainly the most universally used military training aircraft of all time, the North American Harvard was derived from the NA-16 prototype, built at the General Aviation Factory at Dundalk, Maryland to a US Army Air Corps specification for a basic trainer, which was the forerunner of more than 17,000 examples of a number of derivatives.

Originally with open tandem cockpits, the NA-16 was of metal construction with a fabric-covered fuselage, and was provided with fixed tailwheel type landing gear. Powered by a 400-hp (298-kW) Wright R-975 Whirlwind engine, the prototype was flown in April 1935 and, after evaluation at Wright Field, was selected for production for training under the designation BT-9, albeit with some modifications which were to include enclosed cockpits.

Although the prototype, as the NA-18, was so modified and fitted with a 600-hp (447-kW) Pratt & Whitney R-1340 Wasp engine, the R-975-7 Whirlwind was retained for the production BT-9; the first of 42

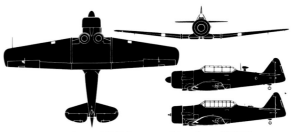

North American AT-6 (upper side view: T-6G)

was flown at Inglewood, Los Angeles (North American's new factory) in April 1936. Forty BT-9As, with one fixed forward-firing gun, and another on a flexible mounting in the rear cockpit, were then built for the USAAC Reserve, followed by 117 BT-9Bs with detail improvements, and the series ended with 67 similarly armed BT-9Cs. The first of the BT-9Cs was, in fact, completed as the sole Y1BT-10, with a 600-hp (447-kW) R-1340 radial engine, and another was designated

North American Texan/SNJ/Harvard

The North American SNJ-5 was the naval equivalent of the USAAF's AT-6D, the example illustrated being seen over Naval Air Station Moffett Field in California.

BT-9B to signify a change, namely the substitution of the more angular outer wing panels and rudder which had, by then, been developed for the BC-1A. The basic BT-9B with similar modifications, a metal-covered fuselage and a 450-hp (336-kW) R-985-25 Wasp Junior engine was designated BT-14 and production of these for the USAAC totalled 251, of which 27 were later fitted with an R-985-11 engine and redesignated BT-1A. The US Navy received 40 NJ-1s, which were basically BT-9s with the R-1340 engine.

Export orders for the fixed landing gear variants included three for Honduras and 85 for China, together with one for Australia as a pattern aircraft for licence production by Commonwealth Aircraft Corporation. Two pattern aircraft each were supplied to licence-holders ASJA of Sweden and Mitsubishi of Japan. France had taken 200 NA-57 basic trainers for the Armée de l'Air and 30 for the Aéronavale; delivery of a similar repeat order for the NA-64 was in progress when France fell before the German advance and 119 undelivered aircraft were acquired by the UK and supplied to the Royal Canadian Air Force, becoming known as the Yale I.

Three years earlier, however, USAAC Circular Proposal 37-220, for a basic combat trainer, had led to the development of the NA-26 with retractable landing gear, a 600-hp (447-kW) R-1340 engine, armament and instrumentation representative of contemporary operational types. Evaluation of the prototype, still with rounded wing tips and rudder and with a fabric covered fuselage, resulted in USAAC orders totalling 180 BC-1s, of which 30 were BC-1I instrument trainers; the last three were completed as BC-2s with R-1340-45 engines, three-blade propellers and the BT-9D/BT-14 wing, rudder and fuselage covering improvements. There followed 83 BC-1As before the USAAC changed the role designations to advanced trainer, and subsequent production carried AT designations, the series commencing with 94 AT-6 Texans which included the last nine of the BC-1A order.

Replacement of the centre-section integral fuel tank by removable tanks and installation of the R-1340-49 engine produced the AT-6A, 517 of which were built at Inglewood before all production was transferred to Dallas, Texas where North American had already established a second line. Dallas-built aircraft to USAAF contracts included 1,330 AT-6As, 400 of the gunnery-trainer AT-6B version, 2,970 AT-6Cs, 3,404 AT-6Ds and 956 AT-9Fs. The AT-6D arose from the need to

North American Texan/SNJ/Harvard

alleviate an anticipated shortage of light alloy and approximately 1,250 lb (567 kg) per aircraft was saved by the use of other light metals, bonded plywood construction for the rear fuselage, and ply-covered tail surfaces. Although the AT-6E marked a return to all-metal construction, it featured also a 24-volt rather than a 12-volt electrical system, and the AT-6F had a redesigned rear fuselage and strengthened wings.

US Navy procurement of these aircraft, which had commenced with the NJ-1, continued with 16 SNJ-1s which were equivalent to the US Army's BC-1 but with a metal-covered fuselage. Some 61 SNJ-2s with a variable-pitch propeller and R-1340-56 engine were built at Inglewood and 150 at Dallas. The AT-6C and -6D were manufactured at Dallas as the SNJ-4 and SNJ-5, production totalling 2,400 and 1,357 respectively. Some SNJ-5s were fitted with an arrester hook for deck-landing training and were designated SNJ-5C. Finally, 931 of the US Army's 956 AT-6Fs were actually procured on behalf of the US Navy, which used them as the SNJ-6.

British interest became evident in June 1938 when the British Purchasing Commission placed an order for 200 BC-1s with British equipment, these being designated Harvard I. On 3 December 1938, the first was delivered to the Aircraft Armament and Experimental Establishment at Martlesham Heath for acceptance testing before the type entered service at the Flying Training Schools, commencing with No. 3 FTS, Grantham. The major part of this first batch was shipped to Southern Rhodesia, to equip the three Service Flying Training Schools established there as part of the British Commonwealth Air Training Plan, and the RAF retained most of the second batch, the original order having been doubled.

Thirty similar aircraft were purchased on behalf of the Royal Canadian Air Force and in 1939 the Purchasing Commission ordered 600 AT-6s, which were known as the Harvard II; 20 were delivered to the RAF, 67 to the Royal New Zealand Air Force and the rest to Canada where the Canadian government had agreed, under the Air Training Plan, to train 20,000 Commonwealth aircrew annually at 74 training schools, at least 14 of which had Harvards on strength. Total deliveries to Commonwealth air forces, mostly under Lend-Lease, exceeded 5,000. In addition to the Mk I and Mk II already mentioned, the Mks IIA, IIB and III were also used, these being equivalent to the AT-6C, AT-16 and AT-6D respectively. AT-16 was the designation given to 2,610 aircraft built by Noorduyn Aviation Ltd of Montreal, for the RAF and RCAF, these corresponding to USAAF AT-6As.

Other overseas purchasers of retractable gear variants included Brazil, China and Venezuela, and many other air arms later received aircraft from surplus USAAF, RAF and RCAF stocks.

From 1949, 2,068 T-6s were remanufactured for the US Air Force and US Navy with a revised cockpit layout, an improved canopy, re-located aerial masts, a square-tipped propeller, F-51 type landing-gear and flap-actuating levers, and steerable tailwheel. Some were converted to LT-6G standard for service in Korea from July 1950, operating with 6147th Tactical Air Control Squadron.

Canadian Car and Foundry, which had taken over Noorduyn in 1946, manufactured 270 T-6G standard Harvard 4s for the RCAF and 285 T-6Js for USAF Mutual Aid programmes.

Specification

Type: two-seat advanced trainer, or close air support aircraft
Powerplant (SNJ-5): one 550-hp (410-kW) Pratt & Whitney R-1340-AN-1 Wasp radial piston engine
Performance: maximum speed 205 mph (330 km/h) at 5,000 ft (1 525 m); cruising speed 170 mph (274 km/h) at 5,000 ft (1 525 m); service ceiling 21,500 ft (6 555 m); range 750 miles (1 207 km)
Weights: empty 4,158 lb (1 886 kg); maximum take-off 5,300 lb (2 404 kg)
Dimensions: span 42 ft 0¼ in (12.8 m); length 29 ft 6 in (8.99 m); height 11 ft 9 in (3.58 m); wing area 253.7 sq ft (23.57 m²)
Armament (AT-6): one fixed forward-firing and one rear cockpit-mounted 0.30-in (7.62-mm) machine-gun, plus underwing pylons for machine-gun pods or light bombs (COIN conversions)
Operators: RAF, RCAF, RN, USAAC/USAAF, USN and many others

Northrop P-61 Black Widow

History and Notes

Achieving the distinction of being the first US aircraft designed specifically as a night fighter, the Northrop Black Widow flew in prototype form as the XP-61 on 21 May 1942.

The design was tailored to meet a specification for a twin-engined aircraft capable of carrying newly developed radar equipment, a heavy offensive armament and a crew of three. Northrop tendered to this specification in November 1940 with the twin-boom P-61 and the US Army Air Corps ordered two

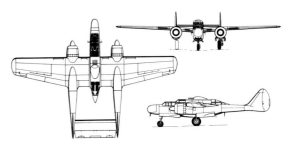

Northrop P-61A Black Widow

Northrop P-61 Black Widow

Northrop P-61A Black Widow of the 6th Night Fighter Squadron, USAAF, based at East Field, Saipan island (Marianas group) in summer 1944.

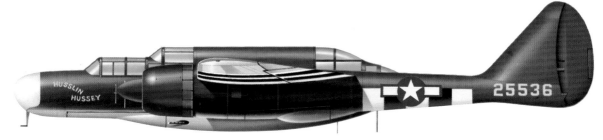

Northrop P-61A Black Widow of the 422nd Night Fighter Squadron, USAAF, based at Scorton (UK) in summer 1944.

Northrop P-61B Black Widow of the 550th Night Fighter Squadron, USAAF, based at Tacloban (Philippine islands) in June 1945.

prototype XP-61s on 11 January 1941. Orders for 13 YP-61s were placed on 10 March 1941, and a further 150 to production standard on 1 September that year, with 410 more following on 12 February 1942.

By September 1942, Northrop had flown the two XP-61s and all 13 YP-61s, while production examples designated P-61A began to appear towards the end of 1943.

As a stop-gap, the USAAF were using the Douglas P-70, a conversion of the A-20 light bomber (known in the RAF as the Boston; the night fighter version was the Havoc). However, the P-70 proved to have a poor performance and unreliable radar, to such an extent that a number of Bristol Beaufighters were lent to the USAAF until the arrival of the P-61s.

Early tests with the new fighter showed buffeting problems with the top turret and this was deleted after the first 37 aircraft had been built as P-61As. From aircraft no. 46 the Pratt & Whitney engines were changed from Srs R-2800-10 to -15, and this variant was the first to enter service, with fighter groups in the South Pacific in the first half of 1944; the first 'kill' was made on 7 July of that year. In the same month

delivery of the next model, the P-61B, began; 450 were to be built and the final 250 had the dorsal turret restored. With this fitted the armament comprised four forward firing 0.50-in (12.7-mm) machine-guns in the remotely-controlled dorsal turret and four 20-mm cannon fixed to fire forward in the lower part of the fuselage. This version was also capable of carrying four 1,600-lb (726-kg) bombs or 300-US gallon (1 136-litre) drop tanks on wing racks.

Black Widows were in service in Europe by July 1944 and shot down four German bombers in their first engagement. Subsequently a number of flying bombs were destroyed during the V-1 offensive against Antwerp, and the P-61 proved itself an extremely versatile aircraft. As might be expected, it had its share of teething troubles including some with the AI (airborne interception) radar in the nose, equipment developed at the Massachusetts Institute of Technology from original British designs.

The last production batch comprised 41 P-61Cs; these were fitted with 2,800-hp (2 088-kW) R-2800-73 engines and had various equipment changes, while further examples built in prototype form included a

Northrop P-61B Black Widow of the 418th Night Fighter Squadron based in the Pacific theatre, 1944.

Northrop P-61 Black Widow

pair of P-61As converted to XP-61Ds as improved night fighters, and two XP-61Es, converted from P-61Bs, as day fighters without radar and top turret and featuring a new two-seat tandem cockpit. By the time Black Widow production was phased out at the end of 1945, 706 had been built, but the type was not quite dead as, following redesign of a P-61A and an XP-61E, 36 examples of a reconnaissance variant, the F-15A Reporter, were built in 1946. With the tandem seating as used on the XP-61E, the Reporter remained in service until 1952, having been redesignated RF-61C in 1948.

The US Navy received 12 ex-USAAF P-61A Black Widows late in 1945, for use as radar-equipped night fighter trainers for crews of the Douglas F3D Skyknights which were on order. USN Black Widows were designated F2T-1.

Specification

Type: three-seat night fighter and intruder
Powerplant (P-61A): two 2,250-hp (1 678-kW) Pratt & Whitney R-2800-65 Double Wasp radial piston engines
Performance: maximum speed 369 mph (594 km/h) at 20,000 ft (6 095 m); service ceiling 33,100 ft (10 090 m); range 1,000 miles (1 609 km) on internal fuel or 1,900 miles (3 058 km) with maximum external fuel
Weights: empty 20,965 lb (9 510 kg); maximum take-off 32,400 lb (14 696 kg)
Dimensions: span 66 ft 0 in (20.12 m); length 48 ft 11 in (14.91 m); height 14 ft 2 in (4.32 m); wing area 664 sq ft (61.69 ^{2})
Armament: four 20-mm cannon in lower forward fuselage, supplemented in some aircraft by four 0.50-in (12.7-mm) machine-guns in dorsal turret
Operators: USAAF, USN

The Northrop P-61 Black Widow was a massive night-fighter, and the early-production P-61A shown here retained the dorsal turret (four 0.50-in/12.7-mm guns) deleted from the 38th aircraft onwards.

Piper L-4 Grasshopper

History and Notes

Evaluated for the role of artillery spotting and front-line liaison, as were the Aeronca L-3 and Taylorcraft L-2, four examples of the Piper Aircraft Corporation's Cub Model J-3C-65 were acquired for this purpose by the US Army Air Corps in mid-1941. These were duly allocated the designation YO-59 and, almost simultaneously, 40 additional examples were ordered as O-59s. These were all delivered quickly enough for the US Army to employ them on a far wider evaluation basis than had been anticipated, using them in the field as if on operational service during annual manoeuvres held at the end of 1941.

There was no doubt at all after this very practical test that the little Cubs were of more value than had been envisaged, and this useful experience made it possible to procure a new version more specifically tailored to the US Army's requirements. This, designated O-59A, was of braced high-wing monoplane

Piper L-4 Grasshopper

configuration and was of composite construction comprising wooden spars, light alloy ribs and fabric covering. The fuselage and braced tail unit had basic structures of welded steel tube and were fabric-covered. Landing gear was of the fixed tailwheel type,

Piper L-4 Grasshopper

and the powerplant of the O-59A comprised a 65-hp (48-kW) Continental O-170-3 flat-four engine. Primary requirement of the O-59A specification was improved accommodation for pilot and observer, which was achieved with a modified enclosure for the tandem cockpits to provide better all-round visibility.

Orders for this version totalled 948, but designation changes resulted in all becoming L-4As, the previously supplied YO-59s and O-59s becoming L-4s. Subsequent procurement covered 980 L-4Bs with reduced radio equipment, 1,801 L-4Hs which had only detail changes, and 1,680 L-4Js which introduced a variable-pitch propeller that made a significant improvement to take-off performance. In addition to the various L-4 Grasshoppers procured specifically for the US Army, more than 100 were impressed from civil sources and designated L-4C (J-3C-65s), L-4D (J-3F-65s), L-4E (J-4Es), L-4F (J-5As), and L-4G (J-5Bs).

In 1942 Piper was requested to develop a training glider from the basic L-4 design, this involving the removal of the powerplant and landing gear. In its modified form it had a simple cross-axle landing gear with hydraulic brakes, and the powerplant was replaced by a new front fuselage to accommodate an instructor, and he and both pupils were provided with full flying controls. A total of 250 was built for the USAAF under the designation TG-8, plus three for evaluation by the US Navy which designated them XLNP-1.

Apart from the three XLNP-1s which the US Navy acquired for evaluation, this service also procured 230 NE-1s, basically similar to the US Army's L-4s, and these were used as primary trainers. Twenty similar aircraft procured at a later date were designated NE-2, and 100 examples of the Piper J-5C Cub which were acquired for ambulance use (carrying one stretcher) were originally HE-1. When, in 1943, the letter H was allocated to identify helicopters, the HE-1s were redesignated AE-1s.

After the war improved versions were built under the designations YL-14, L-18B, L-18C, L-21A, L-21B and TL-21A.

Specification
Type: two-seat lightweight liaison aircraft
Powerplant (L-4): one 65-hp (48-kW) Continental O-170-3 flat-four piston engine
Performance: maximum speed 85 mph (137 km/h); cruising speed 75 mph (121 km/h); service ceiling 9,300 ft (2 835 m); range 190 miles (306 km)
Weights: empty 730 lb (331 kg); maximum take-off 1,220 lb (533 kg)
Dimensions: span 35 ft 3 in (10.74 m); length 22 ft 0 in (6.71 m); height 6 ft 8 in (2.03 m); wing area 179 sq ft (16.63 m²)
Armament: none
Operators: USAAC/USAAF, USN

Republic (Seversky) P-35A

History and Notes
In 1935 the US Army Air Corps initiated a series of design competitions aimed at procuring an advanced monoplane fighter. At an earlier date, however, the Seversky Aircraft Corporation had begun the design of a two-seat aircraft in this category as a private venture, completing its construction in early 1935. This, which had the company designation SEV-2XP, was submitted for competition, but while being flown to Wright Field on 18 June 1935 to participate, was involved in an accident which caused sufficient damage to necessitate its return to the factory for repair. Designed by Alexander Kartveli who, a little later, was to head design of the famous P-47 Thunderbolt, the SEV-2XP had fixed landing gear. During the repair process Kartveli had designed retractable main and tailwheel units which could be adapted to the original airframe without major structural change and, at the same time, changed the accommodation to provide only a single seat. The powerplant comprised an 850-hp (634-kW) Wright R-1820-65 Cyclone 9, and with the changes which had been made the company were confident of attaining a maximum speed in excess of 300 mph (483 km/h).

In this form the redesignated SEV-1XP was delivered to Wright Field in mid-August 1935, but failing to achieve the desired performance was taken back to the Seversky plant where a Pratt & Whitney

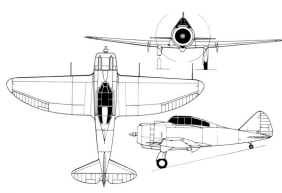

Seversky P-35A

R-1830-9 Twin Wasp engine was installed. Theoretically of the same rating as the Cyclone which it had displaced, it was found during testing that the SEV-7 (as it was then designated) had suffered a decline in performance. Finally, under the designation AP-1 and with an R-1830-9 engine which was rated officially at 950 hp (708 kW), the USAAC ordered 77 of this version as the P-35. With a low-set elliptical wing that was very similar in planform to that of the Supermarine Spitfire, these fighters were armed with one 0.50-in (12.7-mm) and one 0.30-in (7.62-mm) synchronised machine-gun

Republic (Seversky) P-35A

Republic (Seversky) P-35 of the 94th Pursuit Squadron, 1st Pursuit Group, USAAC, based at Selfridge Field (Michigan) and camouflaged for war games in 1939.

Republic (Seversky) P-35A of Lieutenant Boyd Wagner, commander of the 17th Pursuit Squadron, USAAF, based at Nichols Field, Luzon island (Philippines) in 1941.

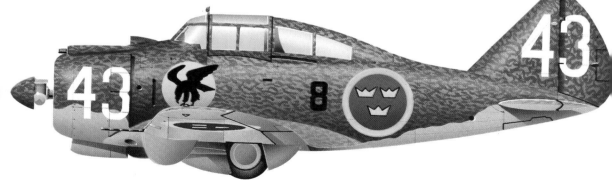

Republic (Seversky) EP-1 of the 3rd Squadron, Flygflottilj 8, Royal Swedish air force, based at Barkarby (Sweden) in 1942-3.

mounted in the close-fitting engine cowling.

P-35s began to enter service in July 1937, with 76 of them being delivered by August 1938. The last of the production batch was completed in changed form, with airframe modifications and the installation of a 1,200-hp (895-kW) Pratt & Whitney R-1830-19 engine with two-stage supercharger, and this was designated XP-41.

In 1939 Seversky adopted the name Republic Aviation Corporation, and in June received an order from the Swedish government for 15 of the P-35s

which, in export form, were identified as EP-1s. This version differed from the USAAC's P-35s in having a more powerful Twin Wasp engine and two additional machine-guns. The initial order, and a following order for 45 more examples, had been completed by mid-June 1940, but the 60 machines which comprised the final order were requisitioned by the US government in October 1940. Allocated the designation P-35A, these had all been delivered by early February 1941, and during the year 48 were despatched to the Philippines where, equipping such units as the

Republic (Seversky) P-35A

USAAC's 3rd, 17th and 20th Pursuit Squadrons, only eight remained serviceable after the first two days of the Japanese attacks on 7/8 December 1941. The 12 which remained in the US were subsequently supplied to Ecuador.

Specification
Type: single-seat fighter/fighter-bomber
Powerplant (P-35A): one 1,050-hp (783-kW) Pratt & Whitney R-1830-45 Twin Wasp radial piston engine
Performance: maximum speed 310 mph (499 km/h) at 14,300 ft (4 360 m); cruising speed 260 mph (418 km/h) at 10,000 ft (3 050 m); service ceiling 31,400 ft (9 570 m); range 950 miles (1 529 km)
Weights: empty 4,575 lb (2 075 kg); maximum take-off 6,723 lb (3 050 kg)
Dimensions: span 36 ft 0 in (10.97 m); length 26 ft 10 in (8.18 m); height 9 ft 9 in (2.97 m); wing area 220 sq ft (20.44 m²)
Armament: two 0.50-in (12.7-mm) synchronised machine-guns in engine cowling and two 0.30-in (7.62-mm) wing-mounted guns, plus up to 350 lb (159 kg) of bombs beneath wings
Operators: Ecuador, Sweden, USAAC

Republic P-43 Lancer

History and Notes
The last production example of the Republic P-35A had been equipped with a 1,200-hp (895-kW) supercharged Pratt & Whitney R-1830-19 engine, instead of the R-1830-45 engine which powered the remainder of the batch acquired by the US Army Air Corps. This re-engined example was designated XP-41, and was one of two projects initiated by the company to produce a developed version of the P-35. The second was a private-venture exercise by the company, identified as the AP-4, involving the same engine and airframe. The difference lay in the powerplant: the engine of the XP-41 had an integral two-stage supercharger, producing its maximum rated power at medium altitude; that of the AP-4 had a separate turbocharger installed in the rear fuselage which enabled the engine to deliver over 90 per cent of its rated take-off power to a much higher altitude. Both of these airframes had one feature which distinguished them easily from the P-35/-35As which had been supplied to the US Army: the narrow-track aft-retracting main landing gear units were replaced by a wide-track inward-retracting layout which housed the retracted main wheels in the undersurface of the fuselage/wing roots, achieving a far superior aerodynamic structure in flight.

Testing of these aircraft resulted in a USAAC contract, awarded on 12 March 1939, for 13 pre-production YP-43s. These had a number of modifications, including installation of the R-1830-35 engine, and resiting of the turbocharger intake from the port wing root to within a new deep oval-shaped cowling. There were also a number of aerodynamic refinements, and armament was increased to four machine-guns. The first of these was delivered by Republic in September 1940 and all were in service by April 1941, and although service trials progressed without problems, it did not take long to discover that the improvement in performance over the P-35s had already been surpassed by new aircraft developed in Europe.

Although Republic had already evolved an improved AP-4J, 80 of which had been ordered under the designation P-44 as early as September 1939, with follow-on contracts bringing the total to 907, these

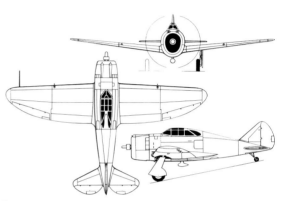

Republic P-43 Lancer

were all cancelled in September 1940 in favour of a much improved design which was to become the P-47 Thunderbolt. This, however, was unlikely to materialise for some months and, as an interim measure, the USAAC ordered 54 P-43s, with the name Lancer and the same powerplant as the YP-43s, plus the 80 P-44s ordered in September 1939 to be completed as P-43As with R-1830-49 engines.

Although it was not in itself very successful, the Republic P-43 Lancer is of historical importance as an evolutionary step towards the magnificent Republic P-47 Thunderbolt heavy fighter.

Republic P-43 Lancer

Delivery of the foregoing Lancers began in 1941, and in June of that year a further 125 similar aircraft were procured for supply to China through Lend-Lease. Designated P-43A-1, these differed by having improved armour and self-sealing tanks, four 0.50-in (12.7-mm) machine-guns and provision for a 50-US gallon (189-litre) drop tank or six 20-lb (9-kg) bombs, and the installation of an R-1830-57 engine. Of this production batch, 107 were delivered to China during 1942.

The P-43s and P-43As supplied to the USAAC were considered unsuitable for combat operations and these, together with the balance of the 18 P-43A-1s built for China, were all converted for use as photo-reconnaissance aircraft. They comprised 150 P-43Bs and two, with a different camera installation, as P-43Cs, and all were supplied to the RAAF under Lend-Lease.

Specification
Type: single seat fighter/fighter-bomber
Powerplant: (P-43A-1): one 1,200-hp (895-kW) Pratt & Whitney R-1830-57 Twin Wasp radial piston engine
Performance: maximum speed 356 mph (573 km/h) at 20,000 ft (6 095 m); cruising speed 280 mph (451 km/h); service ceiling 36,000 ft (10 975 m); range 650 miles (1 046 km)
Weights: empty 5,996 lb (2 720 kg); maximum take-off 8,480 lb (3 846 kg)
Dimensions: span 36 ft 0 in (10.97 m); length 28 ft 6 in (8.69 m); height 14 ft 0 in (4.27 m); wing area 223 sq ft (20.72 m²)
Armament: four 0.50-in (12.7-mm) machine-guns, plus six 20-lb (9-kg) bombs
Operators: China, RAAF, USAAC

Republic P-47 Thunderbolt

History and Notes
The Republic P-47 Thunderbolt really should have been designed and built in Texas, for it was the biggest and heaviest fighter ever to have served with the USAAF when it entered service in 1942; it also belonged to the select, small number of aircraft to be designed, developed, produced and used in action during World War II; it was one of an even more select group, being among the trio of great USAAF fighter aircraft created by the US aviation industry to take part in that war.

And with regard to the foregoing, it is perhaps a little ironic to discover that the P-47 was originated to meet a US Army Air Corps requirement of 1940 for a lightweight interceptor. Its family tree, of course, was clearly defined, from Alexander Seversky's P-35, via the XP-41, P-43 and P-44. The P-35 had been evolved from an earlier Seversky design by the genius of Alexander Kartveli, and it was the design team headed by this man that produced the P-47.

In 1939 the team began development of a new fighter, of lightweight construction, powered by a 1,150-hp (858-kW) Allison V-1710-39 inline engine, and with armament of only two 0.50-in (12.7-mm) machine-guns. A prototype contract for this XP-47 was awarded by the USAAC in November 1939, as was a contract for an even lighter-weight variant designated XP-47A. But with the receipt in early 1940 of information on air combat in Europe, it was immediately clear that neither of these projects was likely to mature into a fighter aircraft that would have any reasonable chance of survival in European skies. The XP-47/-47A prototypes were cancelled, for not even by wearing the rosiest of 'rose-coloured' spectacles could Republic or the USAAC envisage either of the prototypes proving suitable to incorporate the heavy armour, eight-gun firepower and self-sealing fuel tanks that appeared to be the minimum premium for a World War II survival policy.

Kartveli immediately outlined his proposals for a

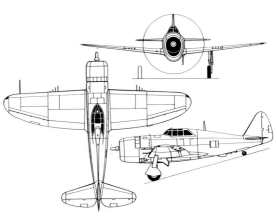

Republic P-47C Thunderbolt

fighter that would meet the requirements, basing it on the use of a turbocharged Pratt & Whitney R-2800 Double Wasp that would have sufficient power to lift a heavy load of arms, ammunition, armour and fuel, and still have something left over to give high performance. Evaluation of the design by the USAAC resulted in a contract for an XP-47B prototype being awarded on 6 September 1940, and this time the question of weight was ignored. From the design point of view this eliminated a major problem, for whilst it was obviously desirable to arrive at the lightest possible structure to provide integrity for the whole machine, it meant that any really worthwhile feature needed no longer be subordinated to weight saving.

The heart of the machine, of course, was its powerplant, comprising a 2,000-hp (1 491-kW) R-2800 Double Wasp engine and its turbocharger. The aircraft was designed around the best possible installation of these two major assemblies, with the aim of having the most direct ducting between them so that ambient air, exhaust gases and pressure air would not be in any way restricted by cramped ducting. This put the turbocharger in the aft fuselage, the low wing

Republic P-47 Thunderbolt

being mounted a slight distance up from the base of the fuselage so that the main air ducting could pass straight beneath the wing spars. To convert this abundant horsepower into tractive effort a four-blade constant-speed propeller 12 ft 2 in (3.71 m) in diameter was necessary, but to provide adequate ground clearance for the propeller tips a taller than average landing gear was needed. These main units, retracting inward, were then too long for the space available in the wing between the inboard machine-gun and wing root, and the solution of this was a specially designed oleo-pneumatic strut that shortened as it retracted.

The rest of the structure was conventional all-metal (the prototype had fabric-covered control surfaces), but on a large scale, and when rolled out for the first time the new aircraft appeared to be nothing short of gigantic for a fighter. However, the first flight, on 6 May 1941, gave a hint of the aircraft's potential, but there were a number of problems to be resolved. Fortunately most of these had been cleared before the XP-47B crashed on 8 August 1942. This occurred before the flight test programme had been completed, but there was no significant hold up on the production which had already got under way to meet the initial contracts, for 773 aircraft, which had been placed very shortly after that for the prototype.

Production P-47Bs, which were given the name Thunderbolt, began to come off the line in March 1942, and in June began to equip the squadrons of the USAAF's 56th Fighter Group. The P-47B differed in only minor respects from the prototype: fabric-covered control surfaces replaced by all-metal structures, and the cockpit canopy aft-sliding instead of side-hinged, and the powerplant having an R-2800-21 production engine. By January 1943 the 56th Fighter Group had joined the 8th Air Force in Britain, soon after reinforced by the 78th Fighter Group, and these groups began operational sorties during April 1943. Initial encounters

Examples of the first operational Thunderbolt version, the Republic P-47B. This went into action for the first time in April 1943, displaying first-class dive performance and structural strength.

with the German fighters showed that the Thunderbolt was lacking in performance and manoeuvrability at low and medium altitudes, and also that range was inadequate, limited as it was by the amount of internal fuel that could be carried.

Production of the P-47B totalled 171, the last off the line being modified by the introduction of a pressurised cockpit under the designation XP-47E. The P-47Cs which followed, from September 1942, were basically similar to the P-47B except for a fuselage extension of 10½ in (0.27 m) and changes in the rudder and elevator balance system to improve manoeuvrability. The powerplant was unchanged for early production aircraft, but the engine model introduced later had a water injection system to provide a combat rating of 2,300 hp (1 715 kW) at 27,000 ft (8 230 m). The most important change in relation to the tactical use of the Thunderbolt was the provision of attachment points for one 200-US gallon (757-litre) drop tank, and from July 1943 P-47Cs could be used deep into German airspace.

A total of 602 P-47Cs had been built before production switched to the most extensively built version, the P-47D. The large numbers of Thunderbolts contracted were far beyond the productive capacity of Republic's Farmingdale, Long Island, factory so the company established a new production line at Evansville, Indiana, and arranged with Curtiss-Wright to begin production at Buffalo, New York. Between them these lines produced no fewer than 12,956 aircraft comprising 6,509 at Farmingdale (3,962 low and 2,547 high block numbers); 6,093 at Evansville (1,461 low and 4,632 high block numbers); and Curtiss-Wright built 354 to P-47D

Republic P-47 Thunderbolt

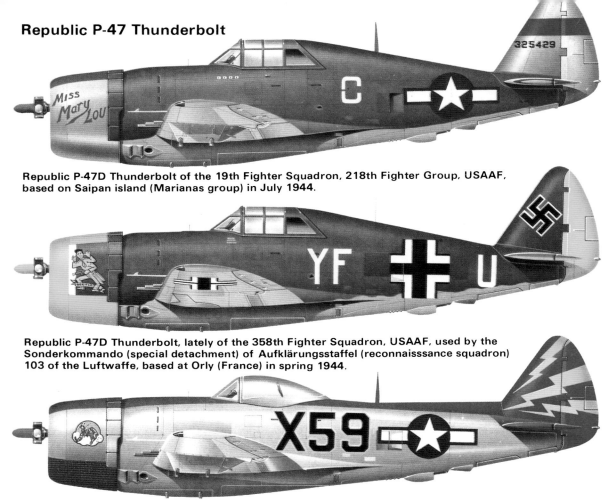

Republic P-47D Thunderbolt of the 19th Fighter Squadron, 218th Fighter Group, USAAF, based on Saipan island (Marianas group) in July 1944.

Republic P-47D Thunderbolt, lately of the 358th Fighter Squadron, USAAF, used by the Sonderkommando (special detachment) of Aufklärungsstaffel (reconnaisssance squadron) 103 of the Luftwaffe, based at Orly (France) in spring 1944.

Republic P-47D Thunderbolt of the 86th Fighter Squadron, 79th Fighter Group, USAAF, based at Fano (Italy) in February 1945.

standard under the designation P-47G.

P-47Ds in the early block numbers varied only slightly from the previous production version, introducing more extensive armour protection for the pilot, slight improvements in the turbocharger exhaust system, and a water injection system as standard for both the R-2800-21 and -59 engines. However, improvements were incorporated throughout the entire production programme, and these included wing strengthening, the provision of underwing pylons to carry additional weapons or fuel and, on the high block numbers, one important change was made to improve rear vision, hitherto restricted by the rear fuselage decking. One P-47D was taken from the line and modified to reduce the height of the rear fuselage, and to replace the framed aft-sliding cockpit canopy by an all-round vision bubble canopy as used on the Hawker Typhoon. Designated XP-47K, this aircraft was tested extensively in July 1943, the improved visibility being acclaimed by all the pilots who flew it, and resulting in immediate adoption of the modification on production aircraft. At a slightly later stage a small dorsal fin was

added to P-47Ds to offset the loss of keel surface which resulted from this change in configuration. Other changes in the later production aircraft included the introduction of the R-2800-63 engine and a paddle-blade propeller, and provision for a total internal plus external fuel capacity of 715 US gallons (2 706 litres) to provide a maximum range of 1,800 miles (2 897 km) under optimum conditions.

The US 8th Air Force began to receive its first P-47Ds towards the end of 1943, and the variant later began to equip units of the 9th and 15th Air Forces in Europe. The 348th Fighter Group in Australia was the first to introduce the type in the Pacific theatre when it received its first aircraft late in 1943. The P-47D was also the first version of the Thunderbolt to be supplied under Lend-Lease, the RAF receiving 240 from the low block numbers with the original framed sliding cockpit canopy, and 585 from the high block numbers with the bubble canopy, under the designations Thunderbolt I and Thunderbolt II respectively. All were used to equip RAF squadrons operating in Burma, the first to receive these aircraft being No. 5.

Republic P-47 Thunderbolt

Republic P-47D Thunderbolt of the 1° Gruppo de Caca, Brazilian air force, based at Tarquinia (Italy) in November 1944.

Republic P-47D Thunderbolt of the 352nd Fighter Squadron, 353rd Fighter Group, USAAF, based at Gaydon (UK) in July 1944.

Republic P-47D Thunderbolt of the 512th Fighter Squadron, 406th Fighter Group, USAAF, based at Nordholz (Germany) in summer 1945.

Other squadrons to be equipped included Nos. 30, 34, 42, 60, 79, 81, 113, 123, 131, 134, 135, 146, 258, 261 and 615. They proved to be a most valuable aircraft as deployed there in a fighter-bomber role, using the 'cab-rank' technique in which patrolling aircraft could be called in by ground controllers to blast Japanese troops with which they were in contact. The armament of up to three 500-lb (227-kg) bombs and the concentrated firepower of their eight machine-guns, combined with high performance at low level, represented a potent weapon against which the Japanese troops had little chance of retaliation. Other nations to be supplied with P-47Ds included Brazil (88), which had one squadron operating with the 12th Air Force in Italy in late 1944; the Free French forces (446); Mexico (25) to equip the 201st Fighter Squadron of the Mexican air force; and the USSR (203).

A small run of P-47Ms followed the -47D on the production line, but it is desirable before discussing this to account for the designation gaps. The XP-47E with pressurised cockpit has been mentioned above, as has the P-47G which was the designation of the

Curtiss-built P-47D; one P-47B was used to test the installation of laminar-flow wings as the XP-47F; and the XP-47H designation covered two P-47Ds in which new 2,300-hp (1-715-kW) Chrysler XIV-2220-1 inline engines were installed to give higher performance, overall length being increased by 3 ft 0¼ in (0.92 m) in comparison with the P-47D. This version did not materialise in production form, however, as the Chrysler engine was not put into production.

Far more radical was the XP-47J (no use was made of an I designation) which introduced a closely-cowled fan-cooled R-2800-61 engine rated at 2,100 hp (1 566 kW), and provided with a wing of lighter weight which incorporated six 0.50-in (12.7-mm) machine-guns. First flown on 26 November 1943, this experimental aircraft was the first piston-engined machine to exceed a speed of 500 mph (805 km/h) in level flight, attaining 504 mph (811 km/h) on 4 August 1944, but despite this sparkling performance production plans were abandoned in favour of an even more advanced project designated XP-72. The XP-47K has been mentioned above, and the XP-47L resulted from structural changes in a P-

The massive Republic P-47 Thunderbolt was designed round a mighty radial engine and its associated turbocharger, and was intended primarily as an interceptor fighter. However, it found its true role as a far-ranging hard-hitting ground-attack aircraft. The type roved the lower altitudes above German and Japanese installations

and transport, bombing, rocketing and machine-gunning anything that looked in the slightest unfriendly. More than 12,000 examples of the classic P-47D model were eventually built and, during their production, a clear-view canopy was introduced as standard. The Thunderbolt's range at first suited the type to escort missions over the European continent and, fitted with drop tanks, missions could be flown as far as Berlin. *Rabbit* was a P-47D-25-RE built by Republic at its Farmingdale, New York, factory. It was then allocated to the 527th Fighter Squadron, 86th Fighter Group, that fought in North Africa, Sicily and Italy.

Republic P-47 Thunderbolt

Republic P-47D Thunderbolt of the 366th Fighter Squadron, 358th Fighter Group, USAAF, based at Toul (France) in winter 1944.

Republic P-47M Thunderbolt of the 63rd Fighter Squadron, 56th Fighter Group, USAAF, based at Boxted (UK) in spring 1945.

47D to provide greater internal fuel capacity.

The P-47M, of which only 130 were built, was preceded by three YP-47Ms. These introduced a similar combination of R-2800 engine and turbocharger as those which had proved so successful in the XP-47J, the engine being able to develop a combat rating of 2,800 hp (2 088 kW) at 32,500 ft (9 905 m). The aim was to produce a high-speed version of the P-47D airframe which could be put into operation quickly to counter the turbine- and rocket-powered fighters in service with the Luftwaffe in Europe, as well as the V-1 flying-bombs which were very vulnerable to a high-speed fighter. Most of these aircraft served after D-Day with USAAF units in Europe, operating initially from Normandy.

Last of this famous line of fighters was the P-47N, of which 1,816 were produced, making the grand total of 15,677 Thunderbolts when production ended soon after VJ-Day. The P-47N utilised the powerplant which had proved so successful in the P-47M, the basic P-47D airframe with an enlarged dorsal fin, strengthened landing gear, and a new strengthened wing of increased span which, for the first time in any version of the P-47, included wing fuel tanks. The resulting maximum internal plus external fuel capacity of 1,146 US gallons (4 338 litres) was sufficient to ensure that the P-47Ns deployed in the Pacific were able to provide an adequate escort service for XXI Bomber Command (VH) Boeing B-29 Superfortresses during long over-water missions.

This enormous fighter, a true juggernaut, indeed the 'Jug' to its friends, proved itself to be robust and

reliable, able to absorb an enormous amount of punishment before being beaten by an enemy, and resulting in the exceptionally low loss rate for this type of 0.7 per cent. The P-47D and P-47N remained in USAF service for a number of years after the war, passing to Air National Guard units before being phased out of service in 1955, by which time they had been redesignated F-47D and F-47N respectively.

Even then, however, Thunderbolts had many more useful years of service to offer, operating with the air forces of Bolivia, Brazil, Chile, Colombia, Dominica, Ecuador, France, Guatemala, Honduras, Iran, Italy, Mexico, Nationalist China, Peru, Turkey and Yugoslavia.

Specification

Type: single-seat escort fighter and fighter-bomber
Powerplant (P-47N): one 2,800-hp (2 088-kW) Pratt & Whitney R-2800-77 radial piston engine
Performance: maximum speed 467 mph (752 km/h) at 32,500 ft (9 905 m); cruising speed 300 mph (483 km/h); service ceiling 43,000 ft (13 105 m); range on internal fuel 800 miles (1 287 km)
Weights: empty 11,000 lb (4 990 kg); maximum take-off 20,700 lb (9 389 kg)
Dimensions: span 42 ft 7 in (12.98 m); length 36 ft 1 in (11.0 m); height 14 ft 7 in (4.44 m); wing area 322 sq ft (29.91 m²)
Armament: six or eight 0.50-in (12.7-mm) machine-guns, plus up to 2,000 lb (907 kg) of bombs or 10 5-in (127-mm) rocket projectiles
Operators: Brazil, FFAF, Mexico, RAF, Soviet Union, USAAF

Ryan FR Fireball

History and Notes

The Ryan FR-1 Fireball looked like a perfectly conventional monoplane fighter unless, of course, one happened to see it in what appeared to be normal flight with the propeller fully feathered. The secret lay in a composite powerplant: a standard Cyclone radial engine just where one would expect to find it, plus one of the new-fangled turbojets installed in the aft fuselage.

The idea had originated towards the end of 1942, when the US Navy drew up a specification for a carrier-based fighter-bomber which would be able to benefit from the higher performance promised by turbojet engines. At the same time, it was appreciated that at the stage of the art there were some problems then inseparable from the use of such an engine, namely high landing speeds and a range-limiting high fuel consumption. The idea of using a conventional piston engine to provide power for landing operations, long-range cruising, and as an auxiliary for high-speed flight seemed perfectly logical and, in addition, represented an insurance policy against the possible failure of the very new turbine engine.

The US Navy's specification was circulated to nine US manufacturers, but of the proposals received Ryan's was by far the most interesting, and three XFR-1 prototypes and 100 production FR-1s were ordered more or less simultaneously in early December 1943. The first of the prototypes was flown initially on 25 June 1944, this flight being made without the turbojet engine installed, and it was not until the following month that a flight was made with both engines operating. By March 1945 the first deliveries of production aircraft began, the FR-1 being powered by a Wright R-1820-72W Cyclone mounted conventionally, and a General Electric J31-GE-3 turbojet in the aft fuselage, with the tailpipe passing below a high-mounted tailplane. Construction was all-metal and, in fact, it was the US Navy's first production aircraft to have a completely flush riveted exterior and metal-covered control surfaces. The low-set monoplane wings folded for carrier stowage and, after the first 14 production aircraft, single-slotted trailing-edge flaps were standard. Landing gear was of the retractable tricycle type, the main units retracting outward into the undersurface of the outer wing panels.

FR-1s began to equip US Navy Squadron VF-66 in March 1945, by which time contracts called for the production of 1,300 aircraft, but VJ-Day cancellations left only 66 to be built, all of them being delivered during 1945. None of these aircraft saw service before the end of the war, but in the immediate postwar years the aircraft were used for carrier trials by Squadron VF-41 (later VF-1E) aboard the USS *Bairoko, Badoeng Strait* and *Wake Island*, finally being withdrawn from service in late 1947.

On the subject of 'insurance policies', it is worth noting that one FR-1 landed on the USS *Wake Island* on 6 November 1945, using the turbojet engine alone, following failure of its piston engine.

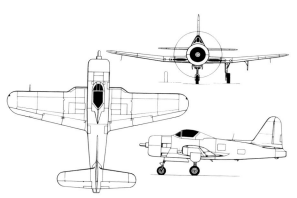

Ryan FR-1 Fireball

Specification

Type: single-seat carrier-based fighter-bomber
Powerplant: one 1,425-hp (1 063-kW) Wright R-1820-72W Cyclone radial piston engine, and one 1,600-lb (726-kg) thrust General Electric J31-GE-3 turbojet engine
Performance: maximum speed (both powerplants) 426 mph (686 km/h) at 18,100 ft (5 515 m); maximum speed (Cyclone only) 295 mph (475 km/h) at 16,500 ft (5 030 m); cruising speed 153 mph (246 km/h); service ceiling 43,100 ft (13 135 m); range 1,030 miles (1 658 km)
Weights: empty 7,915 lb (3 590 kg); maximum take-off 10,595 lb (4 806 kg)
Dimensions: span 40 ft 0 in (12.19 m); length 32 ft 4 in (9.86 m); height 13 ft 7¼ in (4.15 m); wing area 275 sq ft (25.55 m²)
Armament: four wing-mounted 0.50-in (12.7-mm) machine-guns, plus up to 1,000 lb (454 kg) of bombs or eight 5-in (127-mm) rocket projectiles carried externally
Operator: USN

The unusual powerplant of the Ryan FR-1 Fireball is indicated by the feathered propeller, showing that the aircraft is supported by its 1,600-lb (726-kg) thrust General Electric J31 turbojet, located in the rear fuselage.

Ryan PT-16, PT-20, PT-21, PT-22 Recruit

History and Notes

Most readers will recall that it was the Mahoney-Ryan Aircraft Corporation which, in 1927, was responsible for construction of the Ryan NYP monoplane *Spirit of St. Louis*, in which Charles Lindbergh made the first non-stop solo flight across the North Atlantic during 20/21 May 1927. A temporary boom in sales followed this achievement, bringing a merger with Detroit Aircraft Corporation in May 1929, but this company did not survive the depression of 1930-1.

Sensing a new demand for lightplanes, T. Claude Ryan founded the Ryan Aeronautical Company in 1933-4, using as his key to the market a two-seat light monoplane which had the designation S-T. Production of this began in 1934, and it was available in three versions with engines of 95-150 hp (71-112 kW) as S-T, S-T-A or S-T-A Special. When the US Army Air Corps began to evince interest in the procurement of new primary trainers, a single example of the Ryan S-T-A was acquired for evaluation under the designation XPT-16. When delivered in 1939, this represented something of an evolutionary step for the USAAC, for it was the first primary training monoplane to be acquired: all previous USAAC primary trainers had all been biplanes.

Early evaluation resulted in considerable interest in the type, and very soon an additional 15 YPT-16s were procured so that a wider evaluation programme could be completed rather more rapidly. These differed from the XPT-16 only by the addition of an electric starter. Both of these initial versions were powered by the 125-hp (93-kW) Menasco L-365-1 inline engine. An order for 30 production aircraft followed in 1940 under the designation PT-20 and these, delivered in 1941, were generally similar to the YPT-16s except for minor structural changes.

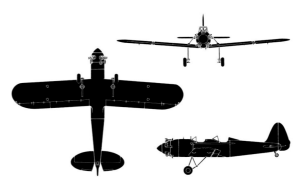

Ryan PT-22

During 1941 the US Army decided that the five-cylinder Kinner R-440 radial engine offered better performance than the Menasco inline, and the 100 PT-21s ordered in 1941 were powered by the 132-hp (98-kW) Kinner R-440-3 engine. The installation of this engine in a streamlined nose fairing, with its five cylinders projecting through the fairing and left uncowled, meant that the PT-21, and subsequent Kinner-engined variants, were easily identified. Their entry into service showed the superiority of this airframe/engine combination, with the result that 14 of the YPT-16s and 27 of the PT-20s were re-engined with 132-hp (98-kW) Kinner R-440-1 engines, and were redesignated PT-16A and PT-20A respectively. Three

The PT-22 Recruit was the ultimate production mark of the USAAF's first monoplane trainer, originally derived from the Ryan S-T civil lightplane. This example was operated in a civilian-run flying school.

Ryan PT-16, PT-20, PT-21, PT-22 Recruit

PT-20s delivered with civil (as opposed to military) versions of the Menasco engine, the D-4, were designated PT-20B.

With the rapid expansion of aircrew training during 1941, Ryan received a contract for 1,023 examples of what was to be the final and most extensively built version, the PT-22 Recruit. This introduced several changes, including deletion of the wheel spats and main landing gear unit fairings which had proved troublesome with aircraft which were making far more take-offs and landings per day than had been envisaged during original design, and by installation of a 160-hp (119-kW) Kinner R-540-1 engine; in other respects these aircraft were similar to the PT-21.

Ryan also built a number of S-Ts for export, and included among these orders was one from the Netherlands for the supply of 25 S-T-3s (basically the same structurally and with the same powerplant as the USAAC PT-22s). By the time these were ready for delivery the Netherlands had already been overrun by German forces and, with the designation PT-22A, they were acquired by the US Army. Used mainly at civilian-operated flying training schools throughout the US, the Ryan trainers gave valuable service before their retirement towards the end of World War II.

This demand for training aircraft was not exclusive to the USAAC, for the US Navy had initiated a similar expansion of its training programme. Following US Army evaluation of this trainer, the US Navy also ordered 100 examples of the Ryan S-T-3 version with Kinner R-440-3 engines on 19 August 1940, these being designated NR-1 Recruit and remaining in service until mid-1944.

Specification
Type: two-seat primary trainer
Powerplant (PT-22): one 160-hp (119-kW) Kinner R-540-1 radial piston engine
Performance: maximum speed 131 mph (211 km/h); cruising speed 123 mph (198 km/h); service ceiling 15,500 ft (4 725 m); range 352 miles (566 km)
Weights: empty 1,313 lb (596 kg); maximum take-off 1,860 lb (844 kg)
Dimensions: span 30 ft 1 in (9.17 m); length 22 ft 5 in (6.83 m); height 6 ft 10 in (2.08 m); wing area 134.25 sq ft (12.47 m²)
Armament: none
Operators: USAAC/USAAF, USN

Sikorsky JRS-1

History and Notes
In 1935 Sikorsky produced a twin-engined amphibian flying boat to accommodate 15 passengers and a crew of three or four. Designated S-43 the type was soon in service with a number of airlines, and because of its amphibious capability appealed to both the US Army Air Corps and US Navy as a utility transport.

Of parasol-wing configuration, the S-43 had a wing which was mounted well above the hull and braced on each side by struts. Wire-braced stabilising floats were strut-mounted beneath the outer panel of each wing. Wing and tailplane had basic metal structures with fabric covering, but the two-step hull was all-metal. All units of the tailwheel type landing gear were hydraulically retractable, and the two Pratt & Whitney Hornet engines were mounted in nacelles at the leading edge of the wing centre-section.

In 1937 the US Navy procured seven of these aircraft under the designation JRS-1 for use as utility transports, obtaining 10 more during the next two years. Generally similar to the civil S-43, they differed by having the R-1690-52 military version of the Hornet radial engine and specifically naval equipment. At about the same time the USAAC acquired five of these aircraft with R-1690-23 Hornets of the same power, and military equipment, and these became designated Y1OA-8. The USAAF commandeered an additional single example of a civil S-43 in 1942, this becoming designated OA-11. Although provided with military equipment, its powerplant differed from that of the Y1OA-8s, comprising two 875-hp (652-kW) R-1690-S2C Hornet radial engines.

Eight of the US Navy's JRS-1s served with Navy Squadron VJ-1 based at San Diego, California, and two were allocated to the US Marines, and Marine Squadrons VMJ-1 and VMJ-2 each operated one for a variety of roles. All of these aircraft in US Army, Marine and Navy use remained in service throughout World War II.

The Sikorsky JRS-1 was a utility amphibian bought by the US Navy, and of the 15 procured by that service, eight (including the example illustrated) were operated by Utility Squadron VJ-1 based at Naval Air Station San Diego, California.

Sikorsky JRS-1

Specification
Type: utility transport amphibian flying boat
Powerplant (JRS-1): two 750-hp (559-kW) Pratt &
Whitney R-1690-52 Hornet radial piston engines
Performance: maximum speed 190 mph (306 km/h)
at 7,000 ft (2 135 m); cruising speed 166 mph
(267 km/h) at 7,000 ft (2 135 m); service ceiling
20,700 ft (6 310 m); range 775 miles (1 247 km)

Weights: empty 12,750 lb (5 783 kg); maximum take-
off 19,096 lb (8 662 kg)
Dimensions: span 86 ft 0 in (26.21 m); length 51 ft 2 in
(15.60 m); height 17 ft 8 in (5.38 m); wing area
780.6 sq ft (72.52 m²)
Armament: none
Operators: USAAF, USMC, USN

Sikorsky R-4

History and Notes
Although Igor Sikorsky could not claim to have
invented the helicopter, he most certainly developed
the basic concept into a practical flying machine which
was to be the foundation of an important sector of the
aviation industry. His successful introduction of the
anti-torque tail rotor overcame the last major control
problem, following months of trials with various
auxiliary rotors fitted to his VS-300 prototype, which
had made its first tethered flight on 14 September 1939.

By the spring of 1941 the VS-300 was achieving free
flight at forward speeds of up to 70 mph (113 km/h) and
the Vought-Sikorsky Division of United Aircraft
received a contract for the development of a two-seat
version which was designated XR-4. It had a fabric-
covered fuselage with an enclosed cockpit and was
powered by a 165-hp (123-kW) Warner R-500 engine
which drove both the main and tail rotors through
gearboxes and driveshafts. The first flight was made
at Stratford, Connecticut on 14 January 1942, and by
April Sikorsky had become sufficiently confident to
demonstrate the helicopter to governmental represen-
tatives and senior military officers.

Over the period from 13 to 18 May the XR-4 was
flown to Wright Field, Ohio for evaluation, accomplish-

Sikorsky R-4

ing the world's first cross-country helicopter delivery
flight. During the 761-mile (1 225-km) journey, which
was completed in 16 hours 10 minutes flying time, the
XR-4 made 16 landings.

**The US Navy's Sikorsky HNS-1 was initially a
USAAF YR-4B received in early 1944. It was
delivered to the Coast Guard Air Station Floyd
Bennett Field, New York, for evaluation in the air-
sea rescue role.**

Sikorsky R-4

Thirty production R-4s were ordered by the USAAF, the first three being designated YR-4A and the rest YR-4B; all were powered by the 180-hp (134-kW) Warner R-550-1 engine and were equipped with a 38-ft (11.58-m) diameter main rotor in place of the 36-ft (10.97-m) diameter rotor fitted to the prototype. Production was completed by 100 R-4Bs with 200-hp (149-kW) R-550-3 engines.

Among the R-4's achievements was the first helicopter landing aboard ship, accomplished on 6 May 1943 when a USAAF YR-4B, flown by Colonel Frank Gregory, landed on the USS *Bunker Hill*, using a deck area only 14 ft (4.27 m) greater than the rotor diameter. The type also recorded the first helicopter rescue, probably that made by a machine from Colonel P. D. Cochran's 1st Air Commando which, in April 1944, rescued the four occupants of a light aircraft which had crashed behind Japanese lines in Burma.

An earlier example of the helicopter's value in a humanitarian role was provided on 3 January 1944 when, in strong winds with sleet and snow, Commander Frank Erickson of the US Coast Guard flew his YR-4B from Floyd Bennett Field, New York to Sandy Hook, New Jersey with plasma for badly-burned survivors of an explosion aboard the destroyer USS *Turner*.

The US Navy had taken a decision to acquire helicopters for evaluation in July 1942 and its first HNS-1 (a YR-4B on loan from the USAAF) was delivered later that year, to be supplemented by two more in March 1943. A total of 25 HNS-1s was eventually to be operated by the US Navy and US Coast Guard.

British observers had been present at Sikorsky's demonstration flight in April 1941 and had been active in the later deck operation trials. Seven YR-4Bs and 45 R-4Bs were flown by Royal Air Force and Fleet Air Arm units under the designation Hoverfly I, entering service with the Helicopter Training Flight at RAF Andover in 1945 and with No. 771 Fleet Requirements Unit at Portland in September of that year.

Specification

Type: two-seat training and rescue helicopter
Powerplant: one 180-hp (134-kW) Warner R-550-1 radial piston engine
Performance: maximum speed 75 mph (121 km/h); service ceiling 8,000 ft (2 440 m); range 130 miles (209 km)
Weights: empty 2,020 lb (916 kg); maximum take-off 2,535 lb (1 150 kg)
Dimensions: main rotor diameter 38 ft 0 in (11.58 m); length 48 ft 2 in (14.68 m); height 12 ft 5 in (3.78 m); disc area 1,134 sq ft (105.35 m²)
Armament: none
Operators: RAF, RN, USAAF, USCG, USN

Sikorsky R-5

History and Notes

Sikorsky's fabric-covered R-4 helicopter was developed into the R-6 which retained the earlier machine's rotor and transmission system, but installed in a streamlined fuselage with a semi-monocoque all-metal tail boom. Powered by a Lycoming O-435 engine of 225 hp (168 kW), the prototype XR-6 was flown on 15 October 1943 and was the forerunner of 228 helicopters which were flown by the USAAF as the R-6A, by the US Navy under the designation HOS-1 and by the Royal Air Force and Fleet Air Arm as the Hoverfly II.

While the R-6 was being developed, work was also in progress on a completely new helicopter, the VS-337 with an all-metal fuselage which provided tandem accommodation for two. Four prototypes, plus one additional at a later date, were ordered by the USAAF, and the first of these, with the designation XR-5 and powered by the 450-hp (336-kW) Pratt & Whitney R-985-AN-5 engine, was flown at Bridgeport, Connecticut on 18 August 1943. Total production of 65 included the remaining four XR-5s (two of which were fitted with British equipment and redesignated XR-5A), 26 YR-5As built for USAAF evaluation and 34 production R-5As. The last could be fitted with a litter carrier on each side of the fuselage and the type was used by the Air Rescue Service. The addition of a nosewheel, a rescue hoist, an external auxiliary fuel tank and a third seat identified the R-5D, 21 of which

Sikorsky R-5A

were modified from R-5As. For training duties, five YR-5As were fitted with dual controls and redesignated YR-5E.

A four-seat version was developed for the civil market as the S-51, the prototype making its first flight on 16 February 1946. Certification was achieved during the following month and initial deliveries were made in August. Los Angeles Airways used the S-51 to open the world's first scheduled helicopter mail service on 1 October 1947.

A total of 379 S-51s was built, many of them for the US armed forces. Eleven were acquired by the USAF

Sikorsky R-5

in 1947 as R-5Fs (redesignated H-5F in June 1948), supplemented by 39 H-5Gs with a rescue hoist, for use by the Air Rescue Service, and USAAF procurement was completed by 16 H-5Hs purchased in 1949.

Following trials with two US Army R-5As, which were redesignated HO2S-1, the US Navy ordered a number of HO3S-1s which were equivalent to the USAF R-5F, deliveries commencing in November 1946. The type were also used by the US Coast Guard as the HO3S-1G. Some 165 S-51s, powered by the 550-hp (410-kW) Alvis Leonides engine, were built under licence by Westland in Britain, operating with the Royal Air Force and Fleet Air Arm as the Dragonfly, and with export and civil customers under the original type number.

Specification
Type: two/four-seat rescue helicopter
Powerplant (R-5B): one 450-hp (336-kW) Pratt & Whitney R-985-AN-5 radial piston engine
Performance: maximum speed 106 mph (171 km/h); cruising speed 85 mph (137 km/h) at 1,000 ft (305 m); service ceiling 14,400 ft (4 390 m); range 360 miles (579 km)
Weights: empty 3,780 lb (1 715 kg); maximum take-off 4,825 lb (2 189 kg)
Dimensions: main rotor diameter 48 ft 0 in (14.63 m); length 57 ft 1 in (17.40 m); height 13 ft 0 in (3.96 m); main rotor disc area 1,810 sq ft (168.15 m²)
Armament: none
Operator: USAAF

Spartan NP-1

History and Notes
In early 1940 the US Navy advised manufacturers of a requirement for a two-seat primary trainer of biplane configuration. A proposal from the Spartan Aircraft Company, based on the company's C-3 three-seat open cockpit biplane of 1927, was selected by the US Navy, and a contract for an XNP-1 prototype, and 200 production aircraft subject to satisfactory prototype testing, was awarded on 10 July 1940.

The first military aircraft to be produced by Spartan, it was an unequal-span single-bay biplane of light alloy construction with fabric covering, except for light alloy side and underfuselage panels forward of rear cockpit. Features included ailerons on both upper and lower wings, and split type landing gear with long-stroke oleo-spring shock-absorbers, and oleo-spring mounted tailwheel. The powerplant, which was uncowled, was a 220-hp (164-kW) Lycoming R-680-8 radial engine driving a two-blade adjustable-pitch propeller. The two open cockpits had dual controls but only basic instrumentation.

Entering service during late 1940, these NP-1s were used to equip newly established US Naval Reserve primary flying training schools based at Atlanta, Georgia; Dallas, Texas; and New Orleans, Louisiana, where they continued to serve for several years.

Specification
Type: two-seat primary trainer
Powerplant: one 220-hp (164-kW) Lycoming R-680-8 radial piston engine
Performance: maximum speed 108 mph (174 km/h); cruising speed 90 mph (145 km/h); service ceiling 13,200 ft (4 025 m); range 315 miles (507 km)
Weights: empty 2,069 (938 kg); maximum take-off 2,775 lb (1 259 kg)
Dimensions: span 33 ft 8½ in (10.27 m); length 24 ft 2¾ in (7.39 m); height 8 ft 4 in (2.54 m); wing area 301.3 sq ft (27.99 m²)
Armament: none
Operator: USN

The NP-1 basic trainer was the US Navy's version of the Spartan C-3.

St Louis PT-15

History and Notes

The long, lean years which stretched from the end of World War I until the late 1930s had, so far as the US Army Air Corps was concerned, done little to expand the force which had been pruned back almost to the roots in the aftermath of the Armistice and the Treaty of Versailles. The outbreak of war in Europe in 1939 brought the realisation that something must be done immediately to train large numbers of pilots to fly aeroplanes that did not then exist, leading to the procurement of training aircraft on an unprecedented scale.

Competing in one of the many prototype competitions for primary trainers in 1939-40 were the St Louis Model PT-1W and the Waco Model UPF-7, both biplane primary trainers which had been built for the civil aircraft market. They were of particular interest to the USAAC because the procurement of an aircraft which was already in production would fill the training gap far more quickly than would the issue of a contract for the design and development of an entirely new trainer, even if it meant accepting something marginally below the desirable standard.

Testing of an example of the St Louis trainer, given the designation XPT-15, resulted in a follow-up order for a small batch of 13 YPT-15 pre-production aircraft. These were single-bay stagger-wing biplanes of clean appearance, with a minimum of struts and bracing wires, and provided with ailerons on the lower wing only. The basic structure was all-metal, with fabric covering on all the aerofoil surfaces and a metal skin on the fuselage. Landing gear was of the non-retractable tailwheel type, the cantilever main units being well faired to reduce drag. Powerplant consisted of a Wright R-760-1 Whirlwind 7 engine, uncowled but faired so that only the cylinder heads and a forward-mounted exhaust ring were exposed. Instructor and pupil were accommodated in tandem open cockpits, and dual controls and full instrumentation were standard.

All of the trainers were delivered in 1940, and though used in the primary training role they were regarded only as 'stop gaps' until aircraft procured to US Army specification became available, and no further orders ensued.

Specification

Type: two-seat biplane primary trainer
Powerplant: one 225-hp (168-kW) Wright R-760-1 Whirlwind 7 radial piston engine
Performance: maximum speed 130 mph (209 km/h); cruising speed 116 mph (187 km/h); service ceiling 15,000 ft (4 570 m); range 350 miles (563 km)
Weights: empty 2,059 lb (934 kg); maximum take-off 2,766 lb (1 255 kg)
Dimensions: span 33 ft 10 in (10.31 m); length 26 ft 5 in (8.05 m); height 9 ft 5 in (2.87 m); wing area 279.9 sq ft (26.00 m²)
Armament: none
Operator: USAAC/USAAF

The enormous expansion schemes implemented by the USAAC during 1939 and 1940 found the service very short of primary trainers, and numerous civil types were thus evaluated for military use. One of these was the St Louis PT-1W, of which 14 were bought as PT-15s.

Stinson AT-19 Reliant

History and Notes

The high-wing cabin monoplane was a popular formula in the USA during the second half of the 1930s, and before World War II the Stinson Reliant had been produced in a number of versions. Indeed, by 1939 the manufacturer was able to offer no less than eight different engine installations ranging from the 245 hp (183 kW) Lycoming to the 450 hp (336 kW) Pratt & Whitney Wasp. In that year All American Aviation Inc. of Morgantown, West Virginia, inaugurated a non-stop air mail delivery and pick up service on two routes covering 56 towns, half of which lacked airport facilities. The service was operated by Reliants with 260-hp (194-kW) Lycoming engines and the aircraft were modified to enable mail to be picked up in flight, a hook beneath the aircraft catching the mail bag suspended

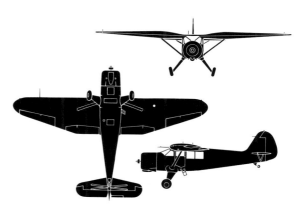

Stinson UC-81 Reliant

Stinson AT-19 Reliant

between two poles.

Production of the Reliant ceased on the USA's entry into the war, but a number of civil machines were impressed for military service where they received the basic designation UC-81, suffix letters up to N indicating the variants.

A few Reliants were in use before World War II with the US Coast Guard as the RQ-1 and the US Navy at Sunnyvale Naval Air Station as the XR3Q-1.

Production was resumed, however, and 500 Reliants, designated AT-19, were eventually supplied to the Royal Navy under Lend-Lease agreements. This model was basically similar to the prewar SR-10 version, but had a fuselage lengthened by 2 ft 2 in (0.66 m) and special interior equipment. Fleet Air Arm Reliants served with Nos. 722 and 733 Squadrons on navigation training and communications duties in the UK and overseas; the first four aircraft in the second batch of 250 supplied went to the Fleet Air Arm in Trinidad.

In addition to the specially built military models, a number of prewar British civil Reliants were impressed for RAF service. They included two SR-5s, one SR-7B, three SR-8S, six SR-9s and three SR-10s. Three were used by the Air Transport Auxiliary for ferry work, and six others served with No. 1 Camouflage Unit at Coventry/Baginton. Others were used on general communications duties and a few of the Fleet Air Arm aircraft eventually found their way either to the RAF or Fleet Air Arm units.

Specification

Type: four-seat navigation/radio trainer and communications aircraft
Powerplant: one 290-hp (216-kW) Lycoming R-680 radial piston engine
Performance: (AT-19): maximum speed 141 mph (227 km/h); cruising speed 130 mph (209 km/h); service ceiling 14,000 ft (4 265 m); range 810 miles (1 303 km)
Weights: empty 2,810 lb (1 275 kg); maximum take-off 4,000 lb (1 814 kg)
Dimensions: span 41 ft 10½ in (12.76 m); length 30 ft 0 in (9.14 m); height 8 ft 7 in (2.62 m); wing area 258.5 sq ft (24.01 m²)
Armament: none
Operators: RAF, RN, USAAF, USMC, USN

The main operator of the Stinson AT-19 Reliant was the Royal Navy, which received 500 Lend-Lease examples.

Stinson L-1 Vigilant

History and Notes
The two-seat light observation aircraft had been an essential adjunct to US Army operations, a concept dating back to World War I when, initially, this was about the only task that an aircraft was considered capable of carrying out effectively. In the years between then and the late 1930s, observation aircraft had, of course, been developed to offer much improved performance, some with high-lift devices which made it possible for them to operate into and out of quite small unprepared areas.

When, in 1940, the US Army Air Corps realised the need to reinforce its aircraft in this category, specifications were circulated and these resulted in contracts for three examples each of Bellanca and Ryan designs, designated YO-50 and YO-51 respectively. Stinson, however, were awarded a contract for 142 of their design under the designation O-49. Built by Vultee after its acquisition of Stinson during the summer of 1940, the O-49 appeared as a braced high-wing monoplane, with an all-metal basic structure, part metal- and part fabric-covered. To provide the essential low-speed and high-lift performance, the whole of the wing leading edge was provided with automatically-operated slats, and the entire trailing edge was occupied by wide span (almost two-thirds) slotted flaps

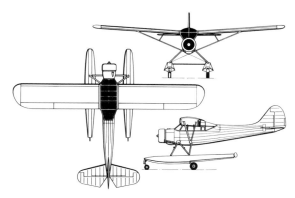

Stinson L-1F Vigilant

and large slotted ailerons which drooped 20° when the flaps were fully down. The non-retractable tailwheel type landing gear was designed specially for operation from unprepared strips. The powerplant consisted of a 285-hp (213-kW) Lycoming R-680-9 radial engine with a two-blade constant-speed propeller. An enclosed cabin seated two in tandem, and the pilot and observer had excellent vision all around, above and below.

A second contract covered the construction of 182 O-49As, which differed by having a slightly longer

Stinson L-1 Vigilant

fuselage, minor equipment changes and detail refinements. Designation changes in 1942 resulted in the O-49 and O-49A becoming the L-1 (liaison) and L-1A respectively. Both versions were supplied to the RAF under Lend-Lease and these were given the British name Vigilant. The designation O-49B (later L-1B) was applied to three O-49s converted to serve in an ambulance role: these could accommodate one stretcher and had a special loading door in the upper surface of the fuselage. Other designations included an L-1C ambulance converted to have a different internal arrangement; 21 L-1Ds modified from L-1As to provide pilot training in glider pick-up techniques; seven L-1s equipped for the ambulance role and provided with amphibious floats as L-1Es; and five similarly-equipped L-1As redesignated L-1Fs.

No further production of new Vigilant aircraft followed, for the type was superseded by the more effective lightweight Grasshopper family. Nevertheless, Vigilants saw quite wide use in both the European and Pacific theatres, the RAF operating many of its aircraft for artillery liaison in Italy, Sicily and Tunisia.

Specification
Type: two-seat light liaison/observation aircraft
Powerplant (L-1A): one 295-hp (220-kW) Lycoming R-680-9 radial piston engine
Performance: maximum speed 122 mph (196 km/h); service ceiling 12,800 ft (3 900 m); range 280 miles (451 km)
Weights: empty 2,670 lb (1 211 kg); maximum take-off 3,400 lb (1 542 kg)
Dimensions: span 50 ft 11 in (15.52 m); length 34 ft 3 in (10.44 m); height 10 ft 2 in (3.10 m); wing area 329 sq ft (30.56 m²)
Armament: none
Operators: RAF, USAAC/USAAF

When fitted with twin Edo amphibious floats, the Stinson L-1A casualty-evacuation ambulance aircraft was redesignated L-1F. Five such conversions were effected.

Stinson L-5 Sentinel

History and Notes
Stinson's 105 Voyager was an attractive three-seat civil lightplane, and in 1941 the US Army acquired six of these civil aircraft which it evaluated for use in a light liaison role under the designation YO-54. Successful testing resulted in an initial order of 1941 for 275 similar aircraft to be powered by the Lycoming O-435-1 flat-four engine, and these were allocated the designation O-62. The following order covered 1,456 similar aircraft, but when delivery of these began in 1942 the designation had been changed to L-5, and the 275 O-62s from the earlier order were also redesignated L-5.

Construction of these L-5s was changed somewhat from the original Voyager design to follow a policy of that particular period which sought to conserve light alloy materials, which it was considered should be reserved for the construction of combat aircraft. Instead of the mixed construction which had been used for the wing and tail unit of the Voyager, those of the L-5 were all-wood, but retained the welded steel-tube fuselage structure. Other changes included rearrangement of the enclosed cabin to seat two in tandem, a reduction in height of the rear fuselage to provide improved rearward vision, and the provision of clear transparent panels in the roof. The original wing design had included leading-edge slots and slotted

Stinson L-5 Sentinel

trailing-edge flaps, and these were retained. The main units of the non-retractable tailwheel type landing gear were modified so that the stroke of the oleo-spring shock-absorbers was almost doubled.

In 1943 688 of the L-5s were reworked to provide a 24-volt instead of 12-volt electrical system, and the streamlined fairings were removed from the main landing gear units: these aircraft were redesignated L-5A. Modifications, which included the installation of an upward-hinged door in the aft fuselage and provision

Stinson L-5 Sentinel

of a stretcher, identified the 679 L-5Bs which followed. The L-5C, of which 200 were built, had provisions for the installation of a K-20 reconnaissance camera. The L-5D designation was cancelled, and the ensuing 558 L-5Es were generally similar to the L-5C except that they introduced ailerons which drooped when the flaps were fully extended. A single XL-5F introduced some minor changes and an O-435-2 engine, leading to the final production version, the L-5G, with the 190-hp (142-kW) O-435-11 engine. Otherwise generally similar to the L-5E, 115 of this last version were built.

In addition to the aircraft procured directly by the US Army, eight commercial Voyagers were commandeered in 1941 to become designated AT-19A (later L-9A), and 12 others as AT-19B (L-9B).

Used extensively by the USAAF throughout World War II, especially in the Pacific theatre, many L-5s were still in use to provide valuable service during the Korean War. The RAF was allocated 100 of these aircraft under Lend-Lease, and these were used widely in Burma for liaison, spotting and air ambulance duties under the name Sentinel. The US Marine Corps acquired a total of 306 L-5s of differing versions, but all were designated OY-1, the Y signifying origin from Consolidated after a merger with Vultee in early 1943. The US Marine Corps deployed its Sentinels for similar missions to those of the RAF and USAAF in support of its operations in the Pacific.

Specification

Type: two-seat light liaison aircraft
Powerplant (L-5): one 185-hp (138-kW) Lycoming O-435-1 flat-four piston engine
Performance: maximum speed 130 mph (209 km/h); service ceiling 15,800 ft (4 815 m); range 420 miles (676 km)
Weights: empty 1,550 lb (703 kg); maximum take-off 2,020 lb (916 kg)
Dimensions: span 34 ft 0 in (10.36 m); length 24 ft 1 in (7.34 m); height 7 ft 11 in (2.41 m); wing area 155 sq ft (14.40 m²)
Armament: none
Operators: RAF, USAAC/USAAF, USMC

Convair OY-1 of the US Marine Corps aboard USS *Petrof Bay* during the Peleliu landings in late 1944.

Taylorcraft L-2 Grasshopper

History and Notes
In 1941 the US Army conducted an operational evaluation with four of each of three types of two-seat light aircraft for use in the artillery spotting and liaison roles, the three types being the Taylorcraft YO-57, the Aeronca YO-58 and the Piper YO-59; all were known as Grasshoppers. The successful use of the aircraft during the US Army's manoeuvres, operating directly with ground forces, resulted in increased production contracts for all three, although the Piper design was to be the most prolific.

The first four Taylorcraft YO-57s were standard civil Taylorcraft Model Ds, powered by the 65-hp (48-kW) Continental YO-170-3 flat-four engine, and were followed by 70 basically-similar O-57s. However, the need to provide all-round vision resulted in modifications to the cabin and rear fuselage and the introduction of trailing-edge cut-outs at the wing roots. Other alterations to fit the aircraft for its specialised tasks included an observer's seat which could be turned around to face the rear, and the installation of radio. In this form the type was designated O-57A and 336 were manufactured.

A further 140 were built under the designation L-2A, US Army aircraft of this class having been reclassified from observation to liaison in 1942. The YO-57s and O-57s became L-2s and the YO-57As were redesignated L-2A. Some 490 aircraft with special equipment, built for service with the field artillery, were designated L-2B and the final variant, with a production run of 900, was the L-2M, identified by the fully cowled engine and the fitting of wing spoilers.

Taylorcraft aircraft were also involved in the training programme for military glider pilots, involving 43 impressed civil machines which were used to provide an initial powered flying course. This total of 43 comprised nine Model DC-65s and nine BC-12-65s, both with Continental A-65 engines and designated L-2C and L-2H respectively; one DL-65, seven BL-65s and four BL-12-65s, with Lycoming O-145s and designated L-2D, L-2F and L-2J; and seven DF-65s, two BFT-65s, three BF-12-65s and one BF-50, all with Franklin

Ordered as the Taylorcraft O-57A, the fully militarised version of the Model D was later redesignated L-2A.

Taylorcraft L-2 Grasshopper

4AC-150s and designated L-2E, L-2G, L-2K and L-2L.

The company also developed a light training glider version which was known as the Taylorcraft ST-100 and given the designation TG-6. The front fuselage was extended and a 'glasshouse' canopy fitted, the landing gear simplified and a skid added under the nose; the lengthened nose necessitated increased fin area. Production totalled 253, including three for US Navy trials.

Specification
Type: two-seat liaison aircraft/training glider

Powerplant: one 65-hp (48-kW) Continental O-170-3 flat-four piston engine
Performance: maximum speed 88 mph (142 km/h); service ceiling 10,000 ft (3 050 m); range 230 miles (370 km)
Weights: empty 875 lb (397 kg); maximum take-off 1,300 lb (590 kg)
Dimensions: span 35 ft 5 in (10.79 m); length 22 ft 9 in (6.93 m); height 8 ft 0 in (2.44 m); wing area 181 sq ft (16.81 m²)
Armament: none
Operators: USAAF, USN

Timm N2T Tutor

History and Notes
The Timm Aircraft Corporation of Van Nuys, California, introduced a method of aircraft construction that involved the use of a plastic-bonded plywood which they themselves had developed. In 1940 the company designed a two-seat light aircraft using this material and, on 2 April 1941 the resulting aeroplane, the Timm S-160-K, became the first of plastic-bonded plywood construction to be awarded an Approved Type Certificate.

The use of construction materials other than the conventional light alloy was of particular interest to the US services, and the US Navy ordered two of these aircraft for evaluation as XN2T-1s, stipulating the installation of a Continental R-670 radial engine. Service trials proved to be satisfactory and resulted in contracts which covered the construction of 260 N2T-1 primary trainers, these being given the name Tutor.

The aircraft presented an external finish that was aerodynamically very clean, free from rivet heads and indifferent fillets or fairings. The configuration was a low-wing cantilever monoplane, with wing, fuselage and tail unit entirely of wooden construction, including extensive use of the plastic-bonded plywood process, which had been patented under the name Aeromold. The fuselage was of particular interest, for most contemporary aircraft of composite construction retained a basic fuselage structure of steel tube. That of the Tutor was, however, all-wood with plastic-bonded veneer circular formers and stringers, all covered with a moulded laminated wood skin. Features included hydraulically operated single-slotted trailing-edge flaps, fixed tailwheel type landing gear with main unit shock-absorption by oleo-pneumatic struts of Timm design, and all controls and control surfaces pivoting on ball or roller bearings. The powerplant comprised the Continental R-670 radial engine specified, with a two-blade variable pitch propeller. This was mounted without any cowlings whatsoever, marring the appearance of what was, for its day, a most attractive training aircraft.

Deliveries of the N2T-1s began in 1943 and, like most other primary trainers in service use, remained in

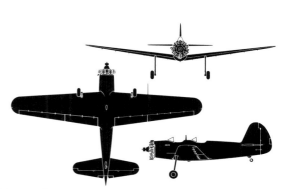

Timm N2T-1 Tutor

employment until towards the end of 1944.

Specification
Type: two-seat primary trainer
Powerplant: one 220-hp (164-kW) Continental R-670-4 radial piston engine
Performance: maximum speed 144 mph (232 km/h); cruising speed 124 mph (200 km/h); service ceiling 16,000 ft (4 875 m); range 400 miles (644 km)
Weights: empty 1,940 lb (880 kg); maximum take-off 2,725 lb (1 236 kg)
Dimensions: span 36 ft 0 in (10.97 m); length 24 ft 10 in (7.57 m); height 10 ft 8 in (3.25 m); wing area 185 sq ft (17.19 m²)
Armament: none
Operator: USN

The clean lines of the Timm N2T-1 Tutor were spoiled completely by the bluff engine installation.

Vought F4U Corsair

History and Notes

While the pundits argue about the merits or demerits of the small number of elite fighters that were developed and used by the combatant nations during World War II, it is pleasant to deal with a far less controversial fighter, the Vought F4U Corsair. Acknowledged universally to be the most outstanding carrier-based fighter to be deployed operationally during the war, Corsair Is and IIs in service with the Royal Navy's Fleet Air Arm were the first to demonstrate the outstanding potential of this design. When used by the US Navy in the Pacific, from April 1944, they were to be credited with no fewer than 2,140 victories for the loss of only 189 of their own number.

This late usage by the US Navy seems hardly possible when it is appreciated that development of this aircraft began as far back as the beginning of 1938, when the US Navy requested proposals for a single-seat carrier-based fighter. To meet this requirement the Vought design team, lead by Tex Beisel, evolved the smallest possible airframe tailored to fit the most powerful engine then available, the Pratt & Whitney XR-2800 Double Wasp. Given the company identification V-166B, the unusual wing configuration of this design resulted from the chosen engine, for with some 2,000 hp (1 491 kW) available, a large-diameter propeller was needed to take full advantage of this power. With a conventional wing layout this would have meant that to give adequate propeller clearance a tall, stalky landing gear would be necessary, far from ideal for carrier operations; so a highly cranked inverted gull-wing was adopted by Beisel. As the main landing gear units were located at the 'pinion joint' of the inverted gull-wing, the landing gear legs were kept as short as possible: at the same time the right-angle junction of wing-root and fuselage was optimum for minimum interference drag at this point.

Of all-metal construction, the remainder of the airframe was conventional, of clean line, and with smooth surfaces which resulted from extensive use of spot welding in the fabrication processes. The wing folded for carrier stowage, and the landing gear was of retractable tailwheel type with all three units retracting aft. Armament comprised a 0.50-in (12.7-mm) and a 0.30-in (7.62-mm) machine-gun in the forward fuselage decking, and one 0.50-in (12.7-mm) machine-gun in each wing. This was the general form of the V-166B design as submitted for US Navy evaluation, and resulted in the award of a contract for a single XF4U-1 prototype on 30 June 1938.

When built the XF4U-1 differed little from this description, except that its XR-2800-4 engine developed only 1,850 hp (1 380 kW) for take-off and 1,460 hp (1 089 kW) at 21,500 ft (6 555 m). The propeller was a Hamilton Standard three-blade constant-speed unit. The XF4U-1 was flown for the first time on 29 May 1940, demonstrating outstanding performance from the very beginning, and during a flight between Stratford and Hartford, Connecticut, on 1 October 1940, a speed of 404 mph (650 km/h) was attained, singling this prototype out as the first US fighter to exceed a speed of 400 mph (644 km/h) in level flight.

By this time, however, reports of combat in Europe

The wartime censor has failed to obliterate the fact that these Vought F4U-1 Corsairs are on the strength of VF-17, the US Navy's first Corsair unit, which became operational on New Georgia in September 1943.

Vought F4U Corsair

Vought Corsair Mk I of No. 1835 Squadron, Fleet Air Arm, based at Brunswick (Nova Scotia) in late 1943.

Goodyear Corsair Mk IV of No. 1850 Squadron, Fleet Air Arm, based aboard HMS *Vengeance* in early 1945.

had been made available to the US forces, and it was considered essential to take full advantage of this information before committing the type to production. First and foremost, much heavier armament was necessary: it was decided to remove the fuselage-mounted guns and to install two additional 0.50-in (12.7-mm) machine-guns in each wing. Unfortunately, to provide sufficient room in the wing for these guns and their ammunition, it meant the deletion of the leading-edge fuel tanks. To compensate for this a fuselage tank had to be provided, and so that the aircraft's trim would not be affected adversely as fuel was consumed, this needed to be as near as possible to the aircraft's centre of gravity. This, in turn, meant that the pilot's cockpit had to be resited some 3 ft (0.91 m) aft. All in all, therefore, the need to include additional guns caused major structural changes. In February 1941 this modified prototype was accepted by the US Navy, and on 30 June following an initial contract was awarded for 584 F4U-1 production aircraft.

The first production example made its initial flight on 25 June 1942, and desirable additions which had been made since acceptance of the modified prototype included an increase in aileron span to provide faster response in the roll axis; the addition of armour, a bullet-proof windscreen, and IFF; an increase of 1 ft 5 in (0.43 m) in fuselage length; and installation of the R-2800-8 engine which provided the original promise of 2,000 hp (1 491 kW). The first of these F4U-1 Corsairs,

as the type had been named, was handed over to the US Navy on 31 July 1942, but it was not until September that sufficient examples had been received to equip the first unit to operate with Corsairs, the US Marine Corps Squadron VMF-124. The first US Navy squadron to receive these aircraft very shortly afterwards was VF-12, but as a result of carrier trials carried out aboard USS *Sangamon* during September, it was considered that the Corsair was unsuitable for carrier operations. This decision was based primarily on the question of forward visibility, but complaints were made also about the landing gear, which was regarded as being too temperamental for the confines of a carrier deck. Remedial action was not long delayed, the cockpit being raised by 7 in (0.18 m) to improve the view forward, and modifications to the landing gear and the addition of a small spoiler near the leading edge of the starboard wing providing the essential directional stability when the wheels touched the deck. Despite this action, it was to be some time before the US Navy took its Corsairs to sea for carrier operations. These modifications were introduced on the production line after 688 F4U-1s had been built, and subsequent aircraft were designated F4U-1A.

Initial operational use, therefore, was confined to land-based units, and the US Marine Corps' VMF-124 was the first to take the Corsair into action, at Guadalcanal on 13 February 1943. The US Navy's first operational squadron, VF-17, was formed in April 1943, and this was the first unit to operate the modified

Vought F4U-1 Corsair

F4U-1As. By that time Vought had received contracts for very large numbers of Corsairs, and to speed deliveries of these aircraft two additional production lines were established: by Brewster at Long Island City, New York, and Goodyear at Akron, Ohio. Brewster-built Corsairs were designated F3A-1 and were generally similar to the Vought F4U-1, the first being delivered in April 1943. Goodyear production had the designation FG-1 and although generally similar to the F4U-1, differed by having fixed instead of folding wings. The first of these was delivered at the end of February 1943.

In June 1943 the Fleet Air Arm began to receive its first Corsairs under Lend-Lease: of the initial versions already mentioned, Britain received 95 F4U-1s which the FAA designated Corsair Is, and 510 F4U-1As as Corsair IIs. The first squadron to receive Corsairs was No. 1830, which formed at the US Navy base at Quonset on 1 June 1943, familiarisation and working up being carried out under US Navy supervision before the squadron was shipped to the UK in an escort carrier. Seven more squadrons were established on the same basis before the end of 1943, working up at NAS Brunswick or NAS Quonset and, in fact, this pattern was followed throughout the war. At its end, 19 squadrons had been formed, the last (No. 1853) collecting its Corsairs from Brunswick in April 1945.

First operational use of Fleet Air Arm Corsair IIs was made by No. 1834 Squadron, flying off HMS *Victorious* and providing a component of the escorting fighters when Fairey Barracudas made their memorable attack on the German battleship *Tirpitz* in Kaafjord, Norway, on 3 April 1944. Corsair Is were put into service virtually as received from the manufacturers, but to make the Corsair IIs more suitable for shipboard operations they were provided with a bulged canopy, so that the pilot could raise his seat to get a better view forward; each wing was clipped by 8 in (0.20 m) to enable the aircraft to be stowed below-deck on the smaller British carriers; provision was made for the carriage of a drop tank; and rocket launch rails were fitted beneath the wings.

When production of these first two versions ended the R-2800-8W engine with water injection had already been introduced. A modified version designated F4U-1C appeared in August 1943 (the designation F4U-1B applied to Lend-Lease deliveries for Britain), this having four wing-mounted 20-mm cannon in place of the standard armament. The F4U-1D which followed reverted to the original armament of six 0.50-in (12.7-mm) guns, but introduced two pylons beneath the wing centre-section each able to accommodate a fuel tank or a 1,000-lb (454-kg) bomb. When F4U-1 production ended Vought had built 4,675, Brewster 735, and Goodyear 4,014.

The designation F4U-2 applied to a night fighter variant which Vought had designed but, because of pressure of work, was unable to transform into hardware apart from a single XF4U-2. Instead, the Naval Aircraft Factory modified 12 standard F4U-1s, deleting two guns and installing AI (airborne interception) radar, and a radome at the starboard wingtip. US Navy Squadrons VFN-75 and VFN-101 each received six of these, and an F4U-2 operated by VFN-75 was to record the first night interception achieved by a radar-carrying single-seat fighter.

The F4U-3 designation had been allocated to a high-altitude interceptor, Vought building three XF4U-3 prototypes with XR-2800-16(C) engines which, with turbochargers, could maintain their rated output of 2,000 hp (1 491 kW) to an altitude of 40,000 ft (12 190 m). Ordered in March 1942, the first was not completed until 1946, by which time the 27 production FG-3s which were to have been built by Goodyear had been reduced to 13. When these eventually entered service with the US Navy after the war they were used for very high altitude test flying.

The final production version of the Corsair to be built during World War II was derived from XF4U-4 prototypes which were flown in 1944. The changes included a redesignated cockpit with a modified canopy, an armoured seat for the pilot, resiting of instruments, further additions of armour, and introduction of the R-2800-18W or -42W engine with a propeller that measured 13 ft 2 in (4.01 m) in diameter. Production F4U-4s began to enter service with the US Navy on 31 October 1944, and although F4U-4Bs were allocated for the Royal Navy no such aircraft were received. Variants of the F4U-4 included the F4U-4C with an armament of four 20-mm cannon, the radar-

Vapour condenses in the tip vortices of an F4U's cranked wings. Though not without its problems, Vought's redoubtable Corsair proved itself perhaps the finest fighter of World War II.

Vought F4U Corsair

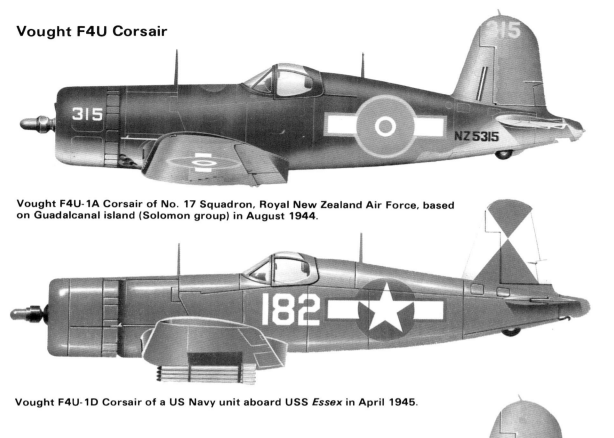

Vought F4U-1A Corsair of No. 17 Squadron, Royal New Zealand Air Force, based on Guadalcanal island (Solomon group) in August 1944.

Vought F4U-1D Corsair of a US Navy unit aboard USS *Essex* in April 1945.

Vought F4U-2 Corsair of VMF(N)-532, US Marine Corps, based on Roi island (Kwajalein group) in 1944.

Vought F4U-4 Corsair of Fuerza Aera Salvadorena (Salvadorean air force), based at San Miguel (El Salvador) in 1958.

Vought F4U Corsair

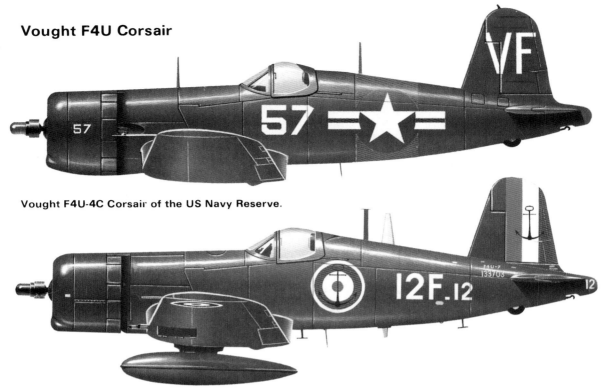

Vought F4U-4C Corsair of the US Navy Reserve.

Vought F4U-7 Corsair of Flottille 12F, Aéronavale (French naval air arm) in 1954.

equipped F4U-4E and F4U-4N, and the camera-carrying tactical reconnaissance F4U-4P. In the postwar years, before production finally ended in December 1952, an XF4U-5 prototype of 1946 with a 2,300-hp (1 715-kW) R-2800-32W engine led to the F4U-5 fighter-bomber, the F4U-5N night-fighter, and the F4U-5P for tactical reconnaissance. These were followed by the prototype XF4U-6 with an R-2800-83W engine, increased capacity for the carriage of underwing weapons and, to enhance pilot and systems protection during the low-altitude operations for which it had been designed, additional armour protection. Under the changed designation AU-1, 110 of these were built and served with the US Marine Corps during the Korean War. Last of the Corsairs to be produced were 90 F4U-7s, generally similar to the AU-1 except for installation of the R-2800-18W engine, and these were procured by the US Navy for supply under MAP to the Aéronavale for operations in Indochina.

In addition to the Corsair Is and IIs which entered service with the Fleet Air Arm from mid-1943, allocations to Britain under Lend-Lease included 430 F3A-1Ds built by Brewster and 977 FG-1Ds built by Goodyear, these being designated Corsair III and IV respectively. The Fleet Air Arm thus received 1,912 of these fighters in total, their major contribution to British naval operations being in the Pacific theatre, where they were deployed from carriers of the East Indies and Pacific Fleets during 1944-5. Their contribution to the attack mounted against targets such as the oil refineries at Palembang, Japanese airfields in the Sakishima Islands, and the Tokyo area

had become a part of Fleet Air Arm history.

The Royal New Zealand Air Force also received small allocations of Corsairs under Lend-Lease, these including 233 F4U-1As, 131 F4U-1Ds, and 61 FG-1Ds, the majority of these being deployed in a close-support role.

When the last F4U-7 came off the line at Dallas, Texas, in December 1952, production had been continuous for just over 10 years, a period in which a total of 12,571 examples of this outstanding fighter had been built by Vought, Brewster and Goodyear.

Specification

Type: single-seat carrier-based fighter/fighter-bomber
Powerplant (F4U-4): one 2,450-hp (1 827-kW), with water injection, Pratt & Whitney R-2800-18W Double Wasp radial piston engine
Performance: maximum speed 446 mph (718 km/h) at 26,200 ft (7 985 m); service ceiling 41,500 ft (12 650 m); normal range 1,005 miles (1 617 km); maximum range 1,560 miles (2 511 km)
Weights: empty 9,205 lb (4 175 kg); maximum take-off 14,670 lb (6 654 kg)
Dimensions: span 40 ft 11 in (12.47 m); length 33 ft 8 in (10.26 m); height 14 ft 9 in (4.50 m); wing area 314 sq ft (29.17 m²)
Armament: six 0.50-in (12.7-mm) machine-guns, plus two 1,000-lb (454-kg) bombs or eight 5-in (127-mm) rocket projectiles
Operators: RN, RNZAF, USMC, USN

Vought SB2U Vindicator

History and Notes

Chance Vought's V-142 design of 1932 was evolved in response to a US Navy requirement for a two-seat fighter. This flew in prototype form as the XF3U-1 in May 1933, but after a period of service testing Vought was requested to modify its design so that, instead, it could serve as the prototype for a scout-bomber under the designation XSBU-1. After the usual process of construction, testing and procurement, aircraft of this latter type began to enter service with the US Navy in November 1935 with the designation SBU-1.

This aircraft was the last biplane produced by Vought for service with the US Navy, and of the 124 which were built examples remained in service with the Naval Reserves until 1941. Long before that (in fact before the first of the SBU-1s had entered service), Vought received a contract from the US Navy to develop the prototype of a monoplane scout-bomber which would offer better performance than the SBU. This was allocated the designation XSB2U-1, and because such an aircraft was regarded as being rather 'way out', four months later Vought received a second contract, in February 1935, which called for an equivalent aircraft of biplane configuration, under the designation XSB3U-1. When completed, this latter aircraft proved to be basically a Vought SBU-1 with aft-retracting main landing gear units.

Both of the aircraft were delivered to and tested by the US Navy in the spring of 1936, and it needed very little time to establish that, without any doubt at all, the performance of the SXB2U-1 monoplane was far superior. On 26 October 1936, Vought received a first production order for 54 SB2U-1s, and the first of these flew in July 1937, with deliveries to US Navy Squadron

VB-3 beginning on 20 December 1937. The type differed considerably from the earlier biplane, being of all-metal basic structure with part fabric, part metal covering. The cantilever monoplane wing was low set, with the outer wing panels folding for carrier stowage; the fuselage was of conventional construction, providing accommodation for a crew of two beneath a long transparent canopy. The main units of the tailwheel type landing gear retracted aft, turning through 90° in the process so that the wheels could lie flush in the undersurface of the wing centre-section. An arrester hook was mounted in the aft fuselage, forward of the tailwheel. The powerplant comprised an 825-hp (615-kW) Pratt & Whitney R-1535-96 Twin Wasp Junior engine driving a two-blade propeller. Armament consisted of two 0.30-in (7.62-mm) machine-guns, one forward-firing and one on a flexible mount, plus a maximum bomb load of 1,000 lb (454 kg).

In early 1938 the US Navy ordered 58 SB2U-2s, which were generally similar to the first production version. The 57 SB2U-3s which followed, ordered on 25 September 1939, introduced the R-1535-02 engine, increased standard fuel capacity and provisions for long-range tanks for ferry flights, improved armour protection, and the two 0.30-in (7.62-mm) guns replaced by two of 0.50-in (12.7-mm) calibre. These aircraft, the first to carry the name Vindicator, began to enter service towards the end of 1940, by which time the

Vought SB2U-1 Vindicators of the US Navy Bombing Squadron VB-3 cross the Sierra Nevada range during 1938. The SB2U was essentially an interim type, serving with seven US Navy and two US Marine Corps squadrons, and was withdrawn after the Battle of Midway.

Vought SB2U Vindicator

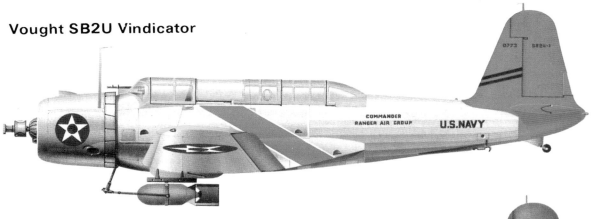

Vought SB2U-1 Vindicator of the Commander Ranger Air Group aboard USS *Ranger* in 1938.

Vought SB2U-1 Vindicator of VS-41, US Navy, based aboard USS *Ranger* in August 1942.

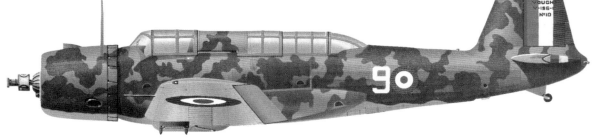

Vought V-156-F of Escadrille AB 3, Aéronavale (French naval air arm) in late 1939.

Vought V-156-B1 Chesapeake of No. 811 Squadron, Fleet Air Arm, based at Lee-on-Solent (UK) in autumn 1941.

Vought SB2U Vindicator

earlier versions were equipping US Navy Squadron VB-2 (aboard USS *Lexington*), VB-3 (USS *Saratoga*), VB-4, VS-41/-42 (USS *Ranger*), and VS-71/-72 (USS *Wasp*). The majority of the SB2U-3s were used to equip the US Marine Corps Squadrons VMSB-131 and VMSB-231. Most of these squadrons saw some action in the Pacific during 1942, including participation in the Battle of Midway, but as more advanced types became available they were soon retired from front-line service, proving too vulnerable to such aircraft as the Mitsubishi A6M Zero.

Vought had also sought export sales for the V-156 design, the initial order coming from France in 1939 for the supply of these aircraft for service with the Aéronautique Maritime. Deliveries had begun before the French capitulation, some of the aircraft falling into German hands. At about this time the UK ordered 50 V-156-B1s, allocated the name Chesapeake I, and these began to arrive in England in early 1941, being assembled at the Burtonwood Repair Depot. First of the Fleet Air Arm's user squadrons was No. 728 based at Arbroath, but No. 811 Squadron was the first to deploy them operationally, and to discover that they were unsuitable for use aboard escort carriers. The Chesapeakes were subsequently relegated for training duties.

Specification

Type: two-seat carrier-based scout/dive-bomber
Powerplant (SB2U-3): one 825-hp (615-kW) Pratt & Whitney R-1535-02 Twin Wasp Junior radial piston engine
Performance: maximum speed 243 mph (391 km/h) at 9,500 ft (2 895 m); cruising speed 152 mph (245 km/h); service ceiling 23,600 ft (7 195 m); range 1,120 miles (1 802 km)
Weights: empty 5,634 lb (2 256 kg); maximum take-off 9,421 lb (4 273 kg)
Dimensions: span 42 ft 0 in (12.80 m); length 34 ft 0 in (10.36 m); height 10 ft 3 in (3.12 m); wing area 305 sq ft (28.33 m²)
Armament: one 0.50-in (12.7-mm) fixed forward-firing machine-gun and one 0.50-in (12.7-mm) gun on flexible mount, plus up to 1,000 lb (454 kg) of bombs
Operators: FNAF, RN, USMC, USN

Vought TBU/Consolidated TBY Sea Wolf

History and Notes

Competing against Grumman in late 1939 to provide the US Navy with a new torpedo-bomber, Vought was awarded a contract on 22 April 1940 to design and develop a prototype to the same requirement. Allocated the designation XTBU-1, this flew for the first time on 22 December 1941, some four months after the first of the Grumman prototypes, and was delivered to the US Navy for evaluation in March 1942.

Vought, of course, was not new to the task of building aircraft for the US Navy, for there had been a steady succession of observation and scouting aircraft stemming from the early 1920s, and the XTBU-1's relationship to other Vought designs was clear to see. Of mid-wing cantilever monoplane configuration, it was of conventional all-metal construction, and provided with retractable tailwheel type landing gear. Accommodation was provided for a crew of three, and a long transparent canopy covered the positions occupied by pilot, radio operator and rear gunner. The powerplant of the prototype comprised a 2,000-hp (1 491-kW) Pratt & Whitney Double Wasp engine,

Designed in competition with the Grumman Avenger, to which it was decidedly superior, the Vought TBU was plagued by production difficulties.

Vought TBU/Consolidated TBY Sea Wolf

driving a three-blade constant-speed propeller. An underfuselage weapon bay provided enclosed accommodation for a torpedo, and armament consisted of a fixed forward-firing 0.50-in (12.7-mm) synchronised machine-gun, a similar gun in a dorsal turret, plus an 0.30-in (7.62-mm) gun in a ventral position aft of the weapon bay.

US Navy testing confirmed that performance of this prototype was considerably superior to that of the Grumman TBF Avenger, and it was decided to procure the type without delay. There was, however, a problem because Vought was unable to produce more aircraft from its limited capacity. This led to the US Navy awarding Consolidated a contract for 1,100 TBY-2 Sea Wolf production aircraft on 6 September 1943, to be built in a new factory at Allentown, Pennsylvania.

Generally similar to the prototype, except that the fixed forward-firing gun had been reinforced by two more 0.50-in (12.7-mm) guns and a radome was mounted beneath the starboard wing, the first of these new aircraft began to enter US Navy service in November 1944. VJ-Day cancellations reduced total construction to 180 examples, the last delivered in September 1945. None of these, however, was used operationally by the US Navy.

Specification
Type: three-seat torpedo-bomber
Powerplant (TBY-2): one 2,000-hp (1 491-kW) Pratt & Whitney R-2800-20 Double Wasp radial piston engine
Performance: maximum speed 306 mph (492 km/h); service ceiling 27,200 ft (8 290 m); range 1,500 miles (2 414 km)
Weight: maximum take-off 18,488 lb (8 386 kg)
Dimensions: span 56 ft 11 in (17.35 m); length 39 ft 2½ in (11.95 m); height 15 ft 6 in (4.72 m)
Armament: three 0.50-in (12.7-mm) fixed forward-firing machine guns, one 0.50-in (12.7-mm) gun in dorsal turret and one 0.30-in (7.62-mm) gun in ventral position, plus torpedo in internal weapon bay
Operator: US Navy

Vought-Sikorsky OS2U-3 Kingfisher

History and Notes
Designed to replace the O3U Corsair biplanes that Vought had built earlier for the US Navy, the VS-310 Kingfisher was the US Navy's first catapult-launched monoplane observation aircraft. It incorporated some novel construction features, including the use of spot welding. A US Navy contract for a prototype XOS2U-1 was awarded on 22 March 1937 and, powered by a 450-hp (336-kW) Pratt & Whitney R-985-4 radial engine, this aircraft was first flown as a landplane on March 1938. Chief test pilot Paul S. Baker was at the controls, as he was for the first flight, on 19 May, of the XOS2U-1 with its Edo floats.

A third, rear support for the central float, and the substitution of an R-985-48 engine, identified the production OS2U-1, 54 of which were built in 1940, the initial example having flown in April. The first delivery was to the battleship USS *Colorado*, others going to the USS *Mississippi*, to Pensacola Naval Air Station, Pearl Harbor Battle Force and the Alameda Naval Air Station Battle Force.

Equipment changes and a later model R-985-50 engine featured in the 158 OS2U-2s which followed in 1941. Of these 45 were delivered as floatplanes, and the rest as landplanes with floats as additional equipment. 46 OS2U-2s were flown from Pensacola and 53 formed the equipment of the Inshore Patrol Squadrons, based at Jacksonville, Florida, which flew anti-submarine missions over the Atlantic and the Gulf of Mexico.

On 17 May 1941 test pilot Boone T. Guyton flew the first OS2U-3, this model having additional wing tanks, 187 lb (85 kg) of crew-protection armour plating and R-985-AN-2 or -8 engines; 368 were delivered during 1941 and another 638 in 1942, when production ended to allow the Vought-Sikorsky factory to concentrate on

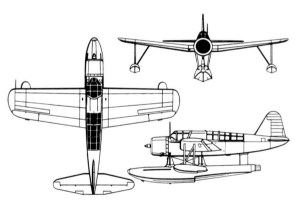

Vought OS2U-3 Kingfisher

the production of the F4U Corsair fighter. The two US Navy OS2U-3 contracts were for 706 and 300 aircraft, the former comprising 175 delivered as seaplanes and 531 as landplanes (with 157 sets of Edo floats), while the latter were all landplanes, most of which were used for training at the Corpus Christi, Jacksonville, Norfolk and Pensacola bases. A further 300 OS2U-3s were built by the Naval Aircraft Factory at Philadelphia, under the designation OS2N-1; these aircraft equipped nine new inshore patrol squadrons formed in 1942.

Most shipboard US Navy Kingfishers were catapulted from the fantails of battleships and cruisers, and three of the surviving OS2Us in the United States are exhibited aboard the preserved battleships USS *Alabama*, USS *Massachusetts* and USS *North Carolina* (the fourth is in the US Naval Aviation Museum at Pensacola). In May 1940, however, the US Navy had

Vought-Sikorsky OS2U-3 Kingfisher

Vought OS2U-3 Kingfisher of the Marine Luchtvaartdienst (Dutch naval air arm), based in Australia during March 1942.

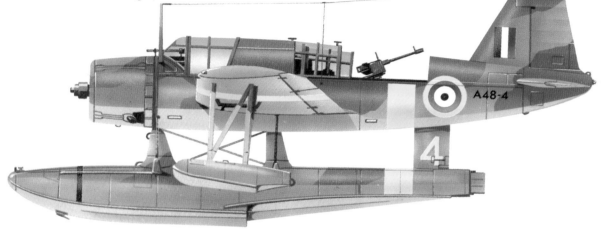

Vought Kingfisher Mk I of the Royal Australian Air Force, based at Rathmines in late 1942.

Vought Kingfisher Mk I of No. 107 Squadron, Royal Australian Air Force, in 1942.

Vought-Sikorsky OS2U-3 Kingfisher

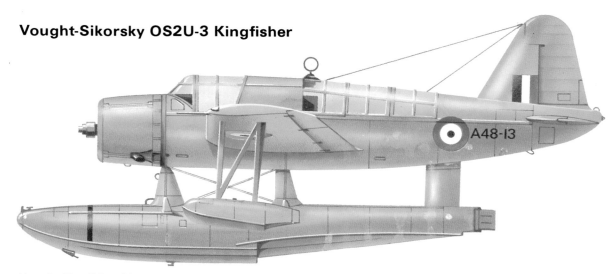

Vought Kingfisher Mk I of the Royal Australian Air Force in 1948.

taken a decision to convert six 'Fletcher' class destroyers to carry Kingfishers, the selected ships being the USS *Halford*, *Hutchins*, *Leutze*, *Pringle*, *Stanley* and *Stevens*. It soon became evident that these ships were too small for effective operation of the floatplanes, and the equipment was removed from at least three of them within months.

The Kingfisher's value as an anti-submarine aircraft, despite its limited bomb load, was demonstrated on 15 May 1943, when one from the Key West Naval Air Station's detachment at Cayo Frances depth-charged *U-176* in the Nicholas Channel, off the Bahamas, the kill being completed by a Cuban submarine chaser. Almost a year earlier, on 15 July 1942, two OS2Us from Squadron VS-9 at the Marine Air Corps Station Cherry Point, had helped gunners aboard the armed merchant-man *Unicol* to sink *U-576* off the Diamond Shoals, North Carolina.

Equally valuable was the Kingfisher's performance as a rescue aircraft, perhaps the best known incident being the rescue of Captain Eddie Rickenbacker, who had ditched in the Pacific in a Boeing B-17 in late October 1942. On 11 November, after an earlier unsuccessful air and sea search, a Kingfisher based on one of the Ellice Islands rescued the pilot and four crewmen. Returning next day, it picked up Ricken-backer and two other members of the crew who had been spotted by other search aircraft. The load being too great for take-off to be achieved, the Kingfisher pilot taxied 40 miles (64 km) to the nearest landfall.

The Kingfisher served with the Royal Navy in both

The Vought OS2U Kingfisher was a useful scouting and observation type. It was stressed for catapult launches from major warships, and taxied up to the crane attachment point after water landings.

Vought-Sikorsky OS2U-3 Kingfisher

landplane and floatplane configuration, 100 having been supplied under Lend-Lease. They were issued to Nos. 763, 764 and 765 Squadrons for seaplane training duties, to No. 726 Squadron at Durban, South Africa, to Nos. 740 and 749 Squadrons for observer training at Arbroath and Piarco, Trinidad, and to No. 703 Squadron. Between May 1942 and March 1944, aircraft from the last unit operated long-range anti-submarine patrols from armed merchant cruisers assigned to the Indian Ocean and the South Atlantic, escorting troop convoys and searching for German blockade runners. These vessels included HMS *Canton*, HMS *Cilicia* and HMS *Corfu*, and some idea of the level of activity can be gained from the achievements of the *Cilicia* and *Corfu* flights which, in 11 months and 10 months, respectively, flew 200 and 142 sorties. Other ships equipped with Kingfishers included the cruisers HMS *Emerald* and HMS *Enterprise*.

The Royal Navy's Kingfisher Mk Is were supplied from the US Navy's contracts for 1,006 OS2U-3s, as were those delivered to Argentina (9), Chile (15), the Dominican Republic (3), Mexico (6) and Uruguay (6). Similarly, the 24 OS2U-3s which had been among the aircraft ordered by the Dutch government for service in the Netherlands East Indies were diverted from the US Navy order. They were being delivered by sea when the islands were overrun by the Japanese, and were docked instead in Australia. Eighteen of them were repainted in Royal Australian Air Force markings for service with No. 3 Operational Training Unit and with No. 107 Squadron, operating in a submarine hunting role from Rathmines on the Australian east coast. The substitution of self-sealing bag fuel tanks in place of the original metal tanks, and the removal of armour plating achieved a weight saving which enabled the RAAF Kingfishers to carry 250-lb (113-kg) depth charges, rather than the 100-lb (45-kg) bombs which were the standard offensive load carried on underwing pylons.

Nine survived the war, including one which, in a high-visibility yellow-overall paint scheme, accompanied the 1947 Deep Freeze Antarctic Expedition aboard the exploration vessel *Wyatt Earp*. The pilot, Squadron Leader R.H.S. Gray, flew the Kingfisher for 55 hours during the expedition, spotting pack ice and ice leads. This was achieved despite limited use as a result of bad weather, and the difficulties of loading and unloading the aircraft which was both longer and of greater span than the ship's beam, necessitating its partial disassembly each time it was stowed.

Specification

Type: two-seat observation aircraft
Powerplant: one 450-hp (336-kW) Pratt & Whitney R-985-AN-2 Wasp Jnr radial piston engine
Performance: maximum speed 164 (264 km/h) at 5,500 ft (1 675 m); cruising speed 119 mph (192 km/h) at 5,000 ft (1 525 m); service ceiling 13,000 ft (3 960 m); range 805 miles (1 296 km)
Weights: empty 4,123 lb (1 870 kg); maximum take-off 6,000 lb (2 722 kg)
Dimensions: span 35 ft 11 in (10.95 m); length 33 ft 10 in (10.31 m); height 15 ft 1½ in (4.61 m); wing area 262 sq ft (24.34 m²)
Armament: one fixed forward-firing 0.30-in (7.62-mm) machine-gun and one rear cockpit flexible 0.30-in (7.62-mm) gun, plus two 100-lb (45-kg) or 325-lb (147-kg) bombs on underwing racks
Operators: Argentina, Chile, Dominican Republic, Mexico, RAAF, RN, Uruguay, USCG, USN

Vultee A-31/A-35 Vengeance

History and Notes

The majority of military aircraft produced by the US industry during the 'between wars' years evolved from requirements of the nation's armed forces and, because of this, enjoyed the benefit of government finance during the development stages. When, in due course, examples became available for export markets, they were costed at a realistic price, one which was not saddled with a share of high development costs. The Vultee V-11/V-12 single-engined light attack bomber was, however, an exception to the foregoing. Developed as a private venture it was sold in the middle and late 1930s, in what were then considered large numbers, to Brazil, China, Turkey and the USSR. When, in late 1938, the US Army Air Corps became interested in the type, only seven were built for service trials under the designation YA-19, but no production order materialised.

Development of the basic design continued, and bearing in mind the interest in dive-bombing techniques which had evolved in the USA and Germany, and been demonstrated as effective in the Spanish Civil War,

The designation A-31 was given to USAAF procurement of the Vultee V-72 Vengeance for supply to the UK under the provisions of the Lend-Lease Act.

Vultee ensured that its new V-72 would have the capability of deployment in such a role. In 1940 a British purchasing mission, which had received plenty

Vultee A-31/A-35 Vengeance

of confirmation of the capabilities of the dive-bomber, placed an order for 700 of these aircraft under the designation Vengeance, and the first of these aircraft for Britain (AN838), flew for the first time in July 1941. A fairly large mid-wing monoplane of all-metal construction, it had hydraulically operated air-brakes on the wings for control in the dive and hydraulically retracted tailwheel type landing gear, and the powerplant comprised one 1,700-hp (1 268-kW) Wright GR-2600-A5 Cyclone 14 twin-row radial engine.

The British order comprised 400 Mk I aircraft and 300 Mk II, built by Northrop and Vultee respectively since the latter had insufficient production capacity. Subsequently, after the introduction of Lend-Lease in 1941, the USAAF ordered 300 more aircraft for Britain, allocating the designation A-31. Northrop-built examples off this latter contract had the designation Vengeance IA in Britain; the Vultee-built aircraft became Vengeance III.

German Junkers Ju 87 dive-bombers, which had been deployed against vital targets in England during the Battle of Britain, had been found to be very vulnerable to high-performance fighters. The RAF realised that its new Vengeance aircraft would be totally unsuited for operations in Europe and, instead, transferred them to equip Nos. 45, 82, 84 and 110 Squadrons operating in Burma where, with Hawker Hurricanes flying escort, they were operated with considerable success against difficult jungle targets.

When the US became involved in the war, at least 243 of the aircraft intended for Britain were commandeered by the USAAF and put into service as V-72s. Further production was initiated for the USAAF, these equipped to US Army standards, and with

armament comprising five 0.50-in (12.7-mm) machine-guns. Vultee built 99 under the designation A-35A; followed by 831 A-35Bs with the Wright R-2600-13 engine and increased armament. Of this total 29 were assigned to Brazil, and the 562 supplied to Britain were known to the RAF as Vengeance IV, of which a small number were transferred to Royal Australian Air Force units. Some of the RAF's aircraft were converted for target towing duties as Vengeance TT.IVs, but almost all of those which served with the USAAF were converted for similar employment. Variants included one XA-31B with a 3,000-hp (2 237-kW) Wright XR-4360-1 Wasp Major radial engine installed for test purposes, and five YA-31Cs with 2,200-hp (1 640-kW) R-3350-13/-37 Cyclone engines installed for development purposes.

Specification

Type: two-seat dive-bomber
Powerplant (A-35B): one 1,700-hp (1 268-kW) Wright R-2600-13 Cyclone 14 radial piston engine
Performance: maximum speed 279 mph (449 km/h) at 13,500 ft (4 115 m); cruising speed 230 mph (370 km/h); service ceiling 22,300 ft (6 800 m); range 2,300 miles (3 701 km)
Weights: empty 10,300 lb (4 672 kg); maximum take-off 16,400 lb (7 439 kg)
Dimensions: span 48 ft 0 in (14.63 m); length 39 ft 9 in (12.12 m); height 15 ft 4 in (4.67 m); wing area 332 sq ft (30.84 m²)
Armament: six 0.50-in (12.7-mm) machine-guns, plus up to 2,000 lb (907 kg) of bombs
Operators: Brazil, RAAF, RAF, USAAF

Vultee BT-13/BT-15/SNV Valiant

History and Notes

A study of the designs initiated by the Vultee Company, especially from the mid-1930s, indicated the introduction of many innovative features. Perhaps one of their most revolutionary design schemes was that for four aircraft to be produced with the same basic tooling and having such assemblies as the wings, aft fuselage and tail unit in common. Variations in powerplant, forward fuselage, landing gear and equipment would ring the necessary changes to provide an advanced, basic or basic combat trainer and a lightweight fighter. While the company probably hoped that such a plan would enable them to get the maximum profit yield from the typical small-scale procurement of the interwar years, it is unlikely that they imagined in their wildest dreams that one aircraft of that quarto would outnumber all other basic trainers produced in the USA during World War II.

It began in a very small way in 1938 when the US Army Air Corps tested Vultee's BC-3 basic combat trainer, evolved from the BC-51 which was one of the four aircraft mentioned above. The BC-3 had retract-

able landing gear and a 600-hp (447-kW) Pratt & Whitney Wasp engine, and while testing showed that it had ideal characteristics for a training role, the complexity of a retractable landing gear was unnecessary for the required basic trainer, as was such a powerful engine. As a result Vultee developed a new Model 74, which had fixed landing gear and a less powerful engine, and US Army testing resulted in an initial contract for 300 aircraft. When placed in September 1939, this was the US Army's largest ever order for basic trainers.

Designated BT-13 by the USAAC, the new basic trainer was a low-wing cantilever monoplane of all-metal structure, with all control surfaces fabric-covered. Landing gear featured oleo-pneumatic shock-struts, a steerable tailwheel, and hydraulic brakes. Accommodation was provided for a crew of two, seated in tandem beneath a continuous transparent canopy, and dual controls and blind-flying instrumentation were standard.

BT-13s were soon established in service, and contracts were placed for very large numbers of these

Vultee BT-13/BT-15/SNV Valiant

aircraft. They were followed on the production line by BT-13As with a different version of the R-985 Wasp Junior engine and detail refinements, and 6,407 of this variant were built. The BT-13B differed only by having a 24-volt electrical system. Production of these airframes was so rapid that the supply of engines dried up, leading to the BT-15 with the 450-hp (336-kW) Wright R-975-11 Whirlwind 9 engine.

The US Navy also had a requirement for basic trainers and, taking the US Army's lead, placed an initial order on 28 August 1940 for a version equivalent to the BT-13A except for the installation of an R-985-AN-1 engine; this had the US Navy designation SNV-1, and 1,350 were built. SNV-2 was the designation of a variant the equivalent of the US Army's BT-13B. In all, well over 11,000 of these trainers were built for service with the USAAF and US Navy in the period 1940-4, but with changing ideas and the introduction of turbine-powered aircraft, all of the aircraft were retired very quickly after the war's end.

The USA's most prolific basic trainer in World War II was the Vultee V-74, produced as the BT-13 and BT-15 for the USAAF, and as the SNV for the US Navy.

Specification

Type: two-seat basic trainer
Powerplant (BT-13A): one 450-hp (336-kW) Pratt & Whitney R-985-AN-1 Wasp Junior radial piston engine
Performance: maximum speed 180 mph (290 km/h); service ceiling 21,650 ft (6 600 m); range 725 miles (1 167 km)
Weights: empty 3,375 lb (1 531 kg); maximum take-off 4,496 lb (2 039 kg)
Dimensions: span 42 ft 0 in (12.80 m); length 28 ft 10 in (8.79 m); height 11 ft 6 in (3.51 m); wing area 239 sq ft (22.20 m²)
Armament: none
Operators: USAAF, USN

Vultee P-66 Vanguard

History and Notes

In the late 1930s, Vultee came up with the idea of developing four different aircraft evolved from the same basic tooling, with wings, aft fuselage and tail unit common to all. These had the initial company designations P-48, BC-51, B-54 and B-54D, the last later being developed as the Vultee BT-13 basic trainer, which was first ordered by the US Army Air Corps in September 1939.

Detail design for a fighter version began at about the same time, during 1938, this deriving from the Vultee's Model 48. Of all-metal construction, except for fabric-covered control surfaces, the configuration was that of a cantilever low-wing monoplane with conventional fuselage and tail surfaces. Features included hydrauli-

cally retractable tailwheel type landing gear with all units retracting into the fuselage; hydraulically actuated slotted trailing-edge flaps; and initially a long streamlined cowling which totally enclosed the 1,200-hp (895-kW) Pratt & Whitney R-1830-S4C4G radial engine. To achieve this it was necessary to adopt a long drive shaft for the propeller, the essential cooling air for the engine being drawn in through a variable-aperture air intake mounted beneath the aircraft's nose, immediately aft of the propeller spinner.

First flown in September 1939, the prototype immediately revealed the fact that the air cooling for the engine was totally inadequate. A number of attempts to resolve this problem proved fruitless, and so it was decided to scrap this feature and install

Vultee P-66 Vanguard

instead a conventional NACA type cowling with adjustable cooling gills on the second prototype. Known as the Model 48X and named Vanguard, this was flown first on 11 February 1940. In addition to the new cowling, the Model 48X introduced compound wing dihedral, increased areas for both horizontal and vertical tail surfaces to overcome a slight instability problem, some improvements in the landing gear and provision for an armament comprising four forward-firing machine-guns.

On February 1940 the company received from the Swedish government an order for 144 of these aircraft, designated Model 48C Vanguard, and immediately initiated their production, completing first a production prototype (NX28300) which made its initial flight on 6 September 1940. Many of the production aircraft were ready for delivery in September 1941, but a US embargo on the supply of aircraft to non-Allied European countries meant these could not go to Sweden. Offered first to Britain, then to Canada, 129 of the total production run eventually went to China under Lend-Lease. The balance were acquired by the USAAF under the designation P-66, the last being delivered in April 1942, and were used by the USAAF in a training role.

Specification

Type: single-seat fighter
Powerplant (P-66): one 1,200-hp (895-kW) Pratt & Whitney R-1830-33 Twin Wasp radial piston engine
Performance: maximum speed 340 mph (547 km/h) at 15,100 ft (4 600 m); cruising speed 290 mph (467 km/h) at 17,000 ft (5 180 m); service ceiling 28,200 ft (8 595 m); range 950 miles (1 529 km)
Weights: empty 5,235 lb (2 375 kg); maximum take-off 7,384 lb (3 349 kg)
Dimensions: span 36 ft 0 in (10.97 m); length 28 ft 5 in (8.66 m); height 9 ft 5 in (2.87 m); wing area 197 sq ft (18.30 m²)
Armament: four 0.30-in (7.62-mm) and two 0.50-in (12.7-mm) forward-firing machine-guns
Operators: China, USAAF

The main operator of the Vultee P-66 Vanguard fighter was the Chinese air force, which received 129 of the 146 built, but Britain did assess the type.

Waco CG-3A

History and Notes

In 1941 the US Army began to initiate the procurement of gliders for troop-carrying, making a first start by requesting proposals for an eight/nine-seat aircraft. Proposals received resulted in the award of contracts for the construction of single prototypes, these going to Bowlus (designation XCG-7), Frankfort (XCG-1), St Louis (XCG-5) and Waco (XCG-3). All of these were completed and flown, except for Frankfort's XCG-1, but it was the Waco XCG-3 which was selected for production.

A high-wing monoplane of large span and external bracing struts, the production CG-3A was of mixed construction with fabric-covered basic structures of wood and steel tube. The fixed tailwheel type landing gear had large low-pressure tyres on the main units, and it was intended that for operational use the main units could be jettisoned if desirable. A skid was mounted beneath the forward fuselage to facilitate landing under wheelless conditions.

Accommodation of the production CG-3As allowed for a total of nine fully armed troops, each carrying 40 lb (18 kg) of equipment, but of these, two men were

Waco CG-3A

required to fulfil the roles of pilot and co-pilot during the flight to an operational deployment of the aircraft.

A total of 300 CG-3As was ordered in 1941, and 100 of these were built by Commonwealth Aircraft Inc. of Kansas City, these being delivered during 1942. The 200 ordered from the Waco Aircraft Company were cancelled, the company's production of the later CG-4A being considered more important. The CG-3As which entered service with the USAAF were used only for the purpose of training.

Specification

Type: troop-carrying glider
Performance: maximum towed speed 120 mph (193 km/h); normal towed speed 100 mph (161 km/h); minimum control speed 38 mph (61 km/h)
Weights: empty 2,044 lb (927 kg); maximum take-off 4,400 lb (1 996 kg)
Dimensions: span 73 ft 1 in (22.28 m); length 43 ft 4 in (13.21 m)
Armament: none

The Waco CG-3 was the USAAF's first production troop-carrying glider, but 200 of the 300 ordered under the designation CG-3A were cancelled, the 100 built being used as trainers for the pilots needed to man the improved CG-4A.

Waco CG-4A

History and Notes

Shortly after four US manufacturers had been contracted in 1941 to build prototypes of eight/nine-seat troop-carrying gliders, the same four manufacturers were each awarded a contract to design and build a single prototype of a 15-seat glider in the same category. These companies were Bowlus (designation XCG-8), Frankfort (XCG-2), St Louis (XCG-6), and Waco (XCG-4). All but the XCG-6 were built and flown, but Waco's XCG-4 was significantly better than its competitors and after early testing a second prototype was ordered.

Of high-wing monoplane configuration, the XCG-4 had wings of all-wood construction, part plywood and completely fabric covered, and rigidly braced by steel tubes with fairings of wood and fabric. Spoilers were mounted on the upper surface of the wings and when deployed these steepened the gliding angle and increased sinking speed. The fuselage was of welded steel tube, fabric-covered over wooden formers and stringers. The conventional tail unit was of wooden construction, plywood- and fabric-covered, with the tailplane wire-braced. Two forms of tailwheel type landing gear were provided: that for training consisted of fixed and well-braced main units each with a long oleo-spring shock strut from half-axle to fuselage/wing spar junction, plus hydraulic brakes; the operational main gear, consisting of a cross axle with brakeless wheels, was jettisonable.

The pilots' compartment was accommodated in the fuselage nose, which was hinged to fold upward to facilitate direct loading into the cabin of heavy equipment. Seats were provided for pilot and co-pilot, side-by-side, and dual controls, mounted on an 'A' frame suspended from the roof, and dual rudder pedals

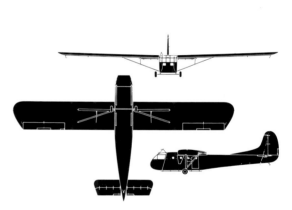

Waco CG-4A

were standard. A door on each side of the rear fuselage gave access to the main cabin which seated 13 fully equipped airborne troops, on benches each side, or could carry varying heavy cargo loaded through the nose. Typical loads included a Jeep with crew of four plus equipment, or a 75-mm howitzer with its guncrew of three, ammunition and supplies. Standard equipment included an intercom system between tug and glider, landing light, and night-flying equipment.

Considered to be very suitable for deployment as a troop/cargo glider, an initial contract was placed with Waco for large-scale production, and arrangements made for licence construction by 15 other manufacturers. These companies (and their production figures) were Babcock (60), Cessna (750), Commonwealth (1,470), Ford Motor Company (4,190), G. & A. Aircraft (627), General Aircraft (1,112), Gibson (1,078), Laister-Kauffman (310), National (1), Northwestern Aero-

Waco CG-4A

nautical (1,509), Pratt & Read (956), Ridgefield (162), Robertson Aircraft (170), Timm Aircraft (433), Ward (7), and of course Waco (1,074) making a grand total of 13,911 including the two prototypes. All of these gliders had the designation CG-4A and the name Haig in USAAF service. Some 694 were supplied to Britain under Lend-Lease, these being designated Hadrian I and II, and the US Navy acquired 13 from the US Army which for experimental purposes under the designation LRW-1. A number of CG-4As were also supplied for use by Canada.

The first examples of the CG-4A began to enter service with the US Army Air Corps during 1941, but it was in 1942-3 that production reached its peak. The CG-4A was first deployed operationally during the invasion of Sicily, on 9 July 1943, when 129 Hadrians and eight British-built Horsas towed by 109 Douglas C-47s of the US Troop Carrier Command, and seven Handley Page Halifax and 21 Armstrong Whitworth Albemarle tugs of the RAF's No.38 Wing were released at night near Cape Murro di Porco. Their target was the Ponte Grande bridge just south of Syracuse, but in gale conditions this initial operation with the Waco CG-4As was near-disastrous. About 50 gliders landed in the sea, and 75 on the island far from the target. Only 12 gliders, one a Horsa, landed near the Ponte Grande, and from them only eight officers and 65 paratroops survived to gain their objective and capture the bridge.

The next operation involving the CG-4A was far more successful when, in early 1944, CG-4A Haigs of the USAAF's 1st Air Commando were used in Burma. As part of the overall plan, it was decided to use Long Range Penetration Groups of Wingate's special force ('the Chindits') to disrupt Japanese communications in the Irrawaddy valley. On 5 March 1944, a total of 54 CG-4As was towed off by C-47s of Troop Carrier Command: 37 landed on target, eight in friendly and nine in enemy territory. There were many on-ground collisions of the 37 gliders which landed on the target area code-named 'Broadway', 30 men being killed and 33 injured. The 490 odd men who survived unharmed created an airstrip within 24 hours, and on the night of 6 March this was able to accept 62 C-47 sorties bringing in men and supplies and carrying out the dead and wounded.

Airborne forces had significant roles to play during

the remainder of the war, and Haigs were used extensively by the USAAF in actions which included the invasions of Normandy and southern France, at Arnhem, and during the Rhine crossings. Apart from the invasion of Sicily, the RAF did not use its Hadrians operationally. The majority were used to equip Nos. 668, 669 and 670 Squadrons of No. 343 Wing and Nos. 671, 672 and 673 Squadrons of No. 344 Wing deployed in South-East Asia Command.

One Hadrian created aeronautical history by becoming the first glider to be flown across the North Atlantic. Towed by an RAF Douglas Dakota, it covered a distance of 3,500 miles (5 633 km) in stages from Montreal to Great Britain, the total flight time being approximately 28 hours.

Variants of the CG-4A included a single all-wood XCG-4B prototype built by the Timm Aircraft Corporation; and with the object of creating a powered glider which could return to base without any payload, two powered prototypes were completed by installing engines on existing CG-4As. These included one XPG-1 converted by Northwestern in 1943 by the installation of two 130-hp (97-kW) Franklin 6AC-298 flat-six engines, and one XPG-2 by Ridgefield with two 175-hp (130-kW) Ranger L-440-1 inverted inline engines. Successful testing of this latter prototype resulted in Northwestern receiving a contract for 10 PG-2As with Ranger L-440 engines, but these were delivered too late in the war to be used operationally.

CG-4As which remained in USAF inventory in mid-1948 were redesignated C-4A, and 35 of these later modified by the addition of a US Navy-developed bar towing device were further redesignated C-4C. The powered PG-2As became G-2As.

Specification

Type: 15-seat cargo-and troop-carrying glider
Performance: maximum towing speed 150 mph (241 km/h); normal towing speed 120 mph (193 km/h)
Weights: empty 3,700 lb (1 678 kg); maximum overload take-off 9,000 lb (4 082 kg)
Dimensions: span 83 ft 8 in (25.50 m); length 48 ft 8 in (14.83 m); height 12 ft 7 in (3.84 m); wing area 852 sq ft (79.15 m²)
Armament: none
Operators: RAF, RCAF, USAAF, USN

Waco CG-13A

History and Notes

In 1942, with the USA well and truly involved in World War II, the US Army began the process of obtaining a larger-capacity troop glider. Although at that time considerable numbers of CG-4As had entered service, none had been used operationally; but it was considered by USAAF planners that for the deployment of both cargo and/or paratroops a glider with double the capacity would have advantages in certain types of

operation. Accordingly, Waco was awarded a contract in 1942 for two XCG-13 prototypes, each of which was required to accommodate a total of 30 troops, two of whom would serve as pilot and co-pilot en route to the target area.

The resulting design by Waco was virtually an enlarged version of the CG-4, the additional capacity gained by a longer and wider fuselage. The wing

Waco CG-13A

differed only by having trailing-edge flaps instead of spoilers, and otherwise its construction was as that of the CG-4. Vertical tail surfaces were increased in area and the tailplane was strut- instead of wire-braced. In other respects layout and equipment were similar to that of its smaller predecessor, except that the cabin could accommodate 28 fully equipped troops.

Delivered and tested successfully in 1943, two additional pre-production prototypes were ordered under the designation YCG-13, one each from Ford Motor Company and Northwestern Aeronautical. Further extensive testing of these four aircraft led to the decision to order production examples, but to introduce in these tricycle type instead of tailwheel type landing gear, leading to the procurement of three YCG-13A pre-production prototypes with such landing gear for service testing before the production contract was finalised. These were far more conventional units than those that equipped the CG-4, the main units being unbraced oleo-pneumatic shock struts with single wheels, the nose unit being braced and carrying twin wheels. This tricycle configuration offered performance advantages at take-off, and simplified the landing problem for the pilot.

Ordered in 1943 in limited quantities from Ford and Northwestern, the former built 48 and the latter 47, and all were delivered during 1944. In that year Ford received a contract for a variant that by changes in seating would accommodate a maximum of 40 troops in the main cabin. A total of 37 was built, these being delivered during 1945, without any changes in

The Waco CG-13A assault glider could carry up to 42 troops or an equivalent weight of cargo.

designation. All those remaining in service in 1948 were redesignated G-13A.

Specification
Type: 30- or 42-seat troop- and cargo-carrying glider
Performance: maximum towing speed 190 mph (306 km/h); landing speed 80 mph (129 km/h)
Weights: empty 8,700 lb (3 946 kg); maximum take-off 18,900 lb (8 573 kg)
Dimensions: span 85 ft 8 in (26.11 m); length 54 ft 4 in (16.56 m); height 20 ft 3 in (6.17 m); wing area 873 sq ft (81.10 m²)
Armament: none
Operator: USAAF

Waco CG-15A

History and Notes
Operational deployment of the CG-4 had highlighted one major shortcoming: the structure of the fuselage nose had very little strength: landings made in forest areas had shown that quite small saplings could cause major damage, and this had resulted in many pilots and co-pilots being killed or seriously injured. In 1944, therefore, a CG-4A was converted by Waco to serve as a prototype of an improved version designated XCG-15. In addition to overcoming the structural weakness of the nose, the opportunity was taken to incorporate several improvements that had been suggested by operational use.

A higher towing speed was desirable, and to achieve this the wing span was clipped by 21 ft 5¾ in (6.55 m), resulting in almost 27% less wing area. As a result the wing upper-surface spoilers were deleted and replaced by trailing-edge flaps, since the latter could not only be used to provide the necessary landing performance, but also to enhance lift for take-off. Because the span was reduced, less drag-inducing struts were required, and while carrying out strengthening of the nose, the opportunity was taken to change its profile to give improved aerodynamic performance. Other changes included the introduction of the same type of cantilever

An improved version of the CG-4A, the Waco CG-15 had revised landing gear and reduced wing span, but could still carry 15 troops, though at higher speed.

oleo-pneumatic struts for the main units of the landing gear as those incorporated in the CG-13A, and standard USAAF cargo tie-down fittings were installed in the cabin.

Testing of this first prototype in 1944 resulted in an order for two additional new-construction prototypes,

Waco CG-15A

designated XCG-15A, and these were delivered in 1945. These proving satisfactory on test, and confirming that the higher maximum speed had been attained, 1,000 examples of the CG-15A were contracted for construction by Waco. After 385 of these had been produced a seating rearrangement provided accommodation for 16 instead of 15, but after only 42 of these had been delivered VJ-Day brought contract cancellation and no further examples were built. Some of the CG-15As were used operationally, deployed alongside CG-4As.

In 1945, with the object of developing a low-cost troop or cargo transport, Waco was awarded a contract to build a powered version of the CG-15A. One of these aircraft was converted by the installation of two designation XPG-3. When in mid-1948 the USAF designations were revised, the CG-15As became G-15A, and the single XPG-3 was changed to G-3.

Specification
Type: 15/16-seat cargo- and troop-carrying glider
Performance: maximum towing speed 180 mph (290 km/h)
Weights: empty 4,000 lb (1 814 kg); maximum take-off 8,035 lb (3 645 kg)
Dimensions: span 62 ft 2¼ in (18.95 m); length 48 ft 10 in (14.88 m); wing area 623 sq ft (57.88 m²)
Armament: none
Operator: USAAF

Waco UC-72

History and Notes
The Waco Aircraft Company had its origin in 1919 when George 'Buck' Weaver established the Weaver Aircraft Company (Waco trademark) at Loraine, Ohio. The name Waco was adopted for the company in 1929, and in the years immediately before World War II this organisation became one of the largest manufacturers of civil aircraft in the United States. Built in a variety of types and, for those days, in quite large numbers, was a series of steadily improving cabin biplanes. In production during the period 1938-40 was the five-seat Model E, which was of mixed construction.

Wings were all-wood, with plywood and fabric covering, and the tailplane and fin were similarly constructed. Fuselage, rudder and elevators all had a basic steel-tube structure, but while the fuselage had wood and fabric covering, the rudder and elevators were fabric-covered only. Frise type ailerons and electrically-operated split trailing-edge flaps were confined to the upper wing. Landing gear was of the non-retractable tailwheel type, the main units being of the divided type and braced in typical Waco manner. Standard accommodation was for five, with pilot and co-pilot/passenger on individual front seats, and three on a bench seat at the rear of the cabin. Dual control facility was provided by means of a 'throw over' control wheel and dual rudder pedals. Powerplants of the Model Es varied from 300-400 hp (224-298 kW).

Immediately after the USA became involved in World War II, the US Army Air Corps began to commandeer a number of Model Es, totalling 15 in all, and subsequently 28 other examples of various earlier models were similarly acquired for use as staff transports and unit ferry aircraft. The first 12 to enter service, all Waco SREs, were allocated the designation UC-72, and subsequent acquisitions were designated UC-72A to -72H, -72J to -72N, and UC-72P. All remained in service throughout the war.

The US Navy also acquired from the Waco company

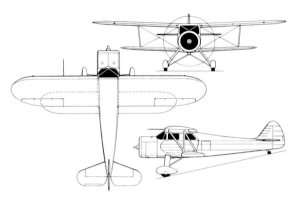

Waco UC-72 (Model E)

in 1936 three examples of an earlier cabin biplane in this series, the EQC-6, which were allocated for use by the US Coast Guard under the designation J2W-1. These were still in service during the war years, their powerplant consisting of 320-hp (239-kW) Wright R-760-E2 Whirlwind 7 engines.

Specification
Type: four/five-seat cabin biplane
Powerplant (UC-72): one 400-hp (298-kW) Pratt & Whitney R-985-33 Wasp Junior radial piston engine
Performance: maximum speed 200 mph (322 km/h); cruising speed 195 mph (314 km/h); service ceiling 23,500 ft (7 165 m)
Weights: empty 2,734 lb (1 240 kg); maximum take-off 4,000 lb (1 814 kg)
Dimensions: span 34 ft 9 in (10.59 m); length 27 ft 9¾ in (8.48 m); height 8 ft 8 in (2.64 m); wing area 285.2 sq ft (26.50 m²)
Armament: none
Operators: USAAC/USAAF, USCG

(**Note:** the X and Y prefixes to aircraft designations, indicating experimental or pre-production status, have been indexed only when the particular aircraft type progressed no further towards full production than either of these two statuses. Thus the XB-29 prototype and YB-29 pre-production variants of the Boeing B-29 Superfortress can be found by reference to the basic B-29 entry, while the sole Grumman XF4F-1 is indexed as such. Additionally, all aircraft designated UC- have been indexed under the C- designation.)

INDEX